Systems & Control: Foundations & Applications

M. I. Zelikin
V. F. Borisov

Theory of Chattering Control

with applications to
Astronautics, Robotics, Economics, and Engineering

Birkhäuser
Boston • Basel • Berlin

Vladimir Borisov
Department of Mathematics
Moscow Technological Institute
Moscow, 141220
Russia

Michail I. Zelikin
Department of Mathematics
Moscow State University
Moscow, 119899
Russia

Library of Congress Cataloging-in-Publication Data

Borisov, V. F. (Vladimir F.), 1961-
    Theory of chattering control with applications to astronautics,
robotics, economics, and engineering / V.F. Borisov, M. I. Zelikin.
          p.   cm. -- (Systems & control)
    Includes bibliographical references and index.
    ISBN 0-8176-3618-8 (alk. paper)
       1.   Chattering control (Control systems) 2. Control theory.
I. Zelikin, M. I. (Mikhail Il'ich)   II. Title.   III. Series
TJ223.C45B67   1994   93-51121
003'.5--dc20   CIP

Printed on acid-free paper
© 1994 Birkhäuser Boston

Birkhäuser   ®

ISBN 0-8176-3618-8
ISBN 3-7643-3618-8
Typeset by the Authors in AMS-TeX
Printed and bound by Quinn-Woodbine, Woodbine, NJ.
Printed in the U.S.A.

9 8 7 6 5 4 3 2 1

CONTENTS

PREFACE

The common experience in solving control problems shows that optimal control as a function of time proves to be piecewise analytic, having a finite number of jumps (called switches) on any finite–time interval.

Meanwhile there exists an old example proposed by A.T. Fuller [1961] in which optimal control has an infinite number of switches on a finite–time interval. This phenomenon is called *chattering*. It has become increasingly clear that chattering is widespread. This book is devoted to its exploration.

Chattering obstructs the direct use of Pontryagin's maximum principle because of the lack of a nonzero–length interval with a continuous control function. That is why the common experience appears misleading. It is the hidden symmetry of Fuller's problem that allows the explicit solution. Namely, there exists a one–parameter group which respects the optimal trajectories of the problem. When published in 1961, Fuller's example incited curiosity, but it was considered only "interesting" and soon was forgotten.

The second wave of attention to chattering was raised about 12 years later when several other examples with optimal chattering trajectories were found.[1] All these examples were two–dimensional with the one–parameter group of symmetries.

In recent years, interest in Fuller's phenomenon has increased due to intense efforts to understand the substance of the concept of regular synthesis. A regular synthesis is a definite system of axioms which pretends to be "typical" for the set of all optimal trajectories with a fixed terminal condition. This system of axioms was proposed by V.G. Boltyanskii [1966] whose intention was to prove sufficient conditions for optimality. However, later on, typical examples of optimal synthesis were published which violate one or another of Boltyanskii's axioms.[2] Among these axioms the following one plays an important part in the proof of optimality: on any finite–time

---

[1] [A.T. Fuller, P.E. Grensted, 1965], [C. Marchal, 1973], [G. Magarill-Il'yaev, 1976], [E. Ryan, 1979], [C. Dorling, E. Ryan, 1981], [Ya.M. Bershchanskii, 1982], [P. Brunovskii, J. Mallet-Paret, 1985], et al.

[2] [P. Brunovskii, 1980], [L. Zelikina, 1982], [H. Sussman, 1990], et al.

interval of a trajectory, there exists a finite number of switches. The question of the ubiquity of Fuller's phenomenon is therefore very much to the point.

This question was solved in the widely known works of I. Kupka [1988, 1990] who proved that in the case of the general position there exists at least one chattering solution to the Hamiltonian system if the dimension of the extended space is greater than or equal to 8. Chattering arcs caused by state constraints were found by A. Milutin [V. Dicoussar, A. Milutin, 1989].

To explain chattering phenomenon, let us consider the Fuller problem. Being subjected to a constrained force $u(t)$, the particle moves along a straight line O_x without friction. Given initial conditions $x(0) = x_0$, $\dot{x}(0) = y_0$, we choose the force to minimize the mean square deviation of the particle from a point O with the coordinate $x = 0$. Therefore, we minimize the functional

$$\int_0^\infty x^2(t)\, dt$$

under the restrictions

$$\begin{cases} \dot{x} = y, \\ \dot{y} = u, \end{cases} \quad |u| \leqslant 1, \quad \begin{cases} x(0) = x_0, \\ y(0) = y_0. \end{cases}$$

It is evident that the most profitable position in the phase space (x, y) is the origin $x = 0$, $y = 0$. This position corresponds to the singular manifold S (see Chapter 1). It may appear that the best strategy is to attain S as soon as possible. However, it is shown in Chapter 2 that this is not exactly true. Let us compare the solution to Fuller's problem with that to the following time–optimal problem

Minimize T subject to

$$\begin{cases} \dot{x} = y, \\ \dot{y} = u, \end{cases} \quad |u| \leqslant 1, \quad \begin{cases} x(0) = x_0, \\ y(0) = y_0, \end{cases} \quad \begin{cases} x(T) = 0, \\ y(T) = 0. \end{cases}$$

Consider, for example, a position $x_0 > 0$, $y_0 = 0$. The optimal control of the time–optimal problem has the only one switch on $[0, T]$. Namely, the force is directed toward the point O (i.e., $u = -1$) on the first time–interval $t \in [0, \tau_0]$. On the second time–interval $t \in [\tau_0, T]$, the force is directed away from O (i.e., $u = 1$). The switching moment $t = \tau_0$ must be chosen in such a way that the particle stops just at the point O.

The optimal strategy of Fuller's problem has infinitely many switches and consists of an infinite number of cycles. On the first cycle, we begin with $u = -1$ and then switch to $u = 1$. The switching moment $t = \tau_0$ causes the particle to stop at a point $x(\tau_1) = -qx_0$ behind the point O. Here $q \approx 0.006$. Since the particle slips past the point O, we say that this is *overregulation.* The cycle is repeated for the new–generated initial position and the particle stops successively at points $-qx_0$, q^2x_0, $-q^3x_0, \ldots$. These points constitute an alternating convergent geometric progression. Thus, we have an infinite number of successive overregulations. Durations of cycles form a convergent geometric progression too, so the whole process takes a finite time. Finally, the particle stops at $\mathcal{O}$ but not in the shortest time.

Consider a two–link manipulator that consists of a rotating vertical cylinder with a horizontal advancing arrow–bar. The robot has four degrees of freedom: an angle of rotation and an angular velocity of the vertical cylinder, and a position and a velocity of the arrow. We have therefore the four–dimensional phase space. The robot has two inputs, $v(t)$ and $u(t)$; v is a torque rotating the cylinder, $|v| \leqslant 1$, u is a force acting on the arrow, $|u| \leqslant 1$. Let us identify the arrow with its center of gravity $\mathcal{A}$ and denote its distance from the axis of the cylinder by x. The robot's motion is described by the nonlinear control system (6.36). We wish to reorientate the robot in the shortest time.

Let us follow the movement of the arrow while the cylinder is rotating. When the point $\mathcal{A}$ lies on the axis of the cylinder, the robot has the minimal moment of inertia. The set of points with the minimal moment of inertia constitutes a manifold S in the phase space. This manifold is two–dimensional because its equations are $x = 0$, $\dot{x} = 0$; the angle of rotation and the angular velocity of the cylinder are arbitrary. The manifold S consists of singular trajectories because one needs the control $u = 0$ to keep the arrow in this position. S is the most profitable set that provides fast rotation of the robot. It may appear that the best strategy is to attain S as soon as possible, disregarding the position of the rotating cylinder. In this case, the control $u(t)$ would have only one switch. In reality, this strategy is nonoptimal. The optimal strategy has an infinite number of switches and consists of an infinite number of cycles. Let us describe this strategy.

Assume that the initial velocity of the arrow is zero. At the first cycle, the force is directed toward the axis and then it switches at a moment t_1 such that the point $\mathcal{A}$ slips past the axis and stops at a point $x(t_1)$ on the opposite side of it. Besides that, we have $|x(t_1)| < |x(0)|$. So we have overregulation again. The cycle is repeated for the new–generated initial position, etc. The sequence of successive maximal deviations of $\mathcal{A}$ from the axis, $x(0)$, $x(t_1)$, $x(t_2)$, $\ldots$, is an alternating convergent asymptotical geometric progression. This means that the successive ratios $x(t_n)/x(t_{n+1})$ are not constant but converge to some $-\gamma$ $(0 < \gamma < 1)$ as $n \to \infty$. Cycles durations constitute a convergent asymptotical geometric progression too, so the whole process takes a finite time. Finally, the arrow stops (but not in the shortest time) at a position corresponding to S.

Chattering is closely related to the existence of singular extremals and their order. To be a little more specific, let us consider an optimal control problem which is affine in the scalar control $u \in [-1, 1]$. Denote the state variable by $x \in \mathbb{R}^n$ and the adjoint variable by ψ. By

$$H_1(\psi, x) = \frac{\partial}{\partial u} H(\psi, x, u),$$

we denote the coefficient of the control u in the Pontryagin's function.

Any extremal $\big(x(t), \psi(t)\big)$, $\quad t \in (t_0, t_1)$, is called singular if

$$H_1\big(\psi(t), x(t)\big) = 0$$

for $t \in (t_0, t_1)$.

The maximum condition completely determines the optimal control u^* along nonsingular arcs (namely, if $H_1 > 0$, then $u^* = 1$, if $H_1 < 0$, then $u^* = -1$) while to find the control on singular arcs, one needs to differentiate the identity $H_1 \equiv 0$.

The order of a singular extremal on (t_0, t_1) is the integer q such that $d^{2q} H_1 / dt^{2q}$ is the lowest of total derivatives of H_1 in which u appears explicitly. It is implicit in this definition that the first appearance of u is an even order derivative of H_1. This is proved by H.M. Robbins [1965] and H.J. Kelley, R.E. Kopp, H.G. Moyer [1967] (see Chapter 2). The theorem of Kelley–Kopp–Moyer also proves that the concatenation of a piecewise smooth nonsingular arc with a singular arc of even order is nonoptimal. Meanwhile the singular manifold S is usually the most profitable subset of the phase space. To design the optimal strategy one needs to reach (if possible) this manifold. If singular solutions lying on S have the second (or any higher) order, one needs the chattering to enter into S.

Thus, we can formulate the first practical device that follows from the theory:

The control with small overregulations gives the
best result in attaining second order singular
manifolds.

As far as we know, overregulation is used in practical engineering for the sake of stability. We claim that this policy relates to optimality as well.

Consider the process of leaving singular manifolds. The optimal escape from a singular manifold of second (and any higher) order is chattering whose frequency of switches accelerates to an infinity in the reverse time current. It is a little difficult to imagine the beginning of this movement because of the lack of an interval of continuity of the control at the initial moment. This behavior can be approximately described as a series of very fast pushes and pulls (like vibration) which gives extremely small alternating deviations from S at the very beginning. Our second device is the following:

use such "vibration" to escape from S.

The principal part of this book is devoted to the case of singular extremals of second order $(q = 2)$. The main reason is the following. In applied mechanical problems, the control u influences the second derivative of the phase coordinate. Generally speaking, it results in the order of singular arcs equaling 2. In the case $q = 2$, our results are complete. We consider also the case of singular extremals of order $q > 2$ but in this case some questions remain.

The following analysis explains the behavior of all chattering arcs and the structure of Lagrangian manifolds which contain these arcs for discontinuous Hamiltonian systems in the vicinity of the manifold of singular trajectories of second order. Denote this manifold by S. In the general case, the codimension of S in $\mathbb{R}^{2n}$ equals 4. We prove that for each point $w \in S$ there exist two mutually tangent two–dimensional manifolds $\mathfrak{N}_w^+$ and $\mathfrak{N}_w^-$ which consist of chattering arcs. The trajectories of $\mathfrak{N}_w^+$ tend to w and have an infinite number of switches on a finite–time interval. Likewise the trajectories of $\mathfrak{N}_w^-$ tend to w in the retrogressive time current. Thus, $\mathfrak{N}^+ = \bigcup_{w \in S} \mathfrak{N}_w^+$ is the stable tendril of S and $\mathfrak{N}^- = \bigcup_{w \in S} \mathfrak{N}_w^-$ is the unstable one. It is worth noting that the set of all switching points of $\mathfrak{N}^+$ and $\mathfrak{N}^-$ are found to be piecewise smooth manifolds. It provides the piecewise smoothness of optimal chattering synthesis. The proof of this theorem is based on the investigation of the Poincaré first return mapping of the switching manifold on itself. This mapping has singularities at points of S, and it is necessary to resolve these singularities by using a generalized blowing up procedure. This means that we introduce the new coordinates such that a projective space is attached instead of the point w. To clarify the necessity of the blowing up procedure, let us consider a very simple example.

Three characters, a pedestrian, a cyclist and a mosquito, start out simultaneously from the same point. The pedestrian and the cyclist travel in the same direction at a velocity of 3 m.p.h. and 6 m.p.h., relatively. The mosquito moves at a velocity of 9 m.p.h. back and forth between the two people, reversing direction immediately upon reaching each person. The question is: where will the mosquito be after an hour? (To state the problem a little more convincingly and to excuse such strange behavior on the part of the third character, we could suggest a puppy rather than a mosquito. We use the mosquito simply to stress his almost infinitely small size and to suggest that our characters are points.)

The answer is somewhat unexpected: the solution is not unique. After an hour the mosquito may be found at any point between the two other characters and may be travelling in either direction. To prove this let us depict the graphs of movement for our characters in the (t, x)–plane (t is the time and x is the distance from the initial point). In Fig. 1, the movement of the pedestrian is shown as the ray $OP = \{x = 3t\}$ and that of the cyclist is shown as the ray $OC = \{x = 6t\}$. The movement of the mosquito is depicted as a polygon with vertices on OP and OC, and with angular coefficients of the sides equaling 9 or -9. The corner points of the polygon are densest near the origin. The reverse of the time current allows

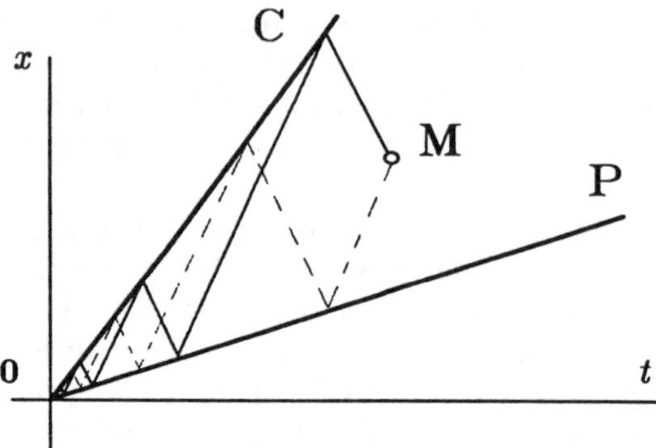

FIG. 1: THE MOSQUITO'S MOTION

us to generate an infinite set of trajectories of the mosquito emanating from the initial point.

It is odd that the same situation takes place inside each manifold $\mathfrak{N}_w^{\pm}$, namely there exists a one–parameter family of chattering trajectories through each point of the singular manifold S. The blowing up procedure provides a way to parameterize these families. To do this we suggest to use a Poincaré mapping of the switching surface on itself. We don't see how to parameterize trajectories of this manifold by infinitesimal values at the point w itself except for the resolution of the singularity of the Poincaré mapping. The Poincaré mapping is the main tool of research in this book.

The manifold $\mathfrak{N}^{\pm}$ is fibrated by two–dimensional piecewise smooth leaves. The asymptotical behavior of the trajectories inside each leaf is the same as in Fuller's problem. Thus, we have two bundles $\pi^{\pm} : \mathfrak{N}^{\pm} \to S$ with base S. The structure of all chattering trajectories of the Hamiltonian system in question in the vicinity of a singular manifold is described by these bundles. The Lagrangian manifolds containing chattering arcs are obtained as the inverse images of Lagrangian submanifolds of S under the projections π^{+} and π^{-}.

The Hamiltonian systems under consideration correspond to affine in control optimal problems. We develop a method to design chattering syntheses for a wide class of such optimal problems. It is sufficient to verify that the projection of a specially chosen Lagrangian manifold to the space of the state variables x is a one–to–one correspondence. The switching manifold of the synthesis is found to be piecewise smooth. Chattering is due to the fact that each concrete optimal trajectory intersects it infinitely.

The developed method provides the possibility of designing the optimal synthesis for a number of examples of applied content, namely, for the problem of a rigid body stabilization, for the resource allocation problem, for controlling the nonlinear, two–link manipulator (robot), for Lowden's problem of the control of rocket motion in the Newtonian gravity field, and other problems.

The problem of stabilization of a rigid body and the problem of optimal control of a two–link manipulator were considered in a number of articles. Their chattering solutions were found by M. Zelikin and V. Borisov. The statement of the resource allocation problem given later in the book is new. Lowden's problem is famous. Articles and monographs have been devoted to its exploration. Most attention has been focused on the question of optimality of the intermediate thrust arcs, i.e., on the optimality of singular trajectories in Lowden's problem. It is no exaggeration to say that this question gave rise to the theory of necessary conditions of higher orders for singular extremals. We describe the optimal synthesis of Lowden's problem, including chattering and intermediate thrust arcs.

At first glance it may seem that the optimal modes which are found in this book are paradoxical from the practical and technological points of view. The surprising thing is that such a large number of nonlinear examples of applied content fall within the realms of Chattering Theory. We think that chattering solutions describe natural situations in accordance with the chosen idealized model. The practical significance of these results lies in the possibility of improving control and estimating the loss in criterion if we use some real suboptimal control. We enlarge the examples to clarify our point of view.

Imagine two parallel walls which move toward each other with constant velocity and an elastic ball which bounces back and forth between these walls. (See Fig. 2.)

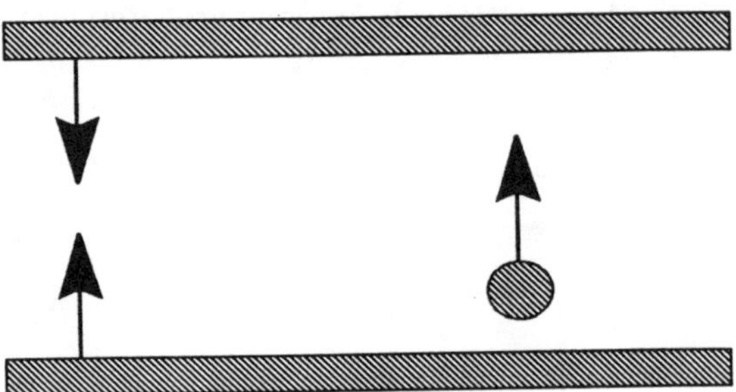

FIG. 2: THE BOUNCING BALL

Suppose that the velocity of the ball changes sign instantaneously at the moment of bounce, remaining the same in absolute value. Then this ball bounces an infinite number of times up to the instant that it is compressed between the colliding walls. This situation is quite familiar to anyone who has tried to stop a ping-pong ball by pressing it against the table with a paddle. A real ball undergoes finite (though very many) bounces because the strike occurs in nonzero time. The shorter the time of strike, the greater the number of bounces. In the theoretical case of zero–time strikes, the number of bounces becomes infinite in a finite–time interval.

We wished to write this book not only to share our results but also to open up other perspectives of this theory. Our theory has a complete form in the case of singular modes of second order for systems with single input. It is natural to develop the theory in two directions. The first one is connected with a study of singular modes of order greater than or equal to 3. For this reason we include a section with conjectures concerning higher order singular modes. The second way of the development is to consider chattering modes for control systems with several inputs. The background of this theory is founded in Chapter 7.

The study of chattering solutions is difficult but interesting. We hope that it is a promising field for those who choose to explore it.

Finally, we wish to express our sincerest gratitude to Birkhäuser Boston, and especially to Professors C.E. Martin and C.I. Byrnes who offered us the opportunity to write this book. We are deeply indebted to Dr. L.Ph. Zelikina whose criticism resulted in the substantial improvement of this book. We are much obliged to Professor E.L. Presman for plenty of useful suggestions. The research was supported by Grant No. 93–011-1731 from the Russian Foundation of Fundamental Research.

Outline

The book is organized as follows. The first chapter is intended as an introduction to chattering theory with special emphasis on a heuristic approach. The objective is to expose the main results of the book and to supply the readers with algorithms for the design of chattering modes.

The second chapter contains a detailed analysis of Fuller's problem in two dimensions. Attention is paid to this particular problem for historical reasons and because the problem can be considered a simple model for creating the whole theory.

The third chapter is the principal one. In this chapter the main theorems of the book are proved, namely, a theorem on chattering bundles, a theorem on Lagrangian manifolds, the chattering optimality theorem, and a theorem on regular projection.

The fourth chapter deals with the ubiquity of Fuller's phenomenon. We outline the results of I. Kupka and prove that in general the codimension of chattering is not greater than 7.

The higher order singular modes are treated in the fifth chapter. The main result is a solution of a class of third order saturating systems with homogeneity.

We devote the sixth chapter to the applications of the chattering theory. We consider designs of optimal syntheses for a number of problems arising in mechanics, space navigation, mathematical economics, etc.

Chapter 7 exhibits perspectives of the study of chattering modes for control systems with several inputs. We present the new results concerning chattering modes of systems with multivariable controls and formulate important research problems.

All chapters can be read independently, except for Chapters 4 and 6 where the results of the basic theorems of Chapter 3 are used. In order to make the volume as self-contained as possible, we include the formulations of the most important concepts upon which some technical details of the proofs are based. In the rest of the book, the comprehension requires a basic acquaintance with the theory of optimal control and differential equations.

The impatient reader who does not want to go deep into mathematical arrangements can read Preface, Introduction, Applications, and Epilogue. The most impatient reader could restrict himself to the title.

Chapter 1

INTRODUCTION

In this chapter we discuss the main results of the book paying attention to heuristic and algorithmic items, without always limiting ourselves to absolutely rigorous mathematical formulations. Exact statements will be given in subsequent chapters. We hope that this facilitates the reading for those readers who don't want to go deeply into the details of mathematical proofs, but who are interested in using the presented results in applied engineering.

1.1 The Subject of the Book

This book deals with the *chattering phenomenon.* In optimization theory this term has two different meanings. The first is related to the situation where optimal control does not exist, i.e., a minimizing sequence of controls does not tend to any limit in the given class of admissible controls. This can be illustrated by the following example of Bolza.

PROBLEM 1.1. *Sliding control.*
Minimize

$$\int_0^{\sqrt{2}} \left(x^2 + (1 - u^2)^2\right) dt$$

subject to

$$\dot{x} = u, \qquad |u| \leqslant 1,$$
$$x(0) = 0, \qquad x(\sqrt{2}) = 0.$$

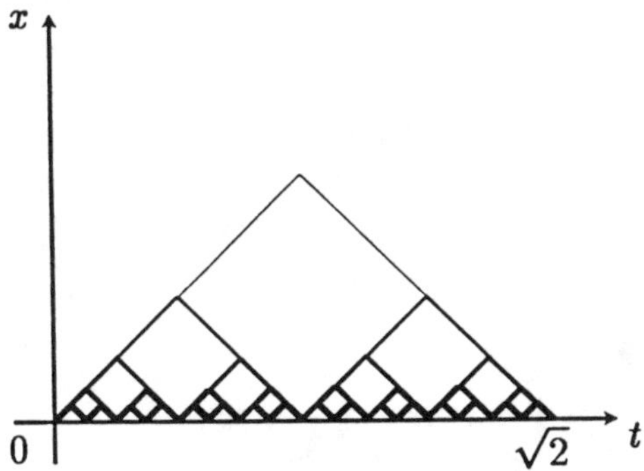

FIG. 3: SLIDING MODE

The minimizing sequence $\left\{x_n(t)\right\}_{n=1}^{\infty}$ consists of polygons where the controls alternate the values 1 and -1 with increasing frequency as $n \to \infty$ (see Fig. 3).

The polygons uniformly tend to the segment $x(t) \equiv 0$, $t \in [0, \sqrt{2}]$; meanwhile the limit of the controls does not exist. Figuratively speaking, the trajectory $x(t) = 0$ corresponds to the control oscillating with an infinite frequency.

We prefer to refer to this as the *sliding mode*. Everywhere below the term *chattering* is reserved for the second situation. Let us consider

PROBLEM 1.2. **Chattering control** (Fuller's problem).
Minimize

$$\int_0^1 x^2 \, dt$$

subject to

$$\dot{x} = y, \quad \dot{y} = u, \quad |u| \leqslant 1,$$
$$x(0) = x_0, \quad y(0) = y_0, \quad x(1) = x_1, \quad y(1) = y_1.$$

It will be proved in Chapter 2 that for any pair of boundary points (x_0, y_0) and (x_1, y_1) in the vicinity of the origin there exists a unique optimal solution $\left(\widehat{x}(t), \widehat{y}(t)\right)$ with a control $\widehat{u}(t)$ containing a countable set of switches (Fig. 4.). The optimal trajectory is composed of three successively adjoined subarcs. The first one is depicted in Fig. 6. The intervals between successive switching points $t_{i+1} - t_i$ constitute a converging geometric progression. This part of the optimal trajectory resembles a kind of spiral, twisting around the origin and attaining the origin in a finite time

$\tau < 1$ with an infinite number of switches. The second arc coincides identically with the origin, $x(t) \equiv y(t) \equiv 0$. The third arc is depicted in Fig. 8. The behavior of this part of the optimal trajectory is similar to that of the first one. The difference is that the switching curve OA_1B_1 coincides with the reflection of the curve OAB in Fig. 6 with respect to the O_x–axis. In particular, the trajectories escape from the origin with an infinite number of switches.

Everywhere in this book the term "chattering control" means optimal control with an infinite number of switches in a finite time interval. A typical behavior of chattering trajectories in a two–dimensional phase space is exhibited in Fig. 6. Namely, there is a one–parameter family of chattering trajectories through the origin. Each trajectory contains an infinite number of corner points corresponding to the switches of the control and these points are densest near the origin. The totality of the corner points of all trajectories fills a smooth one–dimensional manifold. The same is true for the trajectories in Fig. 8 with the difference that they escape from the origin in the direct time current.

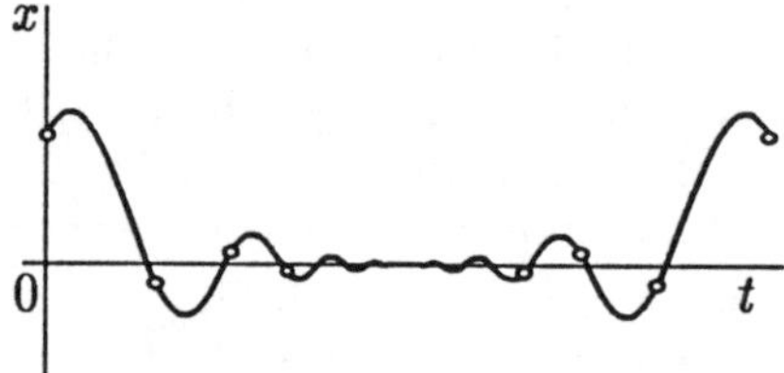

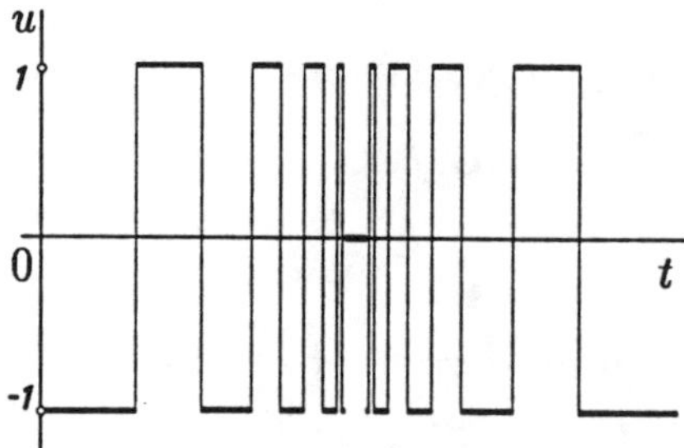

FIG. 4: THE CHATTERING TRAJECTORY AND THE
CORRESPONDING CHATTERING CONTROL

In this book we study the geometry of optimal chattering trajectories for problems, affine in the one–dimensional control u where $|u| \leqslant 1$. As a rule, module of the optimal control equals 1, but there are plenty of interesting physical and mechanical problems where intermediate values of the control $|u| < 1$ are optimal. Trajectories with this peculiarity have been called *singular*. For any singular trajectory some integer q, called the *order of the trajectory*, can be assigned. The exact definition will be given later. In this book we mainly deal with the case $q = 2$. Let us describe a typical example of optimal synthesis in an n–dimensional space. As a rule, singular trajectories of order 2 fill some manifold, call it $\mathcal{L}$, of dimension $n - 2$ (for example, a point in the plane, a curve in the three–dimensional space and so on). There exists one and only one optimal trajectory through each point outside $\mathcal{L}$. On the other hand, there are infinitely many optimal trajectories that enter into each $w \in \mathcal{L}$, and these trajectories establish a two–dimensional manifold $\mathcal{N}_w$. The behavior of the trajectories inside $\mathcal{N}_w$ is shown in Fig. 6. (The point w corresponds to the origin in Fig. 6.) The manifolds $\mathcal{N}_w$ do not intersect each other and fibrate some open neighborhood of $\mathcal{L}$. Further on, there is one and only one singular trajectory emanating from each w. Thus, optimal trajectories go along $\mathcal{N}_w$ with chattering mode, enter into w at finite time, and then proceed along $\mathcal{L}$ with singular mode. The totality of switching points of all trajectories constitute a smooth $(n - 1)$–dimensional manifold.

The three–dimensional case is shown in Fig. 17. Another example of a chattering synthesis can be generated if we use the two–dimensional manifolds with untwisted spirals as in Fig. 8.

The following sections discuss the algorithm of generating a synthesis with chattering arcs. The first step is to define singular extremals.

1.2 Hamiltonian Systems and Singular Extremals

Let us consider

PROBLEM 1.3. *Affine in control problem.*
Minimize

$$\int_0^T \left(f^0(x) + u f^1(x) \right) dt \tag{1.1}$$

subject to

$$\dot{x} = \varphi^0(x) + u\varphi^1(x), \quad |u| \leqslant 1, \tag{1.2}$$

$$x(0) = x_0, \quad x(T) \in \mathcal{M}.$$

Here $x = (x_1, \ldots, x_n) \in \mathbb{R}^n$ is an n–dimensional vector; f^0 and f^1 are scalar functions of x; φ^0 and φ^1 are n–dimensional vector–functions of x; $\mathcal{M}$ is a smooth target manifold in $\mathbb{R}^n$; x_0 is a fixed initial point. The value T in (1.1) is not fixed. The following optimality condition is well known [L.S. Pontryagin, et al., 1970].

Pontryagin's Maximum Principle

Assume that $\big(\widehat{x}(t), \widehat{u}(t)\big)$ is an optimal pair in Problem 1.3. Let $H = H(\psi, x, u)$ be the Pontryagin's function of the problem,

$$H = \sum_{i=1}^{n} \psi_i \left(\varphi_i^0(x) + u\varphi_i^1(x)\right) - \psi_0 \left(f^0(x) + u f^1(x)\right).$$

Then there exists a number $\psi_0 \geqslant 0$ and an absolutely continuous vector–function $\psi(t) = \big(\psi_1(t), \ldots, \psi_n(t)\big)$ (nonzero simultaneously) such that $\big(\widehat{x}(t), \widehat{u}(t)\big)$, and $\psi(t)$ satisfy the following conditions:

(a) **Hamiltonian system**

$$\begin{cases} \dot{\widehat{x}}(t) = \dfrac{\partial H}{\partial \psi}\big(\psi(t), \widehat{x}(t), \widehat{u}(t)\big), \\[2mm] \dot{\psi}(t) = -\dfrac{\partial H}{\partial x}\big(\psi(t), \widehat{x}(t), \widehat{u}(t)\big); \end{cases} \tag{1.3.a}$$

(b) **Maximum condition**

$$\max_{u \in [-1,1]} H\big(\psi(t), \widehat{x}(t), u\big) = H\big(\psi(t), \widehat{x}(t), \widehat{u}(t)\big) \equiv 0; \tag{1.3.b}$$

(c) **Transversality condition**

$$\sum_{i=1}^{n} \psi_i(T)\, \xi_i = 0 \tag{1.3.c}$$

for an arbitrary vector $\xi = (\xi_1, \ldots, \xi_n)$, tangential to the manifold $\mathcal{M}$ at the point $\widehat{x}(T) \in \mathcal{M}$.

Factors ψ, ψ_0 can be multiplied by an arbitrary positive constant. As a rule, we consider the case $\psi_0 \neq 0$, so we can take $\psi_0 = 1$, which will be tacitly assumed below. Everywhere in this book the variable x is referred to as the *state vector* and the n–dimensional space with coordinates x is called the *state* (or *phase*) *space*. The variable ψ is referred to as the *adjoint vector* of the problem. Let us call the $2n$–dimensional vector space with coordinates (ψ, x) as the *extended space*.

Since the function H is affine in u (i.e., contains u in at most the first power), we have

$$H = H_0(\psi, x) + u H_1(\psi, x)$$

where

$$H_0 = \sum_{i=1}^n \psi_i \varphi_i^0 - f^0, \qquad H_1 = \sum_{i=1}^n \psi_i \varphi_i^1 - f^1.$$

To maximize H in (1.3.b) one has to take $u = 1$ if $H_1\big(\psi(t), \widehat{x}(t)\big) > 0$ and $u = -1$ if $H_1\big(\psi(t), \widehat{x}(t)\big) < 0$. If we have

$$H_1\big(\psi(t), \widehat{x}(t)\big) = 0 \tag{1.4}$$

at some interval (t_0, t_1), then any value of $u \in [-1, 1]$ meets condition (1.3.b) and $\psi(t)$, $\widehat{x}(t)$ is called a *singular solution*. To find the value of the optimal control $\widehat{u}(t)$ at (t_0, t_1) one has to differentiate the identity (1.4) with respect to t.

We need a little digression at this point. Let $A = A(\psi, x)$ be an arbitrary smooth function. Let us differentiate it along a solution of system (1.3). We have

$$\frac{dA(\psi(t), \widehat{x}(t))}{dt} = \sum_{i=1}^n \left(\frac{\partial A}{\partial \psi_i} \frac{d\psi_i}{dt} + \frac{\partial A}{\partial x_i} \frac{dx_i}{dt} \right)$$

$$= \sum_{i=1}^n \left(-\frac{\partial A}{\partial \psi_i} \frac{\partial H}{\partial x_i} + \frac{\partial A}{\partial x_i} \frac{\partial H}{\partial \psi_i} \right).$$

The *Poisson bracket* of functions A and B is defined as

$$\{B, A\} \overset{\text{def}}{=} \sum_{i=1}^n \left(\frac{\partial B}{\partial \psi_i} \frac{\partial A}{\partial x_i} - \frac{\partial B}{\partial x_i} \frac{\partial A}{\partial \psi_i} \right).$$

Thus,

$$\frac{dA}{dt} = \{H, A\} \tag{1.5}$$

and the result of the differentiation may be written as a function of (ψ, x).

It is easy to see that the operator $\{\cdot, \cdot\}$ is linear in each of the two arguments $B(\cdot)$ and $A(\cdot)$ and that it is anticommutative,

$$\{B, A\} = -\{A, B\}.$$

Hence, $\{A, A\} \equiv 0$. For any functions A, B, C of (ψ, x) the following *Jacobi's identity* holds:

$$\{A, \{B, C\}\} + \{B, \{C, A\}\} + \{C, \{A, B\}\} \equiv 0. \tag{1.6}$$

Let us return to differentiating H_1 on a singular solution (1.4). We have

$$\frac{dH_1}{dt} = \{H_0 + \widehat{u}H_1, H_1\} = \{H_0, H_1\} + \widehat{u}\{H_1, H_1\} = \{H_0, H_1\}.$$

Since the function dH_1/dt does not depend on u, the equation

$$\{H_0, H_1\} = 0 \tag{1.7}$$

does not give any information about $\widehat{u}(t)$. Let us differentiate (1.7) with respect to t once more. We obtain

$$\frac{d^2 H_1}{dt^2} = \frac{d}{dt}\{H_0, H_1\} = \{H_0, \{H_0, H_1\}\} + \widehat{u}\{H_1, \{H_0, H_1\}\} = 0.$$

If $\{H_1, \{H_0, H_1\}\} \neq 0$, then

$$\widehat{u} = -\frac{\{H_0, \{H_0, H_1\}\}}{\{H_1, \{H_0, H_1\}\}}.$$

In this case $\widehat{x}(t)$ is called a *singular trajectory of order 1*. If we have

$$\{H_1, \{H_0, H_1\}\} \equiv 0, \tag{1.8}$$

for all (ψ, x), then we need to continue differentiating:

$$\frac{d^3 H_1}{dt^3} = \Big\{H_0, \{H_0, \{H_0, H_1\}\}\Big\} + \widehat{u}\Big\{H_1, \{H_0, \{H_0, H_1\}\}\Big\}.$$

It follows from Jacobi's identity (1.6) that

$$\Big\{H_1, \{H_0, \{H_0, H_1\}\}\Big\}$$

$$\equiv -\Big\{H_0, \{\{H_0, H_1\}, H_1\}\Big\} - \Big\{\{H_0, H_1\}, \{H_1, H_0\}\Big\}$$

$$\equiv -\Big\{H_0, \{\{H_0, H_1\}, H_1\}\Big\} \equiv \Big\{H_0, \{H_1, \{H_0, H_1\}\}\Big\}.$$

If condition (1.8) is valid, then

$$\Big\{H_0, \{H_1, \{H_0, H_1\}\}\Big\} = \{H_0, 0\} = 0.$$

In this case the function $d^3 H_1/dt^3$ does not contain the variable u. In general, it is possible to prove that the nonzero coefficient of u in the expression of $d^k H_1/dt^k$ can arise for the first time only when k is even ($k = 2q$). The integer q is called the *order of the singular trajectory*

$\widehat{x}(t)$. In this book we deal mostly with the case of singular trajectories of order 2. The interest in this case can be explained by the fact that the most mechanical optimal control problems belong to it. The reason is that the control variable u is usually included in the expression of the second derivative of state variables.

Thus, we have

$$\frac{d^4 H_1}{dt^4} = \Big\{ H_0, \big\{ H_0, \{ H_0, \{ H_0, H_1 \} \} \big\} \Big\}$$
$$+ \widehat{u} \Big\{ H_1, \big\{ H_0, \{ H_0, \{ H_0, H_1 \} \} \big\} \Big\},$$

and we suppose that the function $\Big\{ H_1, \big\{ H_0, \{ H_0, \{ H_0, H_1 \} \} \big\} \Big\}$ does not equal zero on $(\psi(t), \widehat{x}(t))$. Now the optimal control is equal to

$$\widehat{u}(t) = -\frac{\Big\{ H_0, \big\{ H_0, \{ H_0, \{ H_0, H_1 \} \} \big\} \Big\}}{\Big\{ H_1, \big\{ H_0, \{ H_0, \{ H_0, H_1 \} \} \big\} \Big\}}.$$

The necessary condition for optimality of a singular trajectory (*Kelley's condition*) is the following:

$$(-1)^q \frac{\partial}{\partial u} \frac{d^{2q} H_1}{dt^{2q}} \big(\psi(t), \widehat{x}(t)\big) \leqslant 0.$$

If $q = 2$, this gives

$$\Big\{ H_1, \big\{ H_0, \{ H_0, \{ H_0, H_1 \} \} \big\} \Big\} < 0. \qquad (1.9)$$

To meet the restriction that $|u| \leqslant 1$, one needs the inequality

$$\left| \Big\{ H_0, \big\{ H_0, \{ H_0, \{ H_0, H_1 \} \} \big\} \Big\} \right| \leqslant - \Big\{ H_1, \big\{ H_0, \{ H_0, \{ H_0, H_1 \} \} \big\} \Big\}.$$

Thus, if (1.8) is valid, then second order singular extremals lie in the manifold S that is defined in the extended space by the equations

$$H_1 = 0, \quad \{H_0, H_1\} = 0, \quad \{H_0, \{H_0, H_1\}\} = 0,$$
$$\{H_0, \{H_0, \{H_0, H_1\}\}\} = 0.$$

To investigate optimal trajectories in the vicinity of S, it is convenient to introduce new coordinates. This is the subject of the next section.

1.3 The Semi–Canonical Form of Hamiltonian Systems

We want to establish some standard form of the Hamiltonian system (1.3.a), (1.3.b) in the vicinity of the manifold of second order singular trajectories. Let us introduce a new coordinate system, setting for the first four coordinates

$$z_1 = H_1, \qquad z_3 = \{H_0, \{H_0, H_1\}\},$$
$$z_2 = \{H_0, H_1\}, \qquad z_4 = \Big\{H_0, \{H_0, \{H_0, H_1\}\}\Big\}. \tag{1.10}$$

In view of (1.5) and assumptions (1.8) and (1.9) we have

$$\dot{z}_1 = z_2, \qquad \dot{z}_3 = z_4,$$
$$\dot{z}_2 = z_3, \qquad \dot{z}_4 = \alpha(\psi, x) + u\,\beta(\psi, x), \tag{1.11}$$

where

$$\alpha = \Big\{H_0, \big\{H_0, \{H_0, \{H_0, H_1\}\}\big\}\Big\},$$
$$\beta = \Big\{H_1, \big\{H_0, \{H_0, \{H_0, H_1\}\}\big\}\Big\}.$$

Assume that the functions $(z_1, \ldots, z_4)$ are functionally independent in the vicinity of the trajectory $(\psi(t), \widehat{x}(t))$. We complement the coordinates $z = (z_1, \ldots, z_4)$ by functions $w = (w_1, \ldots, w_{2n-4})$ such that the Jacobi matrix of the mapping $(\psi, x) \rightarrow (z, w)$ is nondegenerate, i.e.,

$$\det \left| \frac{D(z, w)}{D(\psi, x)} \right| \neq 0.$$

Hence, the variables ψ, x can be expressed as functions of z, w. Substitution of these functions in (1.11) and differentiation of w give

$$\dot{z}_1 = z_2, \qquad \dot{z}_3 = z_4,$$
$$\dot{z}_2 = z_3, \qquad \dot{z}_4 = a(z, w) + u\,b(z, w), \tag{1.12.a}$$
$$\dot{w} = F(z, w, u),$$

where

$$u = \begin{cases} 1, & \text{if } z_1 > 0, \\ -1, & \text{if } z_1 < 0, \\ -a/b, & \text{if } z_1 = 0. \end{cases} \tag{1.12.b}$$

For Fuller's problem, the function (1.8) equals zero for all (ψ, x). However, for many applied optimal problems, equation (1.8) holds for the points of the singular trajectory $(\psi(t), \hat{x}(t))$ itself, but not for its neighborhood. In this situation the Hamiltonian system (1.3.a), (1.3.b) being expressed in coordinates (z, w) has somewhat different form. Namely, some additional terms appear on the right–hand side of (1.12.a). We can apply our technique to such a system only in the case where the additional terms are small enough in a certain sense.

To explain that, let us consider the simplest case of system (1.12), when $a \equiv a_0 = \mathrm{Const}$, $b \equiv b_0 = \mathrm{Const}$. Then the first four equations of (1.12.a) can be solved independently and the solutions have the following property. Let $(z_1(t), \ldots, z_4(t))$ be a solution of the system

$$\dot{z}_1 = z_2, \quad \dot{z}_2 = z_3, \quad \dot{z}_3 = z_4,$$

$$\dot{z}_4 = \begin{cases} a_0 + b_0, & \text{if } z_1 > 0 \\ a_0 - b_0, & \text{if } z_1 < 0. \end{cases} \tag{1.13}$$

Then it can readily be verified that for any $\lambda > 0$ the vector–function

$$\left(\lambda^4 z_1(t/\lambda), \quad \lambda^3 z_2(t/\lambda), \quad \lambda^2 z_3(t/\lambda), \quad \lambda z_4(t/\lambda) \right)$$

is also a solution of (1.13). To define the small perturbation of system (1.12) it is natural to assign the following weights $h(z_i)$ to each variable $z_1, \ldots, z_4$:

$$h(z_1) = 4, \quad h(z_2) = 3, \quad h(z_3) = 2, \quad h(z_4) = 1. \tag{1.14}$$

Let us assign the weight $h = 0$ to the variables $w_1, \ldots, w_{2n-4}$. Consider the system

$$\dot{z}_1 = z_2 + f_1(z, w, u), \qquad \dot{z}_3 = z_4 + f_3(z, w, u),$$

$$\dot{z}_2 = z_3 + f_2(z, w, u), \qquad \dot{z}_4 = a(z, w) + u\, b(z, w), \tag{1.15.a}$$

$$\dot{w} = F(z, w, u),$$

where

$$u = \begin{cases} 1, & \text{if } z_1 > 0, \\ -1, & \text{if } z_1 < 0, \\ -a/b, & \text{if } z_1 = 0. \end{cases} \tag{1.15.b}$$

We shall say that system (1.15) is a small perturbation of (1.13) if $f_i = o(z_{i+1})$, i.e.,

$$\lim_{\lambda \to +0} \frac{f_i(\lambda^4 z_1, \lambda^3 z_2, \lambda^2 z_3, \lambda z_4, w, u)}{\lambda^{4-i}} = 0. \tag{1.16}$$

In this case system (1.15) is called *semi–canonical*.

Let us discuss how to reduce Hamiltonian systems to the semi–canonical form. We have the following two cases.

1. If the function $\{H_0, \{H_0, H_1\}\}$ equals zero for all (ψ, x), we say that singular trajectories have the *intrinsic order two*. To reduce system (1.3) to (1.15) we take the function H_1 as a coordinate z_1, and we take its successive derivatives as z_2, z_3, z_4 (see (1.10)).

2. If the function $\{H_0, \{H_0, H_1\}\}$ equals zero on a singular trajectory itself but not in its neighborhood, we say that the singular trajectory has the *local order two*. To reduce (if possible) system (1.3) to (1.15) we can take H_1 and its successive derivatives as the new coordinates again. Sometimes, to obtain simpler equations, it is convenient to choose as new variables some functions which differ from (1.10) in terms of higher order in the sense of graduation (1.14). The choice of (z, w) is not unique and this is the reason why (1.15) is called semi–canonical rather than canonical. Our theory is applicable to systems that can be reduced to the semi–canonical form. Hence, it is necessary to check relations (1.16).

Now we can describe main results of the theory.

1.4 Integral Varieties with Chattering Arcs

Assume that system (1.3) can be reduced to (1.15). If Kelley's condition (1.9) is fulfilled, then we are able to describe the qualitative behavior of all trajectories of the Hamiltonian system (1.3) in the vicinity of the manifold

$$S = \big\{ \psi, x \mid z_1 = z_2 = z_3 = z_4 = 0 \big\}.$$

The strict statement is contained in Theorem 3.1 (the theorem on bundles). Here we explain the intuitive geometric sense of the result. It is found that for any $\sigma = (x_0, \psi_0) \in S$ there are two two–dimensional integral varieties of system (1.3), call them $\mathfrak{N}_\sigma^+$ and $\mathfrak{N}_\sigma^-$, mutually intersecting only at σ, such that the behavior of the solutions inside $\mathfrak{N}_\sigma^+$ and $\mathfrak{N}_\sigma^-$ is similar to the synthesis, exposed in Fig. 6 and in Fig. 8 respectively. The trajectories of $\mathfrak{N}_\sigma^+$ attain σ in a finite time with an infinite number of switches. The same is true for trajectories of $\mathfrak{N}_\sigma^-$ in the reverse time current.

The diagram in Fig. 5 reflects the disposition of the manifolds $\mathfrak{N}_\sigma^+$ and $\mathfrak{N}_\sigma^-$, but two–dimensional manifolds $\mathfrak{N}_\sigma^+$ and $\mathfrak{N}_\sigma^-$ are depicted as one–dimensional lines.

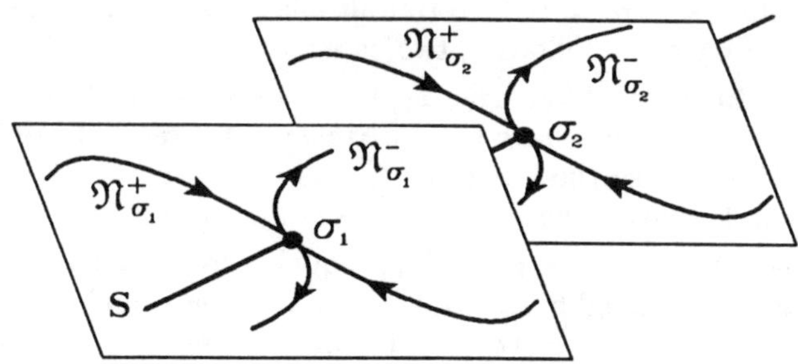

Fig. 5: Chattering bundles

REMARK 1.1. *Being homeomorphic to* $\mathbb{R}^2$, *manifolds* $\mathfrak{N}_\sigma^+$ *and* $\mathfrak{N}_\sigma^-$ *intersect each other at the point* σ *only, like two two-dimensional planes in* $\mathbb{R}^4$. *Since Fig. 5 could make a misleading impression, we call the reader's attention to the fact that the point* σ *does not divide* $\mathfrak{N}_\sigma^+$ *(and* $\mathfrak{N}_\sigma^-$*) into two parts.*

Thus, Theorem 3.1 describes the saddle structure of all solutions in the vicinity of S. According to Pontryagin's maximum principle, the optimal trajectories $x(t)$ in $\mathbb{R}^n$ are the projections of solutions of system (1.3) into x–space. To design an optimal synthesis we select the solutions of (1.3.a), (1.3.b) which satisfy the transversality condition (1.3.c). It is easy to see that the totality of such solutions constitutes some n–dimensional submanifold, call it L, of the space (ψ, x). Whatever the target $\mathcal{M}$, this submanifold has the following property: for any piecewise smooth closed curve $\gamma \subset L$

$$\oint_\gamma \psi \, dx = 0.$$

A submanifold with this property is called a *Lagrangian manifold.* So the manifold consisting of solutions of the Hamiltonian system with transversality condition (1.3.c) is a Lagrangian one. The projection of a smooth n–dimensional manifold is called *regular,* if it is a one–to–one correspondence and the Jacobian of this projection is nondegenerate.

Now we can formulate the following well–known sufficient optimality condition.

Let L be an n–dimensional smooth Lagrangian manifold, generated by solutions of Hamiltonian system (1.3) with $\psi_0 = 1$. Assume that the projection of L into some open neighborhood $U \subset \mathbb{R}^n$ of the target manifold $\mathcal{M}$ is regular. Then projections of the trajectories of L into U are locally optimal for Problem 1.3. This means that if $\big(x(t), u(t)\big)$ is an arbitrary solution of system (1.2) such that $x(t) \in U$ for $t \in [0, T]$ and $x(T) \in \mathcal{M}$, then

$$\int_0^T f\big(x(t), u(t)\big)\, dt \geqslant \int_0^{\widehat{T}} f\big(\widehat{x}(t), \widehat{u}(t)\big)\, dt.$$

Here $f(x, u) = f_0(x) + u f_1(x)$, $\widehat{x}(t)$ is the projection of the trajectory of L–family, $\widehat{x}(0) = x(0)$, $\widehat{x}(\widehat{T}) \in \mathcal{M}$; $\widehat{u}(t)$ is the control on $\widehat{x}(t)$.

Thus, to design the optimal synthesis it is sufficient to generate the corresponding Lagrangian manifold. The usual way to design a Lagrangian manifold is backward integration of the Hamiltonian system from the points of $\mathcal{M}$ (otherwise we cannot specify the initial values of the variable ψ). To do this one needs an interval of continuity of control between the switching points. If we wish to escape from a singular solution and to proceed with a chattering one, this procedure is inapplicable because of the lack of an interval of continuity of control at the junction point of the chattering and singular arcs.

We suggest an explicit procedure for generating a Lagrangian manifold L with chattering trajectories. The exact statement is contained in Theorem 3.2. Its sense can be explained in the following manner. Let L_0 be an $(n-2)$–dimensional Lagrangian submanifold of S. Consider the set of families $\mathfrak{N}_\sigma^+$ for all $\sigma \in L_0$ (i.e., the totality of chattering trajectories passing through the points $\sigma \in L_0$). Denote it by L. The manifold L is an n–dimensional submanifold of $\mathbb{R}^{2n}$, and it is proved in Theorem 3.2 that L is a Lagrangian manifold.

Unfortunately, the above–mentioned sufficient optimal condition is inapplicable to problems in question because the corresponding Lagrangian manifolds are only piecewise smooth. Theorem 3.3 gives a sufficient optimality condition in terms of a properly generalized regular projection. To use this theorem we have to calculate tangent planes to L. This is difficult enough at points of the singular manifold S because the switching points are densest in the vicinity of S. Theorem 3.4 gives an effective sufficient condition which provides the generalized regular projection of L onto $\mathbb{R}^n$.

We are left choosing a Lagrangian submanifold L_0 in S such that the transversality condition (1.3.c) is satisfied. To understand how to proceed, the best way is to examine a concrete problem.

1.5 An Example of Designing a Lagrangian Manifold

In Section 6.5, an economical problem of resource allocation is considered. It is reduced to the following formalization:

$$T \longrightarrow \min$$

subject to

$$\dot{x}_1 = y_1, \quad \dot{y}_1 = u_1 x_1 x_2, \quad \dot{x}_2 = y_2, \quad \dot{y}_2 = u_2 x_1 x_2 \qquad (1.17)$$

with boundary conditions

$$x(0) = (a, b), \quad y(0) = (c, d), \quad \big(x(T), y(T)\big) \in \mathcal{M}.$$

Here $x = (x_1, x_2) \in \mathbb{R}^2$, $y = (y_1, y_2) \in \mathbb{R}^2$ are the state variables and (u_1, u_2) are the control variables subjected to the constraints

$$u_1 \geqslant 0, \quad u_2 \geqslant 0, \quad u_1 + u_2 = 1;$$
$$x_1 \geqslant 0, \quad x_2 \geqslant 0, \quad y_1 \geqslant 0, \quad y_2 \geqslant 0.$$

Let the target manifold $\mathcal{M}$ be given by the equations

$$x_1 = x_2, \quad y_1 = y_2, \quad \frac{y_1 + y_2}{2} = f\left(\frac{x_1 + x_2}{2}\right) \qquad (1.18)$$

f being a C^1–function. The reason that the target manifold $\mathcal{M}$ has been given in this form will become clear later.

Actually, the control (u_1, u_2) is one–dimensional and we reduce the system to (1.2) by means of the following change of variables:

$$x = (x_1 - x_2)/2, \qquad p = (x_1 + x_2)/2,$$
$$y = (y_1 - y_2)/2, \qquad q = (y_1 + y_2)/2,$$
$$u = u_1 - u_2.$$

Now (1.17) is equivalent to the system

$$\dot{x} = y, \quad \dot{y} = u\frac{p^2 - x^2}{2}, \quad \dot{p} = q, \quad \dot{q} = \frac{p^2 - x^2}{2}, \quad u \in [-1, 1].$$

The first step is to find singular trajectories of the Hamiltonian system and to calculate their order. We have

$$H = H_0 + uH_1 \equiv 0,$$

where

$$H_0 = \psi_1 y + \psi_3 q + \psi_4 \frac{p^2 - x^2}{2} - 1, \quad H_1 = \psi_2 \frac{p^2 - x^2}{2}.$$

The Hamiltonian system is

$$
\begin{cases}
\dot{\psi}_1 = u\psi_2 x + \psi_4 x, \\[2mm]
\dot{\psi}_2 = -\psi_1, \\[2mm]
\dot{\psi}_3 = -u\psi_2 p - \psi_4 p, \\[2mm]
\dot{\psi}_4 = -\psi_3,
\end{cases}
\qquad
\begin{cases}
\dot{x} = y, \\[2mm]
\dot{y} = u\dfrac{p^2 - x^2}{2}, \\[2mm]
\dot{p} = q, \\[2mm]
\dot{q} = \dfrac{p^2 - x^2}{2},
\end{cases}
\qquad (1.19.\text{a})
$$

where

$$
u = \begin{cases}
1, & \text{if} \quad \psi_2\,(p^2 - x^2) > 0, \\[2mm]
-1, & \text{if} \quad \psi_2\,(p^2 - x^2) < 0.
\end{cases}
\qquad (1.19.\text{b})
$$

Assume that a trajectory of (1.19) meets the condition $H_1 = \psi_2\,(p^2 - x^2) = 0$ on some interval (t_0, t_1). In many cases we are only interested in finding singular trajectories in the region where (x, y)–variables are near the origin and (p, q)–variables are comparatively large. So we assume that $p^2 - x^2 \neq 0$ and hence $\psi_2 = 0$ at (t_0, t_1). Thus, $d\psi_2/dt = -\psi_1 = 0$, $d^2\psi_2/dt^2 = -u\psi_2 x - \psi_4 x = 0$, hence $\psi_4 x = 0$. If $\psi_4 = 0$, it follows from (1.19.a) that $\psi_3 = 0$, i.e., all variables ψ_1, ψ_2, ψ_3, ψ_4 are identically zero. This contradicts the condition of Pontryagin's maximum principle. It follows that $\psi_4 \neq 0$ on (t_0, t_1), hence, $x = 0$. Now (1.19.a) gives that $y = 0$ and $u = 0$. So in the region $p^2 - x^2 \neq 0$ singular solutions of (1.19) fill the manifold

$$
S = \{x = y = \psi_1 = \psi_2 = 0\}.
$$

According to the scheme given in Section 1.3, we could introduce the new variables as follows $\widetilde{z}_1 = \frac{1}{2}\psi_2\,(p^2 - x^2)$, $\widetilde{z}_2 = d\widetilde{z}_1/dt = -\frac{1}{2}\psi_1\,(p^2 - x^2) + \psi_2\,(pq - xy)$, and so on. Since

$$
\{H_1, \{H_0, H_1\}\} = -\psi_2 x(p^2 - x^2) = 0
$$

on S, singular trajectories have second order (see Section 1.3). However, it is more convenient to put

$$
z_1 = \psi_2, \quad z_2 = dz_1/dt = -\psi_1.
$$

Now $dz_2/dt = -\psi_4 x - u\psi_2 x$, where the term $-u\psi_2 x$ has a higher order (in the sense of (1.14)) than z_1. Besides that, we suppose $\psi_4 \neq 0$ and put

$$
z_3 = -x\psi_4, \quad z_4 = -y\psi_4
$$

i.e., $z_4 = dz_3/dt - x\psi_3$. We put

$$
w_1 = \psi_4, \quad w_2 = \psi_3, \quad w_3 = p, \quad w_4 = q.
$$

Then

$$
\begin{aligned}
\dot{z}_1 &= z_2, & \dot{w}_1 &= -w_2, \\
\dot{z}_2 &= z_3 + u\frac{z_1 z_3}{w_1}, & \dot{w}_2 &= -u z_1 w_3 - w_1 w_3, \\
\dot{z}_3 &= z_4 - \frac{z_3 w_2}{w_1}, & \dot{w}_3 &= w_4, \\
\dot{z}_4 &= \frac{u}{2}\left(\frac{z_3^2}{w_1} - w_1 w_3^2\right) - \frac{z_4 w_2}{w_1}, & \dot{w}_4 &= \frac{1}{2}\left(w_3^2 - \frac{z_3^2}{w_1^2}\right),
\end{aligned}
\tag{1.20}
$$

where

$$
u = \begin{cases}
1, & \text{if} \quad z_1 > 0, \\
-1, & \text{if} \quad z_1 < 0, \\
0, & \text{if} \quad z_1 = 0.
\end{cases}
$$

Let us check whether the additional terms of the right–hand side of (1.20) are of higher order. For $u z_1 z_3 / w_1$ we have

$$
\lim_{\lambda \to 0} \frac{1}{\lambda^2}\left(u\frac{\left(\lambda^4 z_1\right)\left(\lambda^2 z_3\right)}{w_1}\right) = 0
$$

(the sixth order term); for $-z_3 w_2 / w_1$ we have

$$
\lim_{\lambda \to 0} \frac{1}{\lambda}\left(-\frac{\left(\lambda^2 z_3\right) w_2}{w_1}\right) = 0
$$

(the second order term), and so on.

Let us consider the region where Kelley's condition (1.9) is fulfilled. This gives $w_1 > 0$. Now all the conditions of Theorem 3.1 are satisfied and there exist two different families of chattering trajectories of system (1.20) in an appropriate region.

The last step in our construction is choosing a Lagrangian manifold with the required properties. We are first to lift the target manifold $\mathcal{M}$,

$$
\mathcal{M} = \{x, y, p, q \mid x = y = 0, \; q = f(p)\},
$$

into the extended $\mathbb{R}^8$–space to meet the transversality condition. There exists an invariant construction of such a lifting for the given boundary value problem. Let us return to relations (1.18). One can see that the conditions $x_1 = x_2$, $y_1 = y_2$ in (1.18) imply that $\mathcal{M}$ belongs to the projection of the singular manifold S into $\mathbb{R}^4$. That is the reason why

$\mathcal{M}$ is taken in the form (1.18). Assume that the function $f(\cdot)$ in (1.18) has been chosen in such a way that

$$\frac{p^2}{2} - qf'(p) > 0. \tag{1.21}$$

This inequality guarantees that the projection of any singular trajectory on S intersects $\mathcal{M}$ transversally. Indeed, the vectors tangent to $\mathcal{M}$ at a point $(0, 0, p_0, q_0) \in \mathcal{M}$ are directed along the vector

$$v_1 = \big(0, 0, 1, f'(p_0)\big). \tag{1.22}$$

The (x, y, p, q)–component of the velocity on the singular trajectory of (1.19) equals

$$v_2 = \left(0, 0, q_0, \frac{p_0^2}{2}\right).$$

To have the vectors v_1 and v_2 not be proportional, one needs

$$\frac{1}{q_0} \neq \frac{2f'(p_0)}{p_0^2} \qquad \text{or} \qquad \frac{p_0^2}{2} - q_0 f'(p_0) \neq 0.$$

Set $X = (x, y, p, q)$. To lift $\mathcal{M}$ into $\mathbb{R}^8$, we associate the point $X \in \mathcal{M}$ with the point $\big(\psi(X), X\big) \in \mathbb{R}^8$, such that $\psi(X)$ meets the following two conditions. First, the image of $\mathcal{M}$ must belong to the intersection of the singular manifold S and the zero–level surface of the Hamiltonian, i.e.,

$$\psi_1 = \psi_2 = 0,$$
$$\psi_3 q + \psi_4 \frac{p^2}{2} - 1 = 0.$$

Secondly, the transversality condition has to be fulfilled,

$$\sum_{i=1}^{n} \psi_i(X) v_i = 0 \tag{1.23}$$

for any tangent vector v to $\mathcal{M}$ at $X \in \mathcal{M}$.

In view of (1.22), we obtain $\psi_3 + \psi_4 f'(p) = 0$, hence

$$\psi_1 = \psi_2 = 0, \quad \psi_3 = -\frac{f'(p)}{\dfrac{p^2}{2} - qf'(p)}, \quad \psi_4 = \frac{1}{\dfrac{p^2}{2} - qf'(p)}.$$

Now the inequality (1.21) implies Kelley's condition $\psi_4 > 0$.

Let us denote by $\mathcal{M}^*$ the curve $\big(\psi(X), X\big),\;\; X \in \mathcal{M}$. Consider the singular trajectories entering into $\mathcal{M}^*$. Let us denote by L_0 the two–dimensional manifold consisting of the arcs of these trajectories with $t < 0$. The theorem on integral invariant for Hamiltonian systems and the relation (1.23) imply that L_0 is Lagrangian. We have to consider the four–dimensional manifold generated by L_0 as in Theorem 3.2. We are left with verifying that this manifold has a regular projection to X–space. Checking the regularity is the most complicated piece of the construction. It may be shown that for the given problem the manifold does have a regular projection (for details, see Problem 6.5). This provides optimality of the chattering synthesis in the resource allocation problem, as required.

Chapter 2

FULLER'S PROBLEM

Chapter 2 is devoted to the solution of Fuller's problem. Though this problem has been seen as a starting point in the study of the chattering phenomenon and has been treated by many authors, we give here a self–contained presentation that is convenient for our purpose. We study this problem in such detail because our main goal is to investigate the class of problems that are its perturbation. It is customary to apply perturbation theory to linear problems. But Fuller's problem is a nonlinear one because it has a quadratic functional. Nor does it belong to the class of linear–quadratic problems because of the restriction $|u| \leqslant 1$ on the control. Nevertheless, the group of symmetries $\mathfrak{G}$ enables us to investigate systems that are "close" to Fuller's. The sense of the word "close" here is rather wide. We perturb Fuller's control system using terms which are small relative to its former right–hand side in the sense of generalized homogeneity induced by the action of the group $\mathfrak{G}$. At the same time we enlarge the number of state variables, adding new differential equations to the control system. Luckily, it appears that the obtained systems correspond to problems having singular trajectories of the second order.

2.1 Statement of Fuller's Problem

Let us consider the following

PROBLEM 2.1. **Fuller's problem.**

Minimize

$$\int_0^\infty |x|^q \, dt \qquad (q > 1) \tag{2.1}$$

subject to

$$\dot{x} = y, \quad \dot{y} = u, \quad u \in [-1, 1] \tag{2.2}$$

with initial conditions

$$x(0) = x_0, \quad y(0) = y_0.$$

The admissible controls need to be measurable, $u(\cdot) \in L_\infty[0, \infty)$, and the admissible trajectories must be absolutely continuous functions $(x(\cdot), y(\cdot))$.

The existence of the solution is proved in Chapter 5 for much more general situation (Lemma 5.1). Since system (2.2) is linear in x, y, u and integral (2.1) is strictly convex over solutions of system (2.2), a solution to Problem 2.1 is unique for any fixed (x_0, y_0). There is a *homogeneity group* that allows us to determine the optimal solutions to Problem 2.1 in an explicit form.

Let $(x(t), y(t))$ be an admissible solution of system (2.2) under the control $u(t)$. It can readily be seen that for any $\lambda > 0$ the pair

$$\big(x_\lambda(t), \, y_\lambda(t)\big) = \big(\lambda^2 x(t/\lambda), \, \lambda y(t/\lambda)\big) \tag{2.3}$$

is also an admissible solution of system (2.2) under the control $u_\lambda(t) = u(t/\lambda)$. It is convenient to introduce the following one–parameter group $\mathfrak{G} = \{g_\lambda\}$, $\lambda \in \mathbb{R}_+$, of transformations $\mathbb{R}^2 \to \mathbb{R}^2$:

$$g_\lambda(x, y) = (\lambda^2 x, \, \lambda y),$$

for any $(x, y) \in \mathbb{R}^2$.

LEMMA 2.1. [A.T. Fuller, 1961]. *Let* $\big(\hat{x}(t), \, \hat{y}(t), \, \hat{u}(t)\big)$ *be an optimal solution to Problem 2.1 with an initial point* (x_0, y_0). *Then for any* $\lambda > 0$ *the triple* $\big(\hat{x}_\lambda(t), \, \hat{y}_\lambda(t), \, \hat{u}_\lambda(t)\big)$ *is an optimal solution to Problem 2.1 with initial point* $g_\lambda(x_0, y_0)$.

Proof. The statement follows from the invertibility of mapping (2.3):

$$\big(x(t),\, y(t)\big) \;=\; g_{1/\lambda}\,\big(x_\lambda(\lambda t),\, y_\lambda(\lambda t)\big), \quad u(t) \;=\; u_\lambda(\lambda t),$$

and from the relation

$$\int_0^\infty \big|x_\lambda(t)\big|^q dt \;=\; \lambda^{2q+1}\int_0^\infty \big|x(t)\big|^q dt. \qquad \textbf{Q.E.D.}$$

While λ varies from 0 to $+\infty$, the point $g_\lambda(x_0, y_0)$ runs along the branch of the parabola

$$x \;=\; Cy^2, \tag{2.4}$$

where $C = x_0/y_0^2$ (if $y_0 \neq 0$) and y has the same sign as y_0 (either $y > 0$ or $y < 0$).

Since the solution to Problem 2.1 is unique, it follows from Lemma 2.1 that the optimal feedback control to Problem 2.1 is $\mathfrak{G}$–invariant. In particular, the switching set consists of some set of the branches of parabolas (2.4).

2.2 Chattering Arcs

LEMMA 2.2. [A.T. Fuller, 1961].

(1) *For all* $q > 1$ *the curve*

$$x \;=\; -Cy^2 \operatorname{sgn} y \tag{2.5}$$

is the optimal switching set of Problem 2.1. Here $C = C(q) \in (0, 1/2)$.

(2) *Twisting around the origin the optimal trajectories attain the origin in a finite amount of time and intersect the switching curve (2.5) at a countable set of points. The switching instants constitute a convergent geometric progression.*

(3) *The optimal feedback control equals 1 from the left–hand side of the switching curve (2.5) and equals −1 from the right–hand side of it (see Fig. 6).*

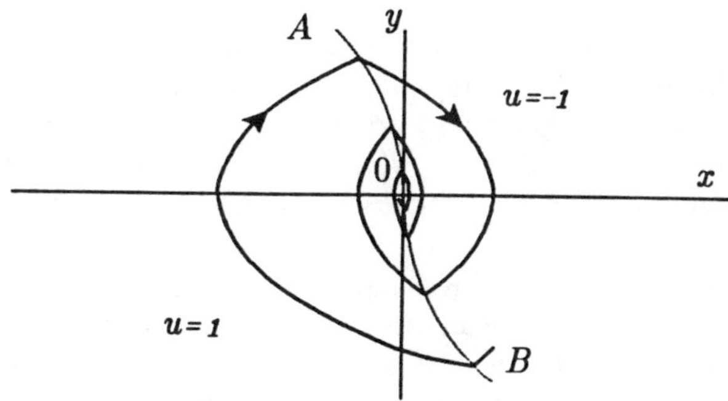

FIG. 6: THE OPTIMAL SYNTHESIS IN FULLER'S PROBLEM

Proof. Assume that $\big(x(t),\, y(t)\big)$ is an admissible solution of (2.2) with the control $u(t)$ such that $y(t) > 0$ while $t \in (t_1, t_2)$. Then $x(t)$ is monotonic and the value $x(t)$ can be taken as a parameter. The curve $\big(x(t),\, y(t)\big)$ can be given as $y = y(x)$. Then

$$\int_{t_1}^{t_2} |x|^q \, dt = \int_{x(t_1)}^{x(t_2)} |x|^q \, \frac{dx}{y}. \tag{2.6}$$

We see that the larger is the function $y = y(x)$, the smaller is the integral on the right–hand side of (2.6). Let $\mathcal{A}, \mathcal{B}$ be two boundary points of the trajectory $\big(x(t),\, y(t)\big)$: $\mathcal{A} = \big(x(t_1),\, y(t_1)\big)$, $\mathcal{B} = \big(x(t_2),\, y(t_2)\big)$. It follows from (2.2) that the larger is $\big|u(t)\big|$, the steeper is the curve $\big(x(t),\, y(t)\big)$ on the $(x,\, y)$–plane. So if either $u(t) \equiv 1$ or $u(t) \equiv -1$ almost everywhere for $t \in (t_1, t_2)$, then an admissible solution through $\mathcal{A}$ and $\mathcal{B}$ is unique in the halfplane $y > 0$. Otherwise, there exists a curve with a single control switch (namely, from $u = 1$ to $u = -1$) which is the "highest" among all admissible solutions of system (2.2) through $\mathcal{A}$ and $\mathcal{B}$ in the halfplane $y > 0$ (see Fig. 7).

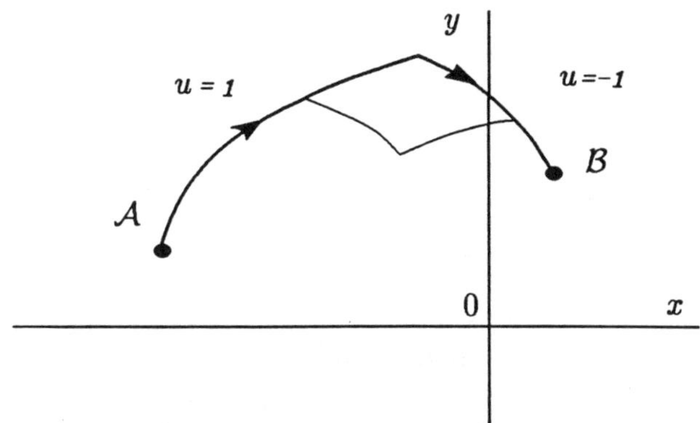

FIG. 7: THE "HIGHEST" TRAJECTORY

Thus, if $\big(x(t),\, y(t)\big)$ is optimal in Problem 2.1, then either there is no control switch while $t \in (t_1, t_2)$ or there is a single switch, namely, from 1 to -1. Symmetry arguments lead to the conclusion that if $\big(x(t),\, y(t)\big)$ is an optimal solution such that $y(t) < 0$ while $t \in (t_1, t_2)$, then the optimal control either has no switches at all or has a single switch from -1 to 1.

It follows from the $\mathfrak{G}$–invariance of the feedback control that the switching set of Problem 2.1 consists of two branches of parabolas, namely, $x = C_1 y^2$ if $y > 0$ and $x = C_2 y^2$ if $y < 0$. The mapping $(x,\, y) \mapsto (-x,\, -y)$ transfers any optimal solution of (2.2) to an optimal one, so $C_1 = -C_2 = C$. The equation $\dot{x} = y$ yields that any admissible trajectory of (2.2) moves to the right if $y > 0$ and moves to the left if $y < 0$. The optimal control switches from $u = 1$ to $u = -1$ at points of the upper branch of the curve (2.5). Consider an optimal trajectory emanating from this branch. On the initial part of it, the control is $u = -1$. If $C > 1/2$ then the trajectory does not intersect the switching curve (2.5) again. Hence, there is no other switch and the integral (2.1) is divergent which is impossible for the optimal trajectory. Thus, $C \leqslant 1/2$ and optimal trajectories cross the upper branch of the curve from the left to the right. Hence, $u = 1$ from the left–hand side of of the curve (2.5) and $u = -1$ from the right–hand side of it. The consideration for the lower branch of the curve (2.5) is quite similar to that of the upper branch. The assertion (3) of the lemma is proved.

If $C < 0$ then trajectories of the corresponding synthesis are represented by untwisted spirals and the functional (2.1) takes an infinite value contrary to the optimality. Thus, we have $0 < C \leqslant 1/2$.

To prove that $C \neq 1/2$ we need the following statement.

Remark 2.1 *In Chapter 1, Pontryagin's Maximum Principle (PMP) is formulated for an integral over a finite time interval. It is easy to prove that PMP holds true (except for the transversality condition) for an integral over an infinite time interval (see* [L.S. Pontryagin, et al., 1970]*).*

Let us demonstrate that $C \neq 1/2$. If $C = 1/2$, then the switching curve (2.5) is a solution of (2.2) with $u = -\operatorname{sgn} y$ and a trajectory of (2.5) moves along it to the origin. Let us show that motion along the curve (2.5) with $C = 1/2$ does not meet the condition of Pontryagin's maximum principle.

Assume that $\big(x(t), y(t)\big)$ is an optimal trajectory to Problem 2.1 with the control $u(t)$. Then there exist an absolutely continuous vector-function $\big(\psi_1(t), \psi_2(t)\big)$ and a number $\psi_0 \geqslant 0$ (simultaneously nonzero) such that

$$
\begin{aligned}
\dot{\psi}_1 &= \psi_0 \, q \, |x|^{q-1} \operatorname{sgn} x, & \dot{x} &= y, \\
\dot{\psi}_2 &= -\psi_1, & \dot{y} &= u(t)
\end{aligned}
\tag{2.7.a}
$$

(the Hamiltonian system), and

$$
\max_{u \in [-1,1]} u\psi_2(t) = u(t)\,\psi_2(t) \quad \text{(a.e.)}
\tag{2.7.b}
$$

(the maximum condition).

Suppose first that $\psi_0 = 0$. Then $\psi_1(t) \equiv \psi_{10}$, $\psi_2(t) = -\psi_{10}t + \psi_{20}$, where either $\psi_{10} \neq 0$ or $\psi_{20} \neq 0$, so there is no more than one switch of $u(t)$. In this case integral (2.1) is diverging, so $\psi_0 \neq 0$ and we can take

$$
\psi_0 = 1/q.
\tag{2.8}
$$

By assumption, the optimal control function $u(t)$ equals -1 at points of the curve $x = -\frac{1}{2}y^2$, $y > 0$. If we integrate (2.2) with $x_0 = -\frac{1}{2}y_0^2$, $y_0 > 0$, we see that the optimal solution $(x(t), y(t))$ meets the relations

$$\begin{cases} x(t) = -\frac{1}{2}y_0^2 + y_0 t - \frac{1}{2}t^2, \\ y(t) = y_0 - t, \end{cases} \quad \text{for} \quad t \in [0, y_0),$$

and

$$x(t) = y(t) \equiv 0, \quad \text{for} \quad t \geqslant y_0.$$

Hence $u \equiv 0$ for $t \geqslant y_0$. Assume that $(\psi_1(t), \psi_2(t))$ is a solution of system (2.7) associated with $(x(t), y(t))$. In view of (2.7), $\psi_1(t) = \psi_2(t) \equiv 0$ for $t \geqslant y_0$. Let $t \in [0, y_0)$. Since $x(t) < 0$, we have $\dot{\psi}_1(t) < 0$. Using the boundary condition $\psi_1(y_0) = 0$ we obtain that $\psi_1(t) > 0$, so $\dot{\psi}_2(t) < 0$ and $\psi_2(t) > 0$. But (2.7.b) implies that $u(t) = \operatorname{sgn} \psi_2(t) = -1$, hence $\psi_2(t) < 0$ while $t \in [0, y_0)$. The contradiction completes the proof of assertion (2).

Remark 2.2 *Let us mention that although the last consideration is almost trivial, its slightly complicated form has been used in a general case to demonstrate that it is impossible to match an optimal nonsingular arc under continuous control directly with a singular arc of second (or any even) order (see [C. Marchal 1971] and the theorem on conjugation in Chapter 3).*

Let us demonstrate that the optimal switching instants constitute a converging geometric progression. Assume that $\mathcal{A}_1$, $\mathcal{A}_2$, $\mathcal{A}_3$ are three points of successive switches of some optimal solution $(x(t), y(t))$ and their coordinates are the following:

$$(-Cy_1^2, y_1), \quad y_1 > 0 \quad \text{for} \quad \mathcal{A}_1,$$

$$(Cy_2^2, y_2), \quad y_2 < 0 \quad \text{for} \quad \mathcal{A}_2,$$

$$(-Cy_3^2, y_3), \quad y_3 > 0 \quad \text{for} \quad \mathcal{A}_3.$$

Integration of (2.2) gives

$$y_2 = y_1 - t_1, \qquad Cy_2^2 = -Cy_1^2 + y_1 t_1 - \frac{1}{2}t_1^2,$$

$$y_3 = y_2 + t_2, \qquad -Cy_3^2 = Cy_2^2 + y_2 t_2 + \frac{1}{2}t_2^2,$$

t_1, t_2 being the intervals between the switching instants. If we eliminate t_1 and t_2 between the equations, we obtain

$$\left(C + \frac{1}{2}\right) y_2^2 = -\left(C - \frac{1}{2}\right) y_1^2,$$

$$-\left(C + \frac{1}{2}\right) y_3^2 = \left(C - \frac{1}{2}\right) y_2^2.$$

Hence,

$$\frac{y_2}{y_1} = \frac{y_3}{y_2} = -\sqrt{\frac{1-2C}{1+2C}}. \tag{2.9}$$

Since $t_2/t_1 = -(y_3 - y_2)/(y_2 - y_1)$, the proportion's property yields

$$\frac{t_2}{t_1} = \sqrt{\frac{1-2C}{1+2C}} < 1. \tag{2.10}$$

It follows from (2.10) that every optimal pair $(x(t), y(t))$ attains the origin in finite time, so it is a chattering arc. **Q.E.D.**

Let us find the dependence of the constant C on q. Since optimal solutions of system (2.7) pass through the origin, all of them belong to the zero–level surface of the Hamiltonian, i.e., in accordance with (2.8),

$$\psi_1 y + |\psi_2| - \frac{|x|^q}{q} = 0. \tag{2.11}$$

It follows from PMP that $\psi_2 = 0$ at points of the switching curve. Hence, the following relations hold at points of the switching curve (2.5):

$$\psi_2 = 0, \quad \psi_1 = \frac{C^q}{q} y^{2q-1}, \quad x = -Cy^2 \operatorname{sgn} y. \tag{2.12}$$

Let the initial conditions ψ_{20}, ψ_{10}, x_0, y_0, $(y_0 > 0)$ of system (2.7.a) meet (2.12). Integrating (2.7.a) with $u(t) \equiv -1$ we obtain

$$\psi_2(t) = -\left(\frac{C^q}{q} y_0^{2q-1} t + \int_0^t (t-s) |x(s)|^{q-1} \operatorname{sgn} x(s)\, ds \right),$$

where $x(s) = -Cy_0^2 + y_0 s - \frac{1}{2}s^2$. Let $\tau > 0$ be the first instant when $\psi_2(\tau) = 0$. It follows from (2.9) that

$$\tau = y_0 \left(1 + \sqrt{\frac{1-2C}{1+2C}} \right). \tag{2.13}$$

The equation $x(s^*) = 0$ implies $s^* = y_0\left(1 - \sqrt{1-2C}\right)$ ($x(s) < 0$ for $0 < s < s^*$ and $x(s) > 0$ for $s^* < s < \tau$). Hence, $\tau > 0$ meets the equation

$$\frac{1}{q} C^q y_0^{2q-1} \tau - \int_0^{y_0(1-\sqrt{1-2C})} (\tau - s)\left(Cy_0^2 - y_0 s + \frac{s^2}{2} \right)^{q-1} ds$$

$$+ \int_{y_0(1-\sqrt{1-2C})}^{\tau} (\tau - s)\left(-Cy_0^2 + y_0 s - \frac{s^2}{2} \right)^{q-1} ds = 0.$$

If we take into account (2.13) and denote

$$\rho = 1 + \sqrt{\frac{1-2C}{1+2C}},$$

we arrive at the equation

$$\frac{1}{q}C^q\rho - \int_0^{1-\sqrt{1-2C}} (\rho - s)\left(C - s + \frac{s^2}{2}\right)^{q-1} ds$$

$$+ \int_{1-\sqrt{1-2C}}^{\rho} (\rho - s)\left(-C + s - \frac{s^2}{2}\right)^{q-1} ds = 0. \qquad (2.14)$$

It was shown in [A.T. Fuller, 1985] that there exists a unique solution $C(q)$ such that $0 < C < 1/2$, and analytical properties of the function $C(q)$ were analyzed. We limit ourselves to the case $q = 2$ as the most important for the following. Let us set

$$\mu = \rho - 1 = \sqrt{\frac{1-2C}{1+2C}}.$$

If $q = 2$, (2.14) gives the reciprocal equation

$$\mu^4 - 3\mu^3 - 4\mu^2 - 3\mu + 1 = 0.$$

The standard substitution $\sigma = \mu + 1/\mu$ yields

$$\sigma^2 - 3\sigma - 6 = 0,$$

hence, $\sigma_1 = \frac{1}{2}(3 + \sqrt{33})$, $\sigma_2 = \frac{1}{2}(3 - \sqrt{33})$. The root $\sigma_2 < 0$ is irrelevant because $\mu > 0$. There are two different values of the variable μ corresponding to σ_1:

$$\mu_1 = \frac{1}{4}\left(3 + \sqrt{33} - \sqrt{26 + 6\sqrt{33}}\right) \in (0, 1),$$

and

$$\mu_2 = \frac{1}{4}\left(3 + \sqrt{33} + \sqrt{26 + 6\sqrt{33}}\right) \in (1, \infty).$$

It follows that $\mu_2 = 1/\mu_1$. We have $C_1 = \frac{1}{2}(1 - \mu_1^2)/(1 + \mu_1^2) \approx 0.444623\ldots$ and

$$C_2 = \frac{1}{2}\frac{1 - \mu_2^2}{1 + \mu_2^2} = -C_1. \qquad (2.15)$$

2.3 Untwisted Chattering Arcs

It appears that the relation (2.15) follows from a general fact. Let us show that if some constant C_1 meets (2.14) for the given $q > 1$, then constant $C_2 = -C_1$ also meets (2.14). Assume that $(x(t), y(t), \psi_1(t), \psi_2(t))$ is a solution of (2.7). A straightforward calculation gives that $(x(-t), -y(-t), -\psi_1(-t), \psi_2(-t))$ is also a solution of (2.7). Hence, if curve (2.12) is the switching set for some family of solutions of system (2.7), then there exists another family of solutions whose switching curve is the following:

$$\psi_2 = 0, \quad \psi_1 = \frac{C^q}{q} y^{2q-1}, \quad x = Cy^2 \operatorname{sgn} y. \tag{2.16}$$

The trajectories of the latter family are called *untwisted chattering arcs*. They can be obtained by reflection of the trajectories of the former family with respect to the plane $y = 0$, $\psi_1 = 0$ and by inversion of the time current (see Fig. 8).

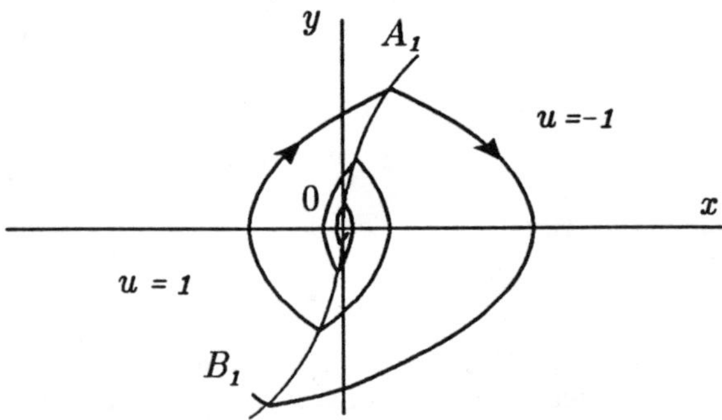

FIG. 8: UNTWISTED CHATTERING ARCS

Denote by $\mathfrak{N}^+$ the family of optimal solutions to Problem 2.1, whose switching set is the curve (2.12), and by $\mathfrak{N}^-$ the family whose switching set is (2.16). The following problem pertains to the $\mathfrak{N}^\pm$–families.

PROBLEM 2.2. *Minimize*

$$\int_0^{T_0} |x|^q \, dt$$

subject to

$$\dot{x} = y, \qquad \dot{y} = u, \qquad u \in [-1, 1]$$

with boundary conditions

$$x(0) = x_0, \quad y(0) = y_0, \quad x(T_0) = x_1, \quad y(T_0) = y_1. \qquad (2.17)$$

The terminal time $T_0 > 0$ is fixed.

LEMMA 2.3. *For any tuple* (x_0, x_1, y_0, y_1) *there exists* $\widetilde{T} > 0$ *such that for any* $T_0 > \widetilde{T}$ *the optimal solution to Problem 2.2 is composed of three successively adjoined segments:*

(1) *a trajectory of the family* $\mathfrak{N}^+$ *with the initial point* (x_0, y_0) *and the endpoint at the origin,*
(2) *the trajectory* $x(t) \equiv y(t) \equiv 0$,
(3) *a trajectory of the family* $\mathfrak{N}^-$ *with the initial point at the origin and the endpoint* (x_1, y_1).

To prove Lemma 2.3 we need the following sufficient condition of optimality for linear control systems with convex functionals. For our purpose it is enough to state it in the simplest form.

Sufficiency theorem. *Consider the problem of minimizing*

$$\int_0^{T_0} f(r,t)\, dt$$

subject to

$$\dot{r} = A(t)r + h(u,t)$$

with boundary conditions

$$r(0) = r_0, \quad r(T_0) = r_1.$$

Here $r \in \mathbb{R}^n$, $u \in [-1, 1]$. *Functions* $A, h, f, \partial f/\partial r$ *are continuous, and the function* f *is convex in* r. *The terminal time* T_0 *is fixed.*

Assume that the triple $\big(\psi(t), r(t), u(t)\big)$ *is an admissible solution of the system*

$$\begin{cases} \dot{\psi} = -\psi A(t) + \dfrac{\partial f}{\partial r}\big(r(t),t\big), \\[2mm] \dot{r} = A(t)r + h\big(u(t),t\big) \end{cases}$$

(ψ is a row–vector). Suppose that the following maximum condition holds:

$$\psi(t)\, h\big(u(t),t\big) = \max_{u \in [-1,1]} \psi(t)\, h(u,t)$$

for almost all $t \in [0, T_0]$. *Then* $r(t)$ *is an optimal solution to the problem.* For the proof of the sufficiency theorem, see [E.B. Lee, L. Markus, 1967; Theorem 5 on p. 341].

Proof of Lemma 2.3. It can be argued in the same way as for Fuller's problem that there exists a unique solution to Problem 2.2 for all sufficiently large $T_0 > 0$. It follows from the sufficiency theorem that every solution of system (2.7) meeting (2.17) gives an optimal solution to Problem 2.2. It has just been demonstrated that an arbitrary $\mathfrak{N}^+$–trajectory starting at (x_0, y_0) attains the origin in finite time. Let us denote this time by $T(x_0, y_0)$. The symmetry arguments yield that the corresponding $\mathfrak{N}^-$–trajectory starting at the origin at $t = 0$ enters into (x_1, y_1) at the instant $t = T(x_1, -y_1)$. Now it is sufficient to take $\widetilde{T} \geqslant T(x_0, y_0) + T(x_1, -y_1)$. **Q.E.D.**

2.4 The Geometry of Trajectories of Hamiltonian Systems

Lemma 2.3 gives the optimal solutions of Problem 2.2 for all boundary conditions (x_0, y_0, x_1, y_1) such that $T_0 \geqslant \widetilde{T}(x_0, y_0, x_1, y_1)$. For other boundary conditions, the optimal solutions lie outside of $\mathfrak{N}^+$ and $\mathfrak{N}^-$. Using the results of [M. Zelikin, V. Borisov, 1991] we give a full description of solutions of system (2.7) on the zero–level surface of the Hamiltonian (2.11) (see (1.3.b)).

As was shown in Section 2.1, Fuller's problem admits a homogeneity group of symmetries $\mathfrak{G} = \{\mathfrak{g}_\lambda\}$. This group induces the homogeneity of the Hamiltonian system (2.7) with respect to the action of the group $\widetilde{\mathfrak{G}} = \{\widetilde{g}_\lambda\}$ of system (2.7),

$$\widetilde{g}_\lambda(\psi_2, \psi_1, x, y) \stackrel{\text{def}}{=} (\lambda^{2q}\psi_2, \lambda^{2q-1}\psi_1, \lambda^2 x, \lambda y), \quad \lambda > 0.$$

Set $\xi = (\psi_2, \psi_1, x, y)$ and write the right–hand side of system (2.7) by $\widetilde{v}(\xi) = (-\psi_1, |x|^{q-1}\operatorname{sgn} x, y, u)$. For any $\xi \neq 0$ the maximum condition (2.7.b) specifies the values of u as follows:

$$u = u(\xi) = \begin{cases} \operatorname{sgn}\psi_2, & \text{if} \quad \psi_2 \neq 0; \\ 1, & \text{if} \quad \psi_2 = 0, \psi_1 < 0 \\ & \text{or} \quad \psi_2 = \psi_1 = 0, x < 0 \\ & \text{or} \quad \psi_2 = \psi_1 = x = 0, y < 0; \\ 0, & \text{if} \quad \psi_2 = \psi_1 = x = y = 0; \\ -1, & \text{otherwise.} \end{cases} \tag{2.18}$$

One can check that if $\xi(t)$ is a solution of system (2.7), then $g_\lambda\big(\xi(t/\lambda)\big)$ is also a solution for any $\lambda > 0$.

There exists a standard procedure of a system's factorization under the homogeneity group action to describe the evolution of orbits of the group.

Let us apply this procedure to system (2.7) and call $\mathfrak{A}$ its result. The *quotient–system* $\mathfrak{A}$ is defined on the orbit–space $(\mathbb{R}^4\backslash\{0\})/\widetilde{\mathfrak{G}}$ within to orbital equivalence. What we wish to do is to define system $\mathfrak{A}$ in an explicit form. Let us take the three–dimensional unit sphere

$$S = \left\{\, \xi \mid \psi_2^2 + \psi_1^2 + x^2 + y^2 = 1 \,\right\}$$

as $(\mathbb{R}^4\backslash\{0\})/\widetilde{\mathfrak{G}}$ (the set of representatives of the coset classes). The intersection of S and a $\widetilde{\mathfrak{G}}$–orbit $\bigcup_{\lambda>0}\widetilde{g}_\lambda(\xi^0)$, $\xi^0 = (\psi_2^0, \psi_1^0, x^0, y^0)$, is given by the equation

$$\lambda^{4q}\left(\psi_2^0\right)^2 + \lambda^{4q-2}\left(\psi_1^0\right)^2 + \lambda^4\left(x^0\right)^2 + \lambda^2\left(y^0\right)^2 = 1. \tag{2.19}$$

If $\lambda > 0$, then the left–hand side of (2.19) is monotonic in λ. Therefore, for any $\xi^0 \neq 0$ there is a unique solution $\lambda = \Lambda(\xi^0)$ of system (2.19). Let us define the mapping

$$\mathbf{g} : \mathbb{R}^4\backslash\{0\} \longrightarrow S$$

as follows

$$\mathbf{g}(\xi) = \widetilde{g}_{\Lambda(\xi)}(\xi) = \left(\Lambda^{2q}\psi_2, \Lambda^{2q-1}\psi_1, \Lambda^2 x, \Lambda y\right).$$

Finally, define the vector field v_S of system $\mathfrak{A}$ on S in the following way:

$$v_S\bigl(\mathbf{g}(\xi)\bigr) = \frac{V(\xi)}{\Lambda(\xi)},$$

where $V(\xi) \overset{\text{def}}{=} D\mathbf{g}\big|_\xi \cdot \widetilde{v}(\xi)$. Here and everywhere below D means the differential of the mapping. Let us demonstrate the correctness of this definition. The differential $D\mathbf{g}$ is represented by the sum of the matrices M_1 and M_2,

$$M_1 = \begin{pmatrix} \Lambda^{2q} & 0 & 0 & 0 \\ 0 & \Lambda^{2q-1} & 0 & 0 \\ 0 & 0 & \Lambda^2 & 0 \\ 0 & 0 & 0 & \Lambda \end{pmatrix},$$

$$M_2 = \begin{pmatrix} 2q\,\psi_2\,\Lambda^{2q-1}\,\mathrm{grad}\,\Lambda \\ (2q-1)\,\psi_1\,\Lambda^{2q-2}\,\mathrm{grad}\,\Lambda \\ 2x\,\Lambda\,\mathrm{grad}\,\Lambda \\ y\,\mathrm{grad}\,\Lambda \end{pmatrix},$$

M_2 being the matrix of rank 1 whose rows are proportional to the co-ordinates of the vector grad Λ. It follows from (2.19) and the implicit function theorem that

$$\text{grad } \Lambda = -\frac{1}{\gamma}(\Lambda^{4q}\psi_2,\ \Lambda^{4q-2}\psi_1,\ \Lambda^4 x,\ \Lambda^2 y),$$

where $\gamma = 2q\Lambda^{4q-1}\psi_2^2 + (2q-1)\Lambda^{4q-3}\psi_1^2 + 2\Lambda^3 x^2 + \Lambda y^2$. So

$$V(\xi) = \Big(-\psi_1\Lambda^{2q} + 2q\psi_2\Lambda^{2q-1}\Delta;\ |x|^{q-1}\text{sgn } x\Lambda^{2q-1}$$
$$+ (2q-1)\psi_1\Lambda^{2q-2}\Delta;\ y\Lambda^2 + 2x\Lambda\Delta;\ u\Lambda + y\Delta \Big),$$

where we put

$$\Delta \overset{\text{def}}{=} \big\langle \text{grad } \Lambda,\ \widetilde{v} \big\rangle =$$
$$-\frac{1}{\gamma}\Big(-\Lambda^{4q}\psi_1\psi_2 + \Lambda^{4q-2}\psi_1|x|^{q-1}\text{sgn } x + \Lambda^4 xy + \Lambda^2 yu \Big).$$

In view of (2.19), for any $\lambda > 0$ we have

$$\Lambda\big(\widetilde{g}_\lambda(\xi)\big) = \frac{\Lambda(\xi)}{\lambda}.$$

A straightforward calculation yields

$$\big\langle \text{grad } \Lambda\big(\widetilde{g}_\lambda(\xi)\big),\ \widetilde{v}\big(\widetilde{g}_\lambda(\xi)\big) \big\rangle = \frac{1}{\lambda^2} \big\langle \text{grad } \Lambda(\xi),\ \widetilde{v}(\xi) \big\rangle,$$

so $V\big(g_\lambda(\xi)\big) = V(\xi)/\lambda$.

We shall call $\mathbf{Orb}\,\xi_0 = \bigcup_{\lambda>0} \widetilde{g}_\lambda(\xi_0)$ the orbit of a point ξ_0 under the $\widetilde{\mathfrak{G}}$–action. Thus, for all $\xi \in \mathbf{Orb}\,\xi_0$ vectors $V(\xi)/\Lambda(\xi)$ coincide and vector v_S is well–defined.

LEMMA 2.4. *The vector–field* v_S *has no fixed points on* S.

Proof. Let us demonstrate that $V(\xi) \neq 0$ for all $\xi \neq 0$. Suppose the contrary: $V(\xi) = 0$ for some $\xi \neq 0$. If $y = 0$, then, because of (2.18), the last coordinate of $V(\xi)$, which is equal to $u\Lambda + y \langle \text{grad } \Lambda, \widetilde{v} \rangle$, is nonzero. Therefore, $y \neq 0$ and $u\Lambda + y\langle \text{grad } \Lambda, \widetilde{v} \rangle = 0$. Hence, $\Delta = -u\Lambda/y$, implying

$$\Lambda^{2q}\Big(-\psi_1 + 2q\psi_2\Big(-\frac{u}{y}\Big)\Big) = 0,$$
$$\Lambda^{2q-1}\Big(|x|^{2q-1}\text{sgn } x + (2q-1)\psi_1\Big(-\frac{u}{y}\Big)\Big) = 0,$$
$$\Lambda^2\Big(y + 2x\Big(-\frac{u}{y}\Big)\Big) = 0.$$

Successive calculations yield

$$x = \frac{y^2}{2u},$$

$$\psi_1 = \frac{y\,|x|^{q-1}\,\mathrm{sgn}\,x}{(2q-1)u} = \frac{y^{2q-1}}{2^{q-1}(2q-1)},$$

$$\psi_2 = -\frac{\psi_1 y}{2qu} = -\frac{y^{2q}}{2^q(2q-1)qu}.$$

Since $\psi_2 \neq 0$, the last equation implies $\mathrm{sgn}\,u = -\mathrm{sgn}\,\psi_2$, which contradicts (2.18). We see that the vector–field v_S has no fixed points on S.

Q.E.D.

Let us denote by **Tor** the intersection of S and zero–level surface of the Hamiltonian (2.11). This manifold is given by the equations

$$\psi_2^2 + \psi_1^2 + x^2 + y^2 = 1,$$
$$\psi_1 y + |\psi_2| - \frac{|x|^q}{q} = 0. \tag{2.20}$$

Let us clear up its topological structure. The manifold **Tor** is non-smooth. We include it into a homotopy of smooth manifolds whose topological structure is known.

LEMMA 2.5. *The manifold **Tor** given by (2.20) is homeomorphic to a two–dimensional torus.*

Proof. Consider the following family of manifolds $T^{\alpha\beta}$:

$$\psi_2^2 + \psi_1^2 + x^2 + y^2 = 1, \tag{2.21.a}$$

$$\psi_1 y + \frac{|\psi_2|^\alpha}{\alpha} - \frac{|x|^\beta}{\beta} = 0. \tag{2.21.b}$$

We assert that if $\alpha > 1$, $\beta > 1$, then $T^{\alpha\beta}$ are all homotopic to **Tor**. Let us take the gradients of the left–hand sides of (2.21),

$$\frac{1}{2}\nabla_1 = (\psi_2, \psi_1, x, y), \quad \nabla_2 = \left(|\psi_2|^{\alpha-1}\mathrm{sgn}\,\psi_2,\ y,\ -|x|^{\beta-1}\mathrm{sgn}\,x,\ \psi_1\right),$$

and check whether they are independent at points of $T^{\alpha\beta}$. Suppose the contrary: $\frac{1}{2}\nabla_1 = k\nabla_2$ for some k. We have

$$\begin{aligned}
\psi_2 &= k|\psi_2|^{\alpha-1}\,\mathrm{sgn}\,\psi_2, \\
\psi_1 &= k\,y, \\
x &= -k\,|x|^{\beta-1}\,\mathrm{sgn}\,x, \\
y &= k\psi_1.
\end{aligned} \tag{2.22}$$

Since both vectors $\frac{1}{2}\nabla_1$ and ∇_2 are nonzero at $T^{\alpha\beta}$, we have $k \neq 0$. The first equation of (2.22) implies that $k > 0$ in the case $\psi_2 \neq 0$ and the third one implies that $k < 0$ in the case $x \neq 0$.

Assume that $\psi_2 = 0$ and $k < 0$. Then the second and the third equations of (2.22) imply $\psi_1 y \leqslant 0$, whereas (2.21.b) implies that $\psi_1 y - |x|^\beta/\beta = 0$; hence $x = 0$, $\psi_1 y = 0$. It follows that $x = y = \psi_1 = \psi_2 = 0$, which contradicts (2.21.a).

Assume that $x = 0$, $k > 0$. Now the second and the third equations of (2.22) give $\psi_1 y \geqslant 0$. Then (2.21.b) implies that $|\psi_2|^\alpha/\alpha + \psi_1 y = 0$ and $x = y = \psi_1 = \psi_2 = 0$. Therefore, systems (2.21) and (2.22) are not compatible. It follows that the manifolds defined by (2.21.a) and (2.21.b) are transversally intersected for all $\alpha > 1$, $\beta > 1$. Hence, the manifolds $T^{\alpha\beta}$ are all homotopic in the domain $\alpha > 1$, $\beta > 1$.

In the case $\alpha = 1$, $\beta > 1$, the previous consideration can be applied to each of two smooth submanifolds of $T^{1\beta}$, namely, to

$$Q_+ \stackrel{\text{def}}{=} \left\{ \psi_1 y + |\psi_2| - \frac{|x|^\beta}{\beta} = 0 \,\middle|\, \psi_2 \geqslant 0 \right\}$$

and to

$$Q_- \stackrel{\text{def}}{=} \left\{ \psi_1 y + |\psi_2| - \frac{|x|^\beta}{\beta} = 0 \,\middle|\, \psi_2 \leqslant 0 \right\}.$$

Manifolds Q_+ and Q_- intersect the surface S transversally, so $Q_+ \cap S$ and $Q_- \cap S$ are both smooth two–dimensional manifolds with common boundary on the section $\psi_2 = 0$. Hence, $T^{\alpha\beta}$ can be extended as the homotopy to the value $\alpha = 1$ with loss of smoothness.

If $\alpha = 1$, $\beta = q$, then $T^{\alpha\beta}$ coincides with **Tor**. If $\alpha = 2$, $\beta = 2$, then (2.21.b) gives the equation $\psi_1 y + \psi_2^2/2 - x^2/2 = 0$, whose left–hand side is represented by the quadratic form with signature 0. This form is associated with a toric cone in $\mathbb{R}^4$ whose intersection with the unit sphere is a two–dimensional torus.   **Q.E.D.**

Let us consider attracting sets of system $\mathfrak{A}$. It was proved above (Lemma 2.4) that there are no fixed points of the system $\mathfrak{A}$ on **Tor**. The manifold $\mathfrak{N}^+$ (and $\mathfrak{N}^-$) is homeomorphic to $\mathbb{R}^2$ since its projection onto (x, y)–space is a one–to–one correspondence. Hence, the image of $\mathfrak{N}^+$ (and $\mathfrak{N}^+$) under the canonical mapping $\mathbb{R}^4 \to S$ is a closed curve. Let us denote it by $\mathcal{Z}^+$ (and $\mathcal{Z}^-$ relatively). The curves $\mathcal{Z}^+$ and $\mathcal{Z}^-$ are cycles of system $\mathfrak{A}$. These cycles consist of two smooth parts. Each part corresponds to an interval where the control is constant (the corresponding orbit maps on itself after two switches). Since the two–dimensional manifolds $\mathfrak{N}^+$ (and $\mathfrak{N}^-$) intersect each other in $\mathbb{R}^4$ only at the origin, the curve $\mathcal{Z}^+$ does not intersect $\mathcal{Z}^-$.

LEMMA 2.6. *There are no* $\mathfrak{A}$*–cycles on* **Tor** *which differ from* $\mathcal{Z}^+$ *and* $\mathcal{Z}^-$.

Proof. Suppose the contrary: there exists an $\mathfrak{A}$–cycle $\mathcal{Z} \subset$ **Tor** which differs from $\mathcal{Z}^+$ and $\mathcal{Z}^-$. Assume that $W(t)$ is a solution of system (2.7) whose starting point $W(0)$ belongs to **Orb** $\mathcal{Z}$. Then $W(t_0) = \widetilde{g}_{\lambda_0}(W(0))$ for some $t_0 > 0$, $\lambda_0 > 0$. Since system (2.7) is $\mathfrak{G}$–homogeneous, the function $\widetilde{W}(t) = \widetilde{g}_{\lambda_0} W(t/\lambda_0)$ is a solution of (2.7) with $\widetilde{W}(0) = W(t_0)$. There is a unique local solution of system (2.7) at all points except for the origin, so

$$W(t + t_0) = \widetilde{g}_{\lambda_0} W(t/\lambda_0). \qquad (2.23)$$

Relation (2.23) means that the solution of system (2.7) with the initial point $W(t_0)$ can be obtained by contracting $W(t)$ along $\mathfrak{G}$–orbits with the coefficient of contraction λ_0. In particular, the next return of $W(t)$ at **Orb** $W(0)$ occurs at the instant $\lambda_0 t_0$ at the point $W(t_0 + \lambda_0 t) = \widetilde{g}_{\lambda_0^2} W(0)$. We have the following sequence of points, where $W(t)$ intersects **Orb** $W(0)$:

$$W(t_0) = \widetilde{g}_{\lambda_0} W(0), \quad W(t_0 + \lambda_0 t_0) = \widetilde{g}_{\lambda_0^2} W(0), \ldots$$

$$\ldots, W(t_0 + \lambda_0 t_0 + \ldots + \lambda_0^{n-1} t_0) = \widetilde{g}_{\lambda_0^n} W(0), \ldots.$$

Let us analyze three different possibilities: $(i)\, 0 < \lambda_0 < 1$, $(ii)\, \lambda_0 = 1$, and $(iii)\, \lambda_0 > 1$. In the first case, we have that $\lim_{n \to \infty} \widetilde{g}_{\lambda_0^n} W(0) = 0$, and the trajectory $W(t)$ enters into the origin at the instant

$$\widetilde{T} = t_0 + \lambda_0 t_0 + \cdots + \lambda_0^n t_0 + \ldots = \frac{t_0}{1 - \lambda_0}.$$

Let us consider Problem 2.2 with $T_0 = \widetilde{T}$, $x_1 = y_1 = 0$ and take as (x_0, y_0) the projection of $W(0)$ into (x, y)–space. Then the sufficiency theorem gives that $W(t)$ is an optimal solution to this problem. In view of Lemma 2.3, $\mathcal{Z}$ coincides with $\mathcal{Z}^+$.

If $\lambda_0 > 1$, then $\lim_{n \to -\infty} \widetilde{g}_{\lambda_0^n} W(0) = 0$ and $W(t)$ escapes from the origin at the negative instant $-\widetilde{\widetilde{T}}$, where

$$\widetilde{\widetilde{T}} = \frac{t_0}{\lambda_0} + \frac{t_0}{\lambda_0^2} + \ldots + \frac{t_0}{\lambda_0^n} + \ldots = \frac{\lambda_0 t_0}{\lambda_0 - 1}.$$

Hence, $W(t - \widetilde{\widetilde{T}})$ is an optimal solution to Problem 2.2 in which we set $T_0 = \widetilde{\widetilde{T}}$, $x_0 = y_0 = 0$ and take the projection of $W(0)$ as (x_1, y_1). In view of Lemma 2.3 $\mathcal{Z}$ coincides with $\mathcal{Z}^+$.

If $\lambda_0 = 1$, then $W(t)$ is an optimal solution to Problem 2.2 in which the boundary points $(x_0, y_0) = (x_1, y_1)$ are the projections of $W(0)$ and $T_0 = nt_0$ (n is any integer). If n is large enough, this contradicts Lemma 2.3.   **Q.E.D.**

COROLLARY 2.1. *α–limit set and ω–limit set of $\mathfrak{A}$–trajectories on* **Tor** *are $\mathcal{Z}^+$ and $\mathcal{Z}^-$.*

Indeed, **Tor** is a smooth manifold outside the curve

$$l \stackrel{\text{def}}{=} \mathbf{Tor} \bigcap \{\psi_2 = 0\}.$$

The vector–field v_S is also smooth at $\mathbf{Tor}\backslash l$. Since all singular solutions of system (2.7) coincide with the origin, any $\mathfrak{A}$–trajectory intersects l at a discrete set. Now the statement follows from the Poincaré–Bendixon theorem [E.A. Coddington, N. Levinson, 1955; Theorem 2.1 on pp. 391–392].

LEMMA 2.7. *$\mathcal{Z}^+$ is a unstable cycle, and $\mathcal{Z}^-$ is a stable one.*

Proof. Let us denote by M_1, M_2 the points where $\mathcal{Z}^+$ intersects l. Denote by $W_0(t)$ the solution of (2.7) with the initial point $W_0(0) \in$ **Orb** M_1. By the definition of the cycle $\mathcal{Z}^+$, the curve $W_0(t)$ returns at **Orb** M_1 with a single switch of the control. Since both $\psi_1(M_1)$ and $\psi_1(M_2)$ are nonzero, solution $W_0(t)$ intersects the plane $\psi_2 = 0$ with nonzero velocities and at nonzero angles. In this case, solutions of ordinary differential equations are smooth in the initial data, so there exists a neighborhood $U \subset l$ of M_1 such that for all $W \in \mathbf{Orb}\, U$ a solution of system (2.7) starting at W returns at $\mathbf{Orb}\, U_1$ with a single control switch ($U_1 \subset l$ being some open neighborhood of M_1). We can take the neighborhood U small enough to guarantee that the trajectories intersect the plane $\psi_2 = 0$ transversally, as $W_0(t)$ does. Assume that $W(t)$ is a solution of system (2.7) starting at $W(0) = W_0 \in U$. Denote by $T(W_0) > 0$ the instant of the first return of $W(t)$ to $\mathbf{Orb}\, U_1$. If U is sufficiently small, then $T(W_0)$ is continuous on U and there exists

$$\sup_{W_0 \in U} T(W_0) \leqslant T^* < +\infty.$$

Let us define the function

$$\Delta(W_0) \stackrel{\text{def}}{=} \Lambda(W_0)/\Lambda\big(W\big(T(W_0)\big)\big)$$

(function $\Lambda(W)$ having been derived from (2.19)). The function $\Delta(W)$ is constant on every $\widetilde{\mathfrak{G}}$–orbit and continuous everywhere in $\mathbf{Orb}\, U_1$. It follows from (2.9) that

$$\Delta(M_1) = \mu^2 = \frac{1 - 2C}{1 + 2C} \in (0, 1).$$

Let us decrease U to obtain the bound $\sup_{W \in U} \Delta(W) \leqslant \delta_0 < 1$ for some $\delta_0 > 0$.

Now we can return to the statement being proved. Assume that there is an $\mathfrak{A}$–trajectory $\widetilde{Z}$ on **Tor** whose ω–limit set coincides with $\mathcal{Z}^+$. Let $\widetilde{W}(t)$ be a solution of system (2.7) with $\widetilde{W}(0) \in \textbf{Orb}\,U$. Let $t_1, t_2, \ldots, t_n, \ldots$ be a sequence of successive instants at which $\widetilde{W}(t_n) \in \textbf{Orb}\,U$ and $\mathcal{Z}_n = \textbf{Orb}\,\widetilde{W}(t_n) \bigcap l$. It follows from Lemma 2.6 that $\lim_{n \to \infty} \mathcal{Z}_n = M_1$. Hence, without loss of generality, we can suppose that there is a unique control switch on $\widetilde{W}(t)$ between t_n and t_{n+1}. By the definition, we have

$$\Delta\big(\widetilde{W}(t_1)\big) \cdots \Delta\big(\widetilde{W}(t_n)\big) = \frac{\Lambda\big(\widetilde{W}(t_1)\big)}{\Lambda\big(\widetilde{W}(t_2)\big)} \cdots \frac{\Lambda\big(\widetilde{W}(t_n)\big)}{\Lambda\big(\widetilde{W}(t_{n+1})\big)},$$

and

$$\Lambda\big(\widetilde{W}(t_{n+1})\big) = \frac{\Lambda\big(\widetilde{W}(t_1)\big)}{\Delta\big(\widetilde{W}(t_1)\big) \cdots \Delta\big(\widetilde{W}(t_n)\big)} \geqslant \frac{\Lambda\big(\widetilde{W}(t_1)\big)}{\delta_0^n}.$$

Hence, $\lim_{n \to \infty} \Lambda\big(\widetilde{W}(t_n)\big) = \infty$ and $\lim_{n \to \infty} \widetilde{W}(t_n) = 0$.

Let us estimate the time of hitting the origin of $\widetilde{W}(t)$. The $\widetilde{\mathfrak{G}}$–homogeneity of system (2.7) yields

$$t_{n+1} - t_n \leqslant \frac{T^*}{\Lambda\big(\widetilde{W}(t_n)\big)} \leqslant \frac{T^* \delta_0^{n-1}}{\Lambda\big(\widetilde{W}(t_1)\big)}.$$

Hence,

$$\sum_{n=1}^{\infty} (t_{n+1} - t_n) \leqslant \frac{T^*}{\Lambda\big(\widetilde{W}(t_1)\big)} \cdot \sum_{n=1}^{\infty} \delta_0^{n-1} = \frac{T^*}{\Lambda\big(\widetilde{W}(t_1)\big)(1 - \delta_0)}.$$

It follows from the sufficiency theorem that $\widetilde{W}(t)$ is an optimal solution to Problem 2.2 where we set $T_0 = T^* \big/ \big(\Lambda\big(\widetilde{W}(t_1)\big)(1 - \delta_0)\big)$, $x_1 = y_1 = 0$ and take the projection of $\widetilde{W}(t_1)$ as (x_0, y_0). This contradicts Lemma 2.3. It follows that $\mathcal{Z}^+$ is the unique α–limit set of $\mathfrak{A}$–trajectories on **Tor**, so $\mathcal{Z}^+$ is an unstable cycle. Respectively, $\mathcal{Z}^-$ is a unique stable cycle on **Tor**.   **Q.E.D.**

Let us define an *angular metric* induced in $\mathbb{R}^4$ by the canonical mapping $\mathbb{R}^4 \to S$. It means that we measure the distance between points of $\mathbb{R}^4$ by the distance between their images in S (the ray's role belongs to $\widetilde{\mathfrak{G}}$–orbits).

Lemmas 2.5–2.7 give a complete description of the behavior of $\mathfrak{A}$–trajectories on **Tor**. Thus, there are two two–dimensional manifolds

$\mathfrak{N}^+$ and $\mathfrak{N}^-$, made up of the optimal chattering solutions to Problem 2.2. The trajectories inside $\mathfrak{N}^+$ attain the origin in finite time with an infinite number of switches at the direct time current (at the retrogressive time current for $\mathfrak{N}^-$, respectively). Any other solution of (2.7) on the zero–level surface of the Hamiltonian approaches in the angular metric the manifold $\mathfrak{N}^-$ as t tends to ∞ and the manifold $\mathfrak{N}^+$ as t tends to $-\infty$.

Chapter 3

SECOND ORDER SINGULAR EXTREMALS AND CHATTERING

3.1 Preliminaries

This chapter is the principal one. Here the main theorems are proved, namely, Theorem 3.1, concerned with the piecewise smooth chattering structure of solutions of discontinuous Hamiltonian systems, Theorem 3.2, which treats the Lagrangian manifolds of these systems, and Theorems 3.3, 3.4, on the optimality of chattering syntheses.

Let us consider

PROBLEM 3.1. **Affine in control problem.**
Minimize

$$\mathcal{J}\big(x(\cdot), u(\cdot)\big) = \int_0^T f_0(x) + u\, f_1(x)\, dt$$

subject to

$$\dot{x} = \varphi_0(x) + u\varphi_1(x), \qquad u \in [-1, 1],$$

with boundary conditions

$$x(0) = x_0, \qquad x(T) \in \mathcal{M}.$$

Here $x \in \mathbb{R}^n$, $f_i : \mathbb{R}^n \to \mathbb{R}$, $\varphi_i : \mathbb{R}^n \to \mathbb{R}^n$ $(i = 0, 1)$, the functions f_i, φ_i are smooth enough. Problem 3.1 is determined by the choice of an initial point x_0 and a smooth target manifold $\mathcal{M} \subset \mathbb{R}^n$. The admissible controls need to be measurable only, the admissible trajectories are assumed to be absolutely continuous.

While the target manifold $\mathcal{M}$ is fixed and the initial point x_0 varies in some open domain $U \subset \mathbb{R}^n$, optimal control becomes a function of the reference position of the trajectory $x(t)$ in U. This function is called *feedback control*. The set of corresponding trajectories gives a synthesis of optimal trajectories in Problem 3.1.

The equations of Pontryagin's maximum principle for Problem 3.1 can be written in the following form:

$$\dot{y} = I \, \mathbf{grad} \; \big(H_0(y) + uH_1(y)\big),$$

$$u = \mathrm{sgn} \; H_1(y),$$

$$(3.1)$$

where $y = (\psi, x) \in \mathbb{R}^{2n}$, $I = \begin{pmatrix} 0 & -E \\ E & 0 \end{pmatrix}$, E being the unit $(n \times n)$–matrix, $H_i(y) = \psi\varphi_i(x) - \psi_0 f_i(x)$, $i = 0, 1$.

Since system (3.1) has a discontinuity of the right–hand side, its solutions need a special definition. Everywhere in this book we use the standard definition proposed by A.F. Filippov [1985, Chapter 2.4, sec. 2, pp. 49–50], which is tacitly assumed below.

Remark 3.1. *Solutions of Differential Equations with Discontinuous Right–Hand Side.* *This book deals with the simplest case of equations with jumps on a smooth hypersurface. Let Σ_0 be a hypersurface,*

$$\Sigma_0 = \big\{r \in \mathbb{R}^d \mid \sigma(r) = 0\big\}$$

for $\sigma : \mathbb{R}^d \to \mathbb{R}^1$ some C^1–function. Consider a point $r^* \in \Sigma_0$ such that $\mathbf{grad} \; \sigma(r^*) \neq 0$. Then a small neighborhood V of r^* is partitioned into two parts:

$$\Sigma_+ = \big\{r \in V \mid \sigma(r) > 0\big\}$$

and

$$\Sigma_- = \big\{r \in V \mid \sigma(r) < 0\big\}.$$

Consider a differential equation

$$\dot{r} = f(r),\qquad\qquad (3.2)$$

where f has a smooth extension from $\mathbb{R}^d \setminus \Sigma_0$ to $\Sigma_0 \bigcup \Sigma_+$ and to $\Sigma_0 \bigcup \Sigma_-$. Assume that r^0 belongs to Σ_0 and put

$$f_+(r^0) \stackrel{\mathrm{def}}{=} \lim_{r \to r^0, \, r \in \Sigma_+} f(r),$$

$$f_-(r^0) \stackrel{\mathrm{def}}{=} \lim_{r \to r^0, \, r \in \Sigma_-} f(r).$$

An absolutely continuous function $r(t)$ *is called a solution of (3.2) iff*

$$\dot{r}(t) \in \mathcal{F}(r),$$

$$\mathcal{F}(r) = \begin{cases} f(r), & \text{if } r \in \Sigma_+ \bigcup \Sigma_-, \\ \displaystyle\bigcup_{0 \le \alpha \le 1} [\alpha f_+(r) + (1-\alpha)f_-(r)], & \text{if } r \in \Sigma_0. \end{cases}$$

Let us return to system (3.1). Since the terminal time T is free, optimal solutions to Problem 3.1 belong to the zero–level surface of the Hamiltonian, i.e.,

$$H_0(y) + |H_1(y)| \equiv 0.$$

We say that a number q is the *order* of a singular arc $y(t)$ (namely, its *intrinsic order*), iff

$$\frac{\partial}{\partial u} \frac{d^k}{dt^k} \bigg|_{(3.1)} H_1(y) = 0, \qquad k = 0, 1, \dots, 2q - 1,$$

$$\frac{\partial}{\partial u} \frac{d^{2q}}{dt^{2q}} \bigg|_{(3.1)} H_1(y) \ne 0 \tag{3.3}$$

for all $y \in \mathbb{R}^{2n}$ in some open neighborhood of the trajectory $y(t)$. If relations (3.3) are valid only at points of the trajectory $y(t)$ itself, then we say that q is a *local order* of the singular arc $y(t)$. For the definitions of order, see [H.J. Kelley, et al., 1967]; [R.M. Lewis, 1980]. It is proved that in both cases the number q is an integer and the following necessary optimality condition for the singular arc $y(t)$ holds,

Kelley's condition:

$$\mathcal{K}(y(t)) = (-1)^q \frac{\partial}{\partial u} \frac{d^{2q}}{dt^{2q}} \bigg|_{(3.1)} H_1(y(t)) \le 0.$$

Theorem on conjugation [H.J. Kelley, R.E. Kopp, H.G. Moyer]. *Assume that the singular solution* $y(t)$ *of system (3.1) at* $t \in (t_0, t_1)$ *is that of second intrinsic order (or that of an arbitrary even intrinsic order). Assume that* $y(t)$ *meets Kelley's condition in the form of the strict inequality* $\mathcal{K}(y(t)) < 0$. *If the singular control is a* C^∞–*function, then* $y(t)$ *cannot be matched directly with a nonsingular piecewise smooth solution of system (3.1) if the control is discontinuous at the conjugation point.*

For the proof, see [J.P. McDonell, W.F. Powers, 1971].

As the immediate consequence of the theorem on conjugation, we obtain that if a singular arc of even order is required to join with a nonsingular arc, then the last one is to contain an infinite number of switches of $u = \operatorname{sgn} H_1$, and the switching points accumulate at the conjugation point.

Denote by

$$ad_{H_0} H_1 = \sum_{i=1}^{n} \left(\frac{\partial H_0}{\partial \psi_i} \frac{\partial H_1}{\partial x_i} - \frac{\partial H_0}{\partial x_i} \frac{\partial H_1}{\partial \psi_i} \right)$$

the *Poisson bracket* of functions H_0, H_1. Assume that there exists a singular arc $y(t)$ of system (3.1) of intrinsic order 2 such that the functions

$$z_i \stackrel{\text{def}}{=} ad_{H_0}^{i-1} H_1 \quad (i = 1, \ldots, 4)$$

are independent on some open neighborhood of $y(t)$. Let us complement the collection $z_1, \ldots, z_4$ by functions $w_1, \ldots, w_{2n-4}$ in such a way that

$$\det \left. \frac{D(z, w)}{Dy} \right|_{y=y(t)} \neq 0,$$

where $z = (z_1, \ldots, z_4) \in \mathbb{R}^4$, $w = (w_1, \ldots, w_{2n-4}) \in \mathbb{R}^{2n-4}$. If we replace y by (z, w) in system (3.1), we obtain

$$\dot{z}_1 = z_2, \quad \dot{z}_2 = z_3, \quad \dot{z}_3 = z_4,$$

$$\dot{z}_4 = \alpha(z, w) + u\beta(z, w), \tag{3.4}$$

$$\dot{w} = \gamma(z, w, u), \quad u = \operatorname{sgn} z_1.$$

In view of the definition of the order of a singular arc, we have $\beta\big|_{y(t)} \neq 0$. Assume that $\beta < 0$, to meet Kelley's condition in its strict form. Assume that $|\alpha| < -\beta$ to meet the restriction that $|u| \leqslant 1$. Then Theorem 3.1 below implies that for any σ in some open region of $\mathbb{R}^{2n-4}$ there exist two families $\mathfrak{N}_\sigma^+$, $\mathfrak{N}_\sigma^-$ of chattering arcs of system (3.4) through points $z = 0$, $w = \sigma$.

3.2 Manifolds with Second Order Singular Trajectories

There are plenty of interesting problems where it is necessary to use chattering controls to hit the singular arcs manifold. These concern both singular arcs of second intrinsic order and those of second local order (in the last case, the theorem on conjugation is inapplicable). To cover the last class of problems we consider the following generalization of system (3.4):

$$\dot{z}_1 = z_2 + f_1(z, w, u), \qquad \dot{w} = F(z, w, u),$$

$$\dot{z}_2 = z_3 + f_2(z, w, u), \qquad u = \operatorname{sgn} z_1.$$

$$\dot{z}_3 = z_4 + f_3(z, w, u), \tag{3.5}$$

$$\dot{z}_4 = \alpha(w) + u\beta(w) + f_4(z, w, u),$$

Here $z = (z_1, z_2, z_3, z_4) \in \mathbb{R}^4$, $w = (w_1, \ldots, w_m) \in \mathbb{R}^m$. Let us consider the domain $|z| < \delta$, $|w - w_0| < \delta$ for some sufficiently small $\delta > 0$ and the functions $f_i \in C^{k_0+5-i}$ $(i = 1, \ldots, 4)$, α, β, $F \in C^{k_0}$ $(1 \leqslant k_0 \leqslant \infty)$.

Assume that $\beta < 0$, $|\alpha| < -\beta$. Functions f_i $(i = 1, \ldots, 4)$ are supposed to be small enough in the following sense. For any $\kappa > 0$, let g_κ be the mapping $\mathbb{R}^4 \to \mathbb{R}^4$ defined as

$$g_\kappa(z) = (\kappa^4 z_1,\ \kappa^3 z_2,\ \kappa^2 z_3,\ \kappa z_4).$$

Everywhere below functions f_i are subject to the following restrictions:

$$\varlimsup_{\kappa \to +0} \frac{|f_i(g_\kappa(z),\, w,\, u)|}{\kappa^{5-i}} < C, \tag{3.6}$$

with C being some positive constant. System (3.5), whose additional terms satisfy (3.6), can be considered a small perturbation of system (3.4) with respect to g_κ–action.

Remark 3.2. *To ascertain that system (3.1) can be reduced to the (3.5)–(3.6) form, we can proceed in the same way as for system (3.4). The function $H_1(\psi, x)$ can be taken as z_1. Since its first derivative along solutions of system (3.1) does not contain the control u, this derivative can be taken as z_2. However, sometimes it is convenient to take as z_2 some function differing from dH_1/dt in a quantity of order less than z_2, and so on.*

It follows from (3.6) that $f_i(0, w, u) = 0$, $i = 1, \ldots, 4$. Hence, for any point $(0, w)$ of the surface $S_0 \stackrel{\text{def}}{=} \{(z, w) \mid z = 0\}$, there exists a singular solution of system (3.5) through $(0, w)$. According to the definition, the velocity vector on the discontinuity hypersurface equals an appropriate linear combination of limits of velocity vectors from both sides of the hypersurface. So the control on the singular solution of system (3.5) is $u = -\alpha(w)/\beta(w)$ and the w–component of the velocity is the following:

$$\dot{w} = \frac{\beta(w) + \alpha(w)}{2\beta(w)} F(0, w, -1) + \frac{\beta(w) - \alpha(w)}{2\beta(w)} F(0, w, 1).$$

The common method of solving Cauchy's problem for a system with a discontinuous right–hand side consists of successive integrations of the system along adjoined segments on which the control is continuous. This method is inapplicable under chattering conditions because chattering arcs do not have a segment with a continuous control adjoined to a singular arc. The method which will now be described allows us to integrate system (3.5) along a chattering arc.

Denote by $S = \{(z, w) \mid z_1 = 0\}$ the switching surface of system (3.5). *Poincaré mapping* $\Phi: S \to S$ joins a point $(z_0, w_0) \in S$ together with a point $(z_1, w_1) \in S$ where the solution of system (3.5) starting at (z_0, w_0) meets S for the first time (at some positive instant $t > 0$). We are looking for a family of solutions which fill some two–dimensional manifold in $\mathbb{R}^{m+4}$ and pass through a fixed point $(0, w) \in S_0$. The intersection of this manifold with the switching surface is invariant with respect to Φ. However, the singularity (degeneration) of mapping Φ at points of S_0 obstructs the use of standard techniques to determine invariant manifolds of diffeomorphisms.

In this book we use the following standard statement of the theory of ordinary differential equations:

Invariant manifold theorem.

(1) Let $\Phi : \mathbb{R}^n \to \mathbb{R}^n$ be a C^{k_0}–diffeomorphism $(k_0 \geqslant 1)$ and let $x_0 \in \mathbb{R}^n$ be a fixed point of Φ. Let the Jacobian $D\Phi(x_0)$ have l_1 eigenvalues less than 1 in absolute value, and l_2 eigenvalues greater than 1 in absolute value $(l_1 + l_2 = n)$. Then there exists an l_1–dimensional C^{k_0}– manifold $\mathfrak{M} \subset \mathbb{R}^n$ $(x_0 \in \mathfrak{M})$ such that $\Phi(\mathfrak{M}) \subset \mathfrak{M}$ and the tangent space $T \overset{\mathrm{def}}{=} T_{x_0}\mathfrak{M}$ is $D\Phi(x_0)$–invariant. The restriction $D_T \overset{\mathrm{def}}{=} D\Phi(x_0)\big|_T$ of the differential $D\Phi(x_0)$ on T has l_1 eigenvalues less than 1 in absolute value.

(2) Let a matrix norm $\|D_T\|$ be less than 1. Then in some neighborhood $U \subset \mathfrak{M}$ of the point x_0 there exists a C^1–change of variables, call it $\phi : (U, x_0) \to (\mathbb{R}^{l_1}, 0)$, such that for any $y \in \mathbb{R}^{l_1}$, we have

$$\phi\Big(\Phi\big(\phi^{-1}(y)\big)\Big) = D_T \cdot y,$$

i.e., the restriction $\Phi\big|_{\mathfrak{M}}$ is C^1–equivalent to its linearization $D\Phi(x_0)\big|_T$ at the fixed point. In this case we shall say that $\mathfrak{M}$ is a contracting invariant manifold.

(3) If the mapping Φ C^{k_0}–depends on a parameter $\sigma \in \mathbb{R}^d$ $(d \geqslant 0)$, then the contracting manifold $\mathfrak{M}$ is C^{k_0} in σ.

For the proof of the theorem, see [P. Hartman, 1964; Theorem 5.1, p. 239] and [M.W. Hirsh, et al., 1977; Theorem 4.1, p. 39].

We resolve the singularity of the mapping Φ at points of S_0 by means of the following *blowing–up procedure*. Let $(0, \sigma) \in S_0$ be a fixed point of Φ. Let us change the variables in the region $S \backslash \{z_4 = 0\}$ as follows:

$$(z, w) \longrightarrow (\kappa, \lambda, \mu, \nu)$$

where

$$z_4 = \kappa, \quad z_3 = \lambda\kappa^2, \quad z_2 = \mu\kappa^3, \quad w = \sigma + \kappa\nu. \qquad (3.7)$$

Here $\kappa, \lambda, \mu \in \mathbb{R}$, $\sigma, \nu \in \mathbb{R}^m$. The values of σ can be considered coordinates on the surface S_0, i.e., $(z, w) \in S_0$ iff $z = 0$, $w = \sigma$. Transformation (3.7) is a one–to–one correspondence if $z_4 \neq 0$ and associates the point $(0, \sigma)$ with the whole $(m + 2)$–dimensional surface $\kappa = 0$. We call $\widehat{\Phi}$ the lifting of the mapping Φ to $(\kappa, \lambda, \mu, \nu)$–space. The mapping $\widehat{\Phi}$ can be continuously prolonged from the region $\kappa \neq 0$ to its boundary $\kappa = 0$. We arrive at the situation described by the invariant manifold theorem. This blowing–up procedure can be applied to any σ in some open region of $\mathbb{R}^m$. It will be proved that the mapping $\widehat{\Phi}^2$ is a diffeomorphism in the vicinity of its fixed point, lying on the surface $\kappa = 0$. Further on, there exists a piecewise smooth $\widehat{\Phi}^2$–invariant contracting manifold. This manifold generates the switching curve of the chattering arcs family, as required.

3.3 The Connection with Fuller's Problem

Consider Pontryagin's system for Fuller's problem (Problem 2.1 in Chapter 2):

$$\begin{aligned}
\dot{\psi}_2 &= -\psi_1, & \dot{x} &= y, \\
\dot{\psi}_1 &= x, & \dot{y} &= \operatorname{sgn} \psi_2.
\end{aligned} \tag{3.8}$$

It was proved in Chapter 2 that system (3.8) is homogeneous under the action of the group $\widetilde{g}_\kappa$, where

$$\widetilde{g}_\kappa(\psi_2, \psi_1, x, y) = (\kappa^4 \psi_2, \kappa^3 \psi_1, \kappa^2 x, \kappa y), \quad \kappa > 0.$$

Denote by $S = \{ \psi_2, \psi_1, x, y \mid \psi_2 = 0 \}$ the switching surface of system (3.8). The Poincaré mapping $\Phi : S \to S$ carries an orbit

$$\boldsymbol{Orb}(\lambda_0, \mu_0) = \bigcup_{\kappa > 0} (0, \kappa^3 \mu_0, \kappa^2 \lambda_0, \kappa)$$

into some orbit $\boldsymbol{Orb}(\lambda_1, \mu_1)$. The mapping $(\lambda_0, \mu_0) \to (\lambda_1, \mu_1)$ can be considered as the residue class of the mapping Φ relative to $\widetilde{g}_\lambda$–action. The mapping $(\psi_1, x, y) \longrightarrow (\mu, \lambda, \kappa)$ transfers mutually tangent curves $\boldsymbol{Orb}(\lambda_0, \mu_0)$ and $\boldsymbol{Orb}(\lambda_1, \mu_1)$ to mutually exclusive straight lines $\lambda = \lambda_i$, $\mu = \mu_i$ $(i = 0, 1)$. System (3.5) is a small perturbation of system (3.8) with respect to $\widetilde{g}_\lambda$–action with an increase in the number of variables. An arbitrary small perturbation of system (3.8) does not preserve the homogeneity under $\widetilde{g}_\lambda$. However, the Poincaré mapping for the perturbed system coincides with the nonperturbed one up to terms of higher order and hence transfers the aforementioned class of tangent curves to mutually exclusive curves. So transformation (3.7) is a sort of resolution of the singularity of the mapping Φ for the perturbed problem at points of a singular manifold.

3.4 Resolution of the Singularity
of the Poincaré Mapping

LEMMA 3.1. *Assume that $(z(t), w(t))$ is a solution of system (3.7) with the initial point $z(0) = g_\kappa(z_0)$, $w(0) = w_0$, defined at $t \in (0, \kappa T)$ (κ is an arbitrary positive parameter). Then for any $T > 0$ there exist constants $\kappa_0 > 0$, $C_1 > 0$ such that*

$$\left| g_{1/\kappa}(z(t)) \right| < C_1$$

for any $0 < \kappa < \kappa_0$.

Proof. Set

$$\Lambda(\tau) = g_{1/\kappa} z(\kappa\tau)$$

and differentiate $\Lambda(\tau)$ with respect to τ along a solution of system (3.5). For the first three coordinates of vector Λ we obtain

$$\frac{d\Lambda_i}{d\tau} = \frac{1}{\kappa^{5-i}} \frac{d}{d\tau}\left(z_i(\kappa\tau)\right)$$
$$= \frac{1}{\kappa^{4-i}}\left(z_{i+1}(\kappa\tau) + f_i\big(z(\kappa\tau), w(\kappa\tau), u(\kappa\tau)\big)\right),$$

($i = 1, 2, 3$). In addition,

$$\frac{d\Lambda_4}{d\tau} = \frac{1}{\kappa} \frac{d}{d\tau}\left(z_4(\kappa\tau)\right)$$
$$= \alpha\big(w(\kappa\tau)\big) + u(\kappa\tau)\beta\big(w(\kappa\tau)\big) + f_4\big(z(\kappa\tau), w(\kappa\tau), u(\kappa\tau)\big).$$

In view of (3.6) and since $z(\kappa\tau) = g_\kappa\big(\Lambda(\tau)\big)$, we have

$$\frac{d\Lambda_i}{d\tau} = \Lambda_{i+1} + \kappa\widetilde{f}_i\big(\Lambda(\tau), w(\kappa\tau), u(\kappa\tau)\big), \quad i = 1, 2, 3,$$
$$\frac{d\Lambda_4}{d\tau} = \alpha\big(w(\kappa\tau)\big) + u(\kappa\tau)\beta\big(w(\kappa\tau)\big) + \kappa\widetilde{f}_4\big(\Lambda(\tau), w(\kappa\tau), u(\kappa\tau)\big),$$
$$\frac{d}{d\tau}\big(w(\kappa\tau)\big) = \kappa\widetilde{F}\big(\Lambda(\tau), w(\kappa\tau), u(\kappa\tau)\big).$$

Here $\widetilde{f}_i(\Lambda, w, u) = f_i\big(g_\kappa(\Lambda), w, u\big)/\kappa^{4-i}$ ($i = 1, 2, 3, 4$), $\widetilde{F}(\Lambda, w, u) = F\big(g_\kappa(\Lambda), w, u\big)$. Functions $\widetilde{f}_i$, $\widetilde{F}$ are continuous in Λ, w, u and uniformly bounded with respect to $\kappa \in [0, \kappa_0]$.

Let us set

$$\Lambda(\tau) = e^{A\tau}\left(z_0 + \xi(\tau) + \int_0^\tau e^{-As}v(s)\,ds\right),$$

where

$$A = \begin{pmatrix} 0 & 1 & 0 & 0 \\ 0 & 0 & 1 & 0 \\ 0 & 0 & 0 & 1 \\ 0 & 0 & 0 & 0 \end{pmatrix}, \quad v = \begin{pmatrix} 0 \\ 0 \\ 0 \\ \alpha(w(\kappa s)) + u(\kappa s)\beta(w(\kappa s)) \end{pmatrix}.$$

It follows that

$$\begin{aligned} \frac{d\xi(\tau)}{d\tau} &= \kappa e^{-A\tau}\widetilde{f}, \\ \frac{dw(\kappa\tau)}{d\tau} &= \kappa\widetilde{F}, \end{aligned} \tag{3.9}$$

where $\widetilde{f} = (\widetilde{f}_1,\ldots,\widetilde{f}_4)$, $\widetilde{F} = \widetilde{F}\left(e^{A\tau}\left(z_0 + \xi(\tau) + \int_0^\tau e^{-As}v(s)ds\right), w(\kappa\tau)\right)$, $\xi_i(0) = 0$ $(i = 1,\ldots,4)$.

We are to prove that for any $T > 0$ there exists $\kappa_0 > 0$ such that the solution $(\xi(\tau), w(\kappa\tau))$ of system (3.9) does not leave a restricted region $U_\lambda = \left\{(\xi, w) \mid \|\xi, w - w(0)\| < \lambda\right\}$ as long as $\tau \in [0, T]$, $\kappa \in (0, \kappa_0)$. By $\|\cdot\|$ we denote the usual Euclidean norm in $\mathbb{R}^{m+4}$.

Since functions $\widetilde{f}_i$ are uniformly bounded with respect to κ, there exists

$$\max\left\{\left\|(e^{-A\tau}\widetilde{f}, \widetilde{F})\right\| : \tau \in [0, T],\right.$$

$$\left.(\xi, w) \in U_\lambda, |u| \leqslant 1\right\} = C_2 > 0.$$

Since $d\|r(\tau)\|/d\tau = \langle dr(\tau)/d\tau, r(\tau)\rangle/\|r(\tau)\|$, we have

$$\left|\frac{d}{d\tau}\left\|(\xi(\tau), w(\tau) - w(0))\right\|\right| \leqslant C_2\kappa,$$

while $(\xi(\tau), w(\tau)) \in U_\lambda$. Hence, the solution $(\xi(t), w(\tau))$ does not leave the region U_λ at $\tau \in [0, T]$, while $\tau C_2\kappa < \lambda$. All that remains is to take $\kappa_0 < \lambda/(TC_2)$.   **Q.E.D.**

Let $(z_0, w_0) \in S$ and $(z_1, w_1) = \Phi(z_0, w_0)$. Denote by $T(z_0, w_0) > 0$ the instant of the first intersection of the solution of system (3.5) with S. Let us write the mapping Φ in terms of the coordinates $(\lambda, \mu, \kappa, \nu)$. The corresponding mapping we denote as $\widehat{\Phi}$. Set $(\mu_1, \lambda_1, \kappa_1, \nu_1) =$

$\widehat{\Phi}(\mu_0, \lambda_0, \kappa_0, \nu_0)$, $T(z_0, w_0) = \kappa_0 \tau_0$. Let us rewrite system (3.5) in the form of integral equations. We start from the equation for z_4–coordinate of the vector (z, w).

$$
\begin{aligned}
\kappa_1 = \kappa_0 + \int_0^{\kappa_0 \tau_0} &\left(\alpha \left(\sigma + \kappa_0 \nu_0 + \int_0^\tau F(z(s), w(s), u)\, ds \right) \right. \\
&+ u\beta \left(\sigma + \kappa_0 \nu_0 + \int_0^\tau F(z(s), w(s), u)\, ds \right) \\
&\left. + u f_4(z(\tau), w(\tau), u) \right) d\tau \\
= \kappa_0 &\left(1 + (\alpha(\sigma) + u\beta(\sigma))\tau_0 \right) + \kappa_0 Q_1,
\end{aligned}
$$

where

$$
\begin{aligned}
Q_1 = \int_0^{\tau_0} &\left(\alpha \left(\sigma + \kappa_0 \nu_0 + \int_0^{\tau \kappa_0} F(s)ds \right) - \alpha(\sigma) \right. \\
&\left. + u(\tau) \left(\beta \left(\sigma + \kappa_0 \nu_0 + \int_0^{\tau_0 \kappa_0} F(s)ds \right) - \beta(\sigma) \right) + u f_4(\kappa_0 \tau) \right) d\tau,
\end{aligned}
$$

$$
F(s) = F(z(s), w(s), u), \quad f_4(s) = f_4(z(s), w(s), u).
$$

In our case the solutions of differential equations are smooth in the initial data, so $(z(s), w(s))$ are C^{k_0}–functions of $(\mu_0, \lambda_0, \kappa_0, \nu_0, \tau_0, u)$, where the number k_0 is given by the smoothness of the right–hand side of system (3.5). Hence, function Q_1 is C^{k_0} in all variables. Let us prove that

$$
\varlimsup_{\kappa \to +0} \frac{1}{\kappa} |Q_1| < C_3,
$$

where the constant C_3 does not depend on the arguments of function Q_1. It follows from Lemma 3.1 that for any $T_0 > 0$, $\tau \in [0, T_0]$, we have

$$
\varlimsup_{\kappa \to +0} \frac{1}{\kappa} \left| f_4(z(\kappa \tau), w(\kappa \tau), u) \right| < C_4,
$$

where the constant C_4 can be chosen depending on T_0 only. In view of the mean value theorem, we obtain

$$
\left| \alpha \left(\sigma + \kappa_0 \nu_0 + \int_0^{\tau \kappa_0} F(s)ds \right) - \alpha(\sigma) \right|
$$

$$
\leqslant \|\alpha'\|_{C^0} \left| \nu_0 + \int_0^\tau F(s\kappa_0)ds \right| \kappa_0,
$$

$$\left| \beta\left(\sigma + \kappa_0\nu_0 + \int_0^{\tau\kappa_0} F(s)ds\right) - \beta(\sigma) \right|$$

$$\leq \|\beta'\|_{C^0} \left| \nu_0 + \int_0^{\tau} F(s\kappa_0)ds \right| \kappa_0.$$

Here, and everywhere below, $\Theta_N(\kappa)$ denotes a C^{k_0+N-1}-function of $(\mu, \lambda, \kappa, \tau, \nu, u)$ such that

$$\varlimsup_{\kappa\to+0} \kappa^{-N}\left|\Theta_N(\kappa)\right| < C,$$

C being some constant. For example, in this notation any C^{k_0-1}-function is $\Theta_0(\kappa)$. Hence, it has been proved that $Q_1 = \Theta_1(\kappa)$.

If we integrate the equation $\dot{z}_3 = z_4 + f_3(z, w, u)$, we obtain

$$\lambda_1\kappa_1^2 = \left(\lambda_0 + \tau_0 + (\alpha(\sigma) + u\beta(\sigma))\frac{\tau_0^2}{2}\right)\kappa_0^2 + Q_2\kappa_0^2 + Q_3\kappa_0$$

where

$$Q_2 = \int_0^{\tau_0} \int_0^{\tau} \left(\alpha\left(\sigma + \nu_0\kappa_0 + \int_0^{\tau'\kappa_0} F(s)ds\right) - \alpha(\sigma)\right.$$

$$\left. + u\left(\beta\left(\sigma + \nu_0\kappa_0 + \int_0^{\tau'\kappa_0} F(s)ds\right) - \beta(\sigma)\right) + f_4(\kappa_0\tau')\right)d\tau'd\tau,$$

$$Q_3 = \int_0^{\tau_0} f_3\bigl(z(\kappa_0\tau), w(\kappa_0\tau), u\bigr) d\tau.$$

It is easy to see that $Q_2 = \Theta_1(\kappa_0)$, $Q_3 = \Theta_2(\kappa_0)$. If we continue the integration, we obtain

$$\mu_1\kappa_1^3 = \left(\mu + \lambda_0\tau_0 + \frac{\tau_0^2}{2} + \left(\alpha(\sigma) + u\beta(\sigma)\right)\frac{\tau_0^3}{6}\right)\kappa_0^3$$

$$+ Q_4\kappa_0^3 + Q_5\kappa_0^2 + Q_6\kappa_0,$$

$$\nu_1\kappa_1 = \bigl(\nu_0 + \tau_0 F(0, \sigma, u)\bigr)\kappa_0 + Q_7\kappa_0.$$

The following equation implies that $\kappa_0\tau_0$ is the switching instant on a solution of system (3.5), i.e., $z_1(\kappa_0\tau_0) = 0$:

$$\kappa_0^4\tau_0\left(\mu_0 + \frac{\lambda_0\tau_0}{2} + \frac{\tau_0^2}{6} + (\alpha(\sigma) + u\beta(\sigma))\frac{\tau_0^3}{24}\right)$$

$$+ Q_8\kappa_0^4 + Q_9\kappa_0^3 + Q_{10}\kappa_0^2 + Q_{11}\kappa_0 = 0.$$

By Q_1, Q_2, Q_4, Q_7, Q_8 are denoted the functions of $\Theta_1(\kappa_0)$-form; by Q_3, Q_5, Q_9 are denoted $\Theta_2(\kappa_0)$-functions; by Q_6, Q_{10} are denoted $\Theta_3(\kappa_0)$-functions, and by Q_{11} is denoted some $\Theta_4(\kappa_0)$-function.

LEMMA 3.2. *Let* $G(\kappa, \Sigma) = \sum_{i=1}^{m} \kappa^{m-i+1} \phi_i(\kappa, \Sigma)$ *and* $\phi_i = \Theta_i(\kappa)$ *(for brevity, Σ denotes the tuple μ, λ, ν, τ). Then*

$$\kappa^{-m} G = \Theta_1(\kappa).$$

Proof. It can readily be seen that $\varliminf_{\kappa \to +0} \kappa^{-(m+1)} |G| < \infty$. We are left with proving that if $\phi = \Theta_{l+1}(\kappa)$, then $\kappa^{-l} \phi \in C^{k_0}$ for any $l \in \mathbb{N}$. By the definition of the $\Theta_{l+1}(\kappa)$–class, we have $\phi \in C^{k_0 + l}$, so Taylor's formula yields

$$\phi(\kappa, \Sigma) = \sum_{i=0}^{l-1} \frac{\partial^i}{\partial \kappa^i} \phi(0, \Sigma) \frac{\kappa^i}{i!} + \int_0^\kappa \frac{(\kappa - \Theta)^{l-1}}{(l-1)!} \frac{\partial^l}{\partial \kappa^l} \phi(\Theta, \Sigma) \, d\Theta$$

$$= \kappa^l \int_0^1 \frac{(1 - \Theta)^{l-1}}{(l-1)!} \frac{\partial^l}{\partial \kappa^l} \phi(\Theta \kappa, \Sigma) \, d\Theta.$$

(We use that the first $(l-1)$–derivatives of the function ϕ at $\kappa = 0$ are zero because $\kappa^{-l+1} \phi(\kappa, \Sigma) \to 0$ as $\kappa \to 0$.) Since $\partial^l \phi / \partial \kappa^l \in C^{k_0}$, we see that $\phi(\kappa, \Sigma) = \kappa^l \phi_1(\kappa, \Sigma)$ where $\phi_1 \in C^{k_0}$.

Q.E.D.

In view of Lemma 3.2 the mapping $\widehat{\Phi}$ can be rewritten in the following form:

$$\kappa_1 = \kappa_0(1 + A\tau_0) + \kappa_0 \Theta_1(\kappa_0),$$

$$\lambda_1 = \left(\lambda_0 + \tau_0 + \frac{A}{2} \tau_0^2 \right) \Big/ (1 + A\tau_0)^2 + \Theta_1(\kappa_0),$$

$$\mu_1 = \left(\mu_0 + \lambda_0 \tau_0 + \frac{1}{2} \tau_0^2 + \frac{A}{6} \tau_0^3 \right) \Big/ (1 + A\tau_0)^3 + \Theta_1(\kappa_0),$$

$$\nu_1 = \left(\nu_0 + \tau_0 F(0, \sigma, u) \right) \Big/ (1 + A\tau_0) + \Theta_1(\kappa_0),$$

$$0 = \mu_0 + \frac{1}{2} \lambda_0 \tau_0 + \frac{1}{6} \tau_0^2 + \frac{A}{24} \tau_0^3 + \Theta_1(\kappa_0),$$

where $A = \alpha(\sigma) + u\beta(\sigma)$.

Let $(\mu_2, \lambda_2, \kappa_2, \nu_2) = \widehat{\Phi}(\mu_1, \lambda_1, \kappa_1, \nu_1)$, $T(z_1, w_1) = \kappa_1 \tau_1$ (i.e., we consider the second iteration of the mapping $\widehat{\Phi}$). Consider the initial conditions $(\mu_0 \, \lambda_0, \kappa_0, \nu_0)$ such that $u = -1$ at $t \in (0, \kappa_0 \tau_0)$ and $u = 1$ at $t \in (\kappa_0 \tau_0, \kappa_0 \tau_0 + \kappa_1 \tau_1)$ for the corresponding solution of system (3.5). To obtain such a succession of switches it is necessary to require that $T(z_0, w_0) > 0$, $T(z_1, w_1) > 0$, and besides that, $\dot{z}_1(+0) < 0$ and $\dot{z}_1(\tau_0 \kappa_0 + 0) > 0$. Then for any sufficiently small κ_0, we have

$$\begin{aligned} \tau_0 \kappa_0 &> 0, & \mu_0 \tau_0 &< 0, \\ \tau_1 \kappa_1 &> 0, & \mu_1 \tau_1 &> 0. \end{aligned} \tag{3.10}$$

Now the mapping $\widehat{\Phi}^2$ is determined by the following system:

$$\kappa_{i+1} = \kappa_i(1 + A_i\tau_i) + \kappa_i\Theta_1(\kappa_i), \tag{3.11.a}$$

$$\lambda_{i+1} = \left(\lambda_i + \tau_i + \frac{A_i}{2}\tau_i^2\right)\Big/(1 + A_i\tau_i)^2 + \Theta_1(\kappa_i),$$

$$\mu_{i+1} = \left(\mu_i + \lambda_i\tau_i + \frac{1}{2}\tau_i^2 + \frac{A_i}{6}\tau_i^3\right)\Big/(1 + A_i\tau_i)^3 + \Theta_1(\kappa_i),$$

$$\nu_{i+1} = \left(\nu_i + \tau_i F(0, \sigma, u_i)\right)/(1 + A_i\tau_i) + \Theta_1(\kappa_i),$$

where $i = 0, 1$ and the functions $\tau_i = \tau_i(\mu_i, \lambda_i, \kappa_i, \nu_i)$ meet the equations

$$\mu_i + \frac{1}{2}\lambda_i\tau_i + \frac{1}{6}\tau_i^2 + \frac{A_i}{24}\tau_i^3 + \Theta_1(\kappa_i) = 0. \tag{3.11.b}$$

Here $u_0 = -1$, $u_1 = 1$, $A_0 = \alpha(\sigma) - \beta(\sigma)$, $A_1 = \alpha(\sigma) + \beta(\sigma)$. In view of the assumption that $|\alpha| < -\beta$, we have $A_0 > 0$, $A_1 < 0$. Everywhere in (3.11) we can set $A_0 = 1$, $A_1 = b < 0$ because it can be obtained by means of the following change of variables:

$$\mu' = A_0^2\mu, \quad \lambda' = A_0\nu, \quad \nu' = A_0\nu, \quad \tau' = A_0\tau, \quad b = A_1/A_0. \tag{3.12}$$

The mapping $\widehat{\Phi}^2$ is piecewise smooth. Let us consider the smooth mapping $\widetilde{\Phi}^2$ defined directly by equations (3.11). The mapping $\widetilde{\Phi}^2$ coincides with $\widehat{\Phi}^2$ at points meeting both (3.10) and the following supplementary condition: each of the equations (3.11.b) has no odd–order roots lying between 0 and τ_0 or 0 and τ_1 respectively.

We would like to find all fixed points of the mapping $\widetilde{\Phi}^2$ on the surface $\kappa = 0$. The substitution $\kappa = 0$, $\lambda_2 = \lambda_0$, $\mu_2 = \mu_0$, $\nu_2 = \nu_0$ into (3.11) gives

$$\lambda_1 = \frac{\lambda_0 + \tau_0 + \frac{1}{2}\tau_0^2}{(1 + \tau_0)^2}, \qquad \lambda_0 = \frac{\lambda_1 + \tau_1 + \frac{b}{2}\tau_1^2}{(1 + b\tau_1)^2}, \tag{3.13.a}$$

$$\mu_1 = \frac{\mu_0 + \lambda_0\tau_0 + \frac{1}{2}\tau_0^2 + \frac{1}{6}\tau_0^3}{(1 + \tau_0)^3}, \; \mu_0 = \frac{\mu_1 + \lambda_1\tau_1 + \frac{1}{2}\tau_1^2 + \frac{b}{6}\tau_1^3}{(1 + b\tau_0)^3},$$

$$\nu_1 = \frac{\nu_0 + \tau_0 F(0,\sigma,-1)}{1+\tau_0}, \qquad \nu_0 = \frac{\nu_1 + \tau_1 F(0,\sigma,1)}{1+b\tau_1},$$

where the functions $\tau_0 = \tau_0(\lambda_0,\mu_0)$ and $\tau_1 = \tau_1(\lambda_1,\mu_1)$ are solutions of the equations

$$\mu_0 + \frac{1}{2}\lambda_0\tau_0 + \frac{1}{6}\tau_0^2 + \frac{1}{24}\tau_0^3 = 0, \qquad (3.13.b)$$

$$\mu_1 + \frac{1}{2}\lambda_1\tau_1 + \frac{1}{6}\tau_1^2 + \frac{b}{24}\tau_1^3 = 0.$$

Since (3.13.b) does not contain the variable ν, the last two equations of (3.13.a) don't influence the (λ,μ)–coordinates of a fixed point and hence can be temporarily excluded.

3.5 The Connection with the Problem of C. Marchal

The solution of system (3.13) is associated with the solution to the following problem of A.T. Fuller – C. Marchal.

PROBLEM 3.2. *Minimize*

$$\int_0^\infty x^2\,dt$$

subject to

$$\dot{x} = y, \quad \dot{y} = u, \quad u \in [b,1],$$

$$x(0) = x_0, \quad y(0) = y_0.$$

Here $b < 0$ is an arbitrary real parameter. Problem 3.2 was treated first in [C. Marchal, 1973]. It appears that Problem 3.2 can be considered modelling for system (3.5) in the following sense: optimal solutions to Problem 3.2 determine the asymptotic behavior of chattering solutions of system (3.5) at their approach to the singular manifold S_0. A more significant question is the dependence of solutions to Problem 3.2 on the parameter b.

Pontryagin's maximum principle for Problem 3.2 leads to the equation

$$\dot{\psi}_1 = x, \quad \dot{\psi}_2 = -\psi_1, \quad \dot{x} = y, \quad \dot{y} = u, \qquad (3.14)$$

where

$$u = \begin{cases} 1, & \text{if } \psi_2 > 0, \\ b, & \text{if } \psi_2 < 0. \end{cases}$$

Let us set $z_1 = -\psi_2$, $z_2 = \psi_1$, $z_3 = x$, $z_4 = y$ and replace z_i, $i = 1, \ldots, 4$ with λ, μ, κ by means of change (3.7). It can be demonstrated by the same way as in Lemma 2.2 that the optimal switching curve in Problem 3.2 has the form

$$\psi_2 = 0, \quad \psi_1 = \begin{cases} \mu_0 y^3, & y < 0, \\ \mu_1 y^3, & y > 0, \end{cases} \quad x = \begin{cases} \lambda_0 y^2, & y < 0, \\ \lambda_1 y^2, & y > 0. \end{cases}$$

A straightforward calculation gives that the tuple $(\lambda_0, \lambda_1, \mu_0, \mu_1)$ meets system (3.13) in which the ν–variable is omitted.

In view of (3.10) we have that the least positive root $\tau_0 > 0$ and the lowest absolute value root $\tau_1 < 0$ of equations (3.11.b) correspond to the branch of the switching curve in the domain $\kappa_0 > 0$, $\kappa_1 < 0$. The inverse distribution of signs $\tau_0 < 0$, $\tau_1 > 0$ corresponds to the case $\kappa_0 < 0$, $\kappa_1 > 0$. Consider the mapping

$$\phi_+ : (\lambda_0, \mu_0) \rightarrow (\lambda_2, \mu_2),$$

having been specified by system (3.11.a) where $\kappa_i = 0$ and the functions $\tau_0 > 0$, $\tau_1 < 0$ are the lowest absolute value roots of (3.11.b). Denote by

$$\phi_- : (\lambda_0, \mu_0) \rightarrow (\lambda_2, \mu_2)$$

the mapping specified by system (3.11.a) where $\kappa_i = 0$ and the functions $\tau_0 < 0$, $\tau_1 > 0$ are the lowest absolute value roots of (3.11.b). The relation between ϕ_+ and ϕ_- can be explained as follows.

Let $Z(t) = \big(\psi_2(t), \psi_1(t), x(t), y(t)\big)$ be a solution of system (3.14). Then it can readily be proved that $\widetilde{Z}(t) = \big(\psi_2(-t), -\psi_1(-t), x(-t), -y(-t)\big)$ is a solution of (3.14) too. Assume that $t = 0$, $t = \tau_1 > 0$, $t = \tau_2 > \tau_1$ are three successive switching instants of the former solution $Z(t)$. Assume also that the switching points meet the relations

$$\psi_2 = 0, \quad \psi_1 = \mu_0 y^3, \quad x = \lambda_0 y^2, \quad y > 0$$

at the instant $t = 0$; the relations

$$\psi_2 = 0, \quad \psi_1 = \mu_1 y^3, \quad x = \lambda_1 y^2, \quad y < 0$$

at the instant $t = \tau_1$; and the relations

$$\psi_2 = 0, \quad \psi_1 = \mu_2 y^3, \quad x = \lambda_2 y^2, \quad y > 0$$

at the instant $t = \tau_2$. Then the solution $\widetilde{Z}(t)$ intersects these curves in reverse succession; namely, it meets the relations

$$\psi_2 = 0, \quad \psi_1 = \mu_2 y^3, \quad x = \lambda_2 y^2, \quad y < 0$$

at the instant $t = -\tau_2$; the relations

$$\psi_2 = 0, \quad \psi_1 = \mu_1 y^3, \quad x = \lambda_1 y^2, \quad y > 0$$

at the instant $t = -\tau_1$; and the relations

$$\psi_2 = 0, \quad \psi_1 = \mu_0 y^3, \quad x = \lambda_0 y^2, \quad y < 0$$

at the instant $t = 0$. It follows that

$$\phi_+(\lambda_0, \mu_0) = (\lambda_2, \mu_2)$$

(the Poincaré mapping along the solution $Z(t)$), and

$$\phi_-(\lambda_2, \mu_2) = (\lambda_0, \mu_0)$$

(the Poincaré mapping along the solution $\widetilde{Z}(t)$), i.e.,

$$\phi_+(\lambda, \mu) = \left(\phi_-\right)^{-1}(\lambda, \mu) \tag{3.15}$$

in an appropriate open domain.

3.6 Fixed Points of the Quotient Mapping

Denote by Ξ the tuple $(\lambda_0, \mu_0, \nu_0, \lambda_1, \mu_1, \nu_1)$.

LEMMA 3.3.

(1) *There exist at least two different solutions of system (3.13), call them* Ξ^+ *and* Ξ^-, *such that the following relations hold:*

$$\lambda_0^+ \in (0, \tfrac{1}{2}), \ \lambda_1^+ \in (\tfrac{1}{2b}, 0), \ \mu_i^+ = \tfrac{1}{2}(\lambda_i^+)^2 \quad (i = 0, 1)$$

for Ξ^+, *and*

$$\lambda_0^- \in (\tfrac{1}{2b}, 0), \ \lambda_1^- \in (0, \tfrac{1}{2}), \ \mu_i^- = \tfrac{1}{2}(\lambda_i^-)^2 \quad (i = 0, 1)$$

for Ξ^-.

(2) *Functions* $\tau_i(\lambda, \mu)$ $(i = 0, 1)$, *being defined by (3.13.b), are* C^{k_0} *at points* $(\lambda_i^\epsilon, \mu_i^\epsilon)$ $(\epsilon = $ "+" *or* "−"$)$, *and meet the inclusions*

$$\tau_0^+ \in (-2, -1), \quad \tau_1^+ \in \left(-1/b, -2/b\right),$$
$$\tau_0^- \in (-\infty, -2), \quad \tau_1^- \in (-2/b, \infty),$$

where $\tau_i^\epsilon = \tau_i(\lambda_i^\epsilon, \mu_i^\epsilon)$.

Proof. We look for solutions of system (3.13) on the zero–level surface of the Hamiltonian of system (3.14), i.e.,

$$\psi_1 y + |\psi_2| - \frac{1}{2}x^2 = 0. \tag{3.16}$$

The substitution of (3.7) into (3.16) gives $\mu_0 = \frac{1}{2}\lambda_0^2$, $\mu_1 = \frac{1}{2}\lambda_1^2$. Now (3.13) implies

$$\lambda_1 = \frac{\lambda_0 + \tau_0 + \frac{1}{2}\tau_0^2}{(1+\tau_0)^2},$$

$$12\lambda_0^2 + 12\lambda_0\tau_0 + 4\tau_0^2 + \tau_0^3 = 0, \tag{3.17.a}$$

and

$$\lambda_0 = \frac{\lambda_1 + \tau_1 + \frac{b}{2}\tau_1^2}{(1+b\tau_1)^2},$$

$$12\lambda_1^2 + 12\lambda_1\tau_1 + 4\tau_1^2 + b\tau_1^3 = 0. \tag{3.17.b}$$

Set

$$\zeta = 1 + b\tau_1, \quad \alpha = 2b\lambda_0 - 1, \quad \beta = 2b\lambda_1 - 1. \tag{3.18}$$

Equations (3.17.b) give

$$\zeta^2 = \beta/\alpha,$$
$$\zeta^3 + \zeta^2 + (1+6\beta)\zeta + 3\beta^2 = 0.$$

If we eliminate the parameter ζ, we get

$$\alpha\beta(1+3\alpha\beta)^2 = (6\alpha\beta + \alpha + \beta)^2. \tag{3.19}$$

If we put $b = 1$, $\tau_1 = \tau_0'$, $\lambda_1 = \lambda_0'$, $\lambda_0 = \lambda_1'$ in (3.18), then (3.17.a) gives

$$\alpha'\beta'(1+3\alpha'\beta')^2 = (6\alpha'\beta' + \alpha' + \beta')^2$$

where

$$\alpha' = 2\lambda_1 - 1 = \frac{\beta + 1 - b}{b},$$

$$\beta' = 2\lambda_0 - 1 = \frac{\alpha + 1 - b}{b}. \tag{3.20}$$

Set

$$X = \alpha\beta, \quad Y = \alpha + \beta. \tag{3.21}$$

The curve (3.19) on the (X, Y)–plane is given by the equation

$$X(1+3X)^2 = (6X + Y)^2. \tag{3.22}$$

System (3.21) is resolvable in α, β in the region

$$X \leqslant Y^2. \tag{3.23}$$

Transformation (3.20) can be rewritten in terms of X, Y as follows:

$$X' = \frac{1}{b^2}X + \frac{1-b}{b^2} + \left(\frac{1-b}{b}\right)^2,$$
$$Y' = \frac{1}{b}Y + 2\frac{1-b}{b}. \tag{3.24}$$

Hence, solutions of system (3.17) are associated with those points of the curve (3.22) that are transferred by mapping (3.24) to the curve (3.22) itself. The mapping (3.24) is affine. By means of the change

$$\xi = X + Y + 1, \quad \eta = Y + 2, \tag{3.25}$$

(3.24) can be rewritten as

$$\xi' = \frac{\xi}{b^2}, \quad \eta' = \frac{\eta}{b}. \tag{3.26}$$

On the (ξ, η)–plane, curve (3.22) is given by the equation

$$(\xi - \eta + 1)\left(3(\xi - \eta + 1) + 1\right)^2 = \left(6(\xi - \eta + 1) + \eta - 2\right)^2.$$

If we take $t^2 = \xi - \eta + 1$ as a parameter on the curve, we get its parametric representation,

$$\xi = 3t^3 - 5t^2 + t + 1,$$
$$\eta = 3t^3 - 6t^2 + t + 2. \tag{3.27}$$

Since the mapping (3.24) respects parabolas $\xi = C\eta^2$, we are left with locating the points (ξ', η') and (ξ, η) on the different sides of the O_x–axis lying in the intersection of (3.27) and some parabola $\xi = C\eta^2$. The parameter $b < 0$ and the coefficient C of the parabola are functions of the parameter t on the curve (3.27). Besides that, from (3.23) we obtain the restriction

$$4\xi \leqslant \eta^2.$$

One can check that

$$\frac{d\eta}{d\xi} = \frac{9t^2 - 12t + 1}{(9t - 1)(t - 1)},$$
$$\frac{d^2\eta}{d\xi^2} = \frac{2(9t^2 - 1)}{\left((9t - 1)(t - 1)\right)^3}.$$

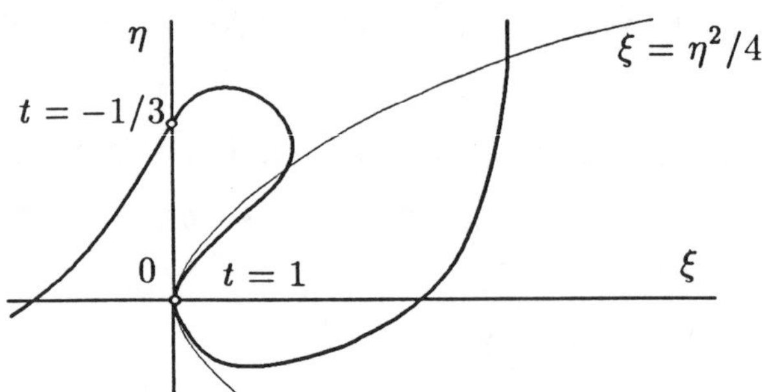

FIG. 9: TO THE PROOF OF SMOOTH DEPENDENCE
OF SOLUTIONS ON THE PARAMETER b

The graph of function $\eta(\xi)$ is shown in Fig. 9.

The values of the variable t at which the curve (3.27) intersects the parabola $\xi = \eta^2/4$ meet the equation

$$3t^2(t-1)^2(3t^2 - 6t - 1) = 0.$$

Hence, $t = 0,\ 1,\ -0.1547\ldots,\ 2.1541\ldots$. Therefore, for any $C < 1/4$, $C \neq 0$, parabolas $\xi = C\eta^2$ intersect the curve at exactly two points outside of the origin. If $0 < C < 1/4$, then the intersection points lie in the same half–plane $\eta > 0$. Hence, the points of interest lie in the half–plane $\xi < 0$. Let us demonstrate that for any $C < 0$ the parabola $\xi = C\eta^2$ intersects the curve (3.27) transversally (at points outside the origin). If the parabola $\xi = C\eta^2$ intersects (3.27) with tangency, then the equation $\xi(t) = C\eta^2(t)$ has multiple roots, i.e., $\xi'(t) = 2C\eta(t)\eta'(t)$. We have

$$3t^3 - 5t^2 + t + 1 = C(3t^3 - 6t^2 + t + 2)^2$$

$$9t^2 - 10t + 1 = 2C(3t^3 - 6t^2 + t + 2)(9t^2 - 12t + 1).$$

If we exclude C, we obtain $t(t-1)^3(9t+1) = 0$. If $t < -1/3$, then there are no multiple intersection points. Now the implicit function theorem says that functions $C(t),\ b(t),\ \tau(t)$ are C^∞. At $t < -1/3$ the functions $C(t),\ b(t),\ \tau(t)$ are uniquely defined by the following relations:

$$\eta(\tau) = \frac{\eta(t)}{b(t)}, \quad \xi(\tau) = \frac{\xi(t)}{b^2(t)}, \quad C(t) = \frac{\xi(t)}{\eta^2(t)}, \tag{3.28}$$

where $\big(\xi(t), \eta(t)\big)$ and $\big(\xi(\tau), \eta(\tau)\big)$ are the pairs of points of the curve (3.27) lying on the parabola $\xi = C\eta^2$. Let us demonstrate that the mapping $t \to b(t)$ is a one–to–one correspondence at $t < -1/3$. Since $d\eta/d\xi > 0$ at $t < -1/3$, the function $\eta(\xi)$ is monotonically increasing. Assume that there are two different values $t_1,\ t_2$ such that $t_1 < t_2 < -1/3$

and $b(t_1) = b(t_2)$. Without loss of generality we can assume $\eta(t_1) > 0$. Then it follows from (3.26) that

$$\xi(t_1) < \xi(t_2) < 0, \quad 0 < \eta(t_1) < \eta(t_2)$$

(the monotonicity of $\eta(\xi)$), and

$$\xi(\tau(t_1)) = \frac{1}{b^2}\,\xi(t_1) < \frac{1}{b^2}\,\xi(t_2) = \xi(\tau(t_2)),$$

$$\eta(\tau(t_1)) = \frac{1}{b}\,\eta(t_1) < \frac{1}{b}\,\eta(t_2) = \eta(\tau(t_2)),$$

where the function $\tau(t)$ is derived from (3.28). The last two inequalities contradict the increasing of $\eta(\xi)$ in the half–plane $\xi < 0$. Hence, the function $b(t) \in C^\infty$ is monotonically increasing from $-\infty$ to 0. The value t, at which $\eta(t) = 0$, corresponds to $b = -1$ $(t = -0.4574\ldots)$.

Let us demonstrate that solutions of (3.17) satisfy one of the following pairs of inequalities:
Either

$$0 < \lambda_0 < \frac{1}{2}, \qquad \frac{1}{2b} < \lambda_1 < 0 \qquad\qquad (3.29.\text{a})$$

or

$$\frac{1}{2b} < \lambda_0 < 0, \qquad 0 < \lambda_1 < \frac{1}{2}. \qquad\qquad (3.29.\text{b})$$

Parameters α, β appear in expression (3.19) in a symmetrical way. Hence, if the pair (λ_0, λ_1) meets (3.17), then the pair (λ_1, λ_0) also meets (3.17) at the same value of b. System (3.29) is equivalent to the following one:
Either

$$b - 1 < \alpha < -1, \qquad -1 < \beta < 0 \qquad\qquad (3.30.\text{a})$$

or

$$-1 < \alpha < 0, \qquad b - 1 < \beta < -1. \qquad\qquad (3.30.\text{b})$$

It was proved in Chapter 2 that if $b = -1$, then the relations (3.30) are valid (in this case we have $\lambda_0 = -\lambda_1 = 0.444623\ldots$). The function $b(t)$ being continuous, it remains to prove that solutions of (3.21) differ from each of the following three values: $\alpha = 0$, $\alpha = -1$, $\alpha = b - 1$. If $\alpha = 0$, then $X = 0$ (see (3.21)). But if $t < -1/3$, then $X = \xi - \eta + 1 = t^2 > 0$ (see (3.25)). If $\alpha = -1$, it follows from (3.21) that $\beta = -X$ and

$X + Y + 1 = 0$. Now (3.25) implies that $\xi = 0$. But if $t < -1/3$, then $\xi < 0$. Finally, consider the case $\alpha = b - 1$. If we substitute $\alpha = b - 1$ into (3.21) and take (3.25) into account, we obtain

$$(b - 1)^2 + (2 - \eta)(b - 1) + \xi - \eta + 1 = 0.$$

Since $b < 0$ and $\xi < 0$, we have

$$b = \frac{\eta}{2} - \sqrt{\frac{\eta^2}{4} - \xi}. \tag{3.31}$$

Assume that there are two values $t_2 < t_1 < -1/3$ such that $t_2 = \tau(t_1)$ and the values $\xi = \xi(t_1)$, $\eta = \eta(t_1)$, $b = b(t_1)$, being defined by (3.28), satisfy (3.31). We have $\eta(t_2) = \eta(t_1)/b(t_1)$ and (3.31) implies

$$b(t_1) = \frac{1}{2}\eta(t_1)\left(1 - \sqrt{1 - 4C(t_1)}\right).$$

Let $\rho = \eta(t_2)$; then $C(t_1) = (\rho - 1)/\rho^2$. Since

$$\frac{\xi(t_2)}{\eta^2(t_2)} = \frac{\xi(t_1)}{\eta^2(t_1)} = C(t_1),$$

it follows from (3.27) that

$$3t_2^3 - 5t_2^2 + t_2 + 1 = \rho - 1,$$

$$3t_2^3 - 6t_2^2 + t_2 + 2 = \rho.$$

Eliminating ρ between the equations, we obtain $t_2^2 = 0$. Hence, the function $b(t)$ does not meet (3.31) at $t < -1/3$. This proves (3.30).

Let $(\lambda_0, \lambda_1, \tau_0, \tau_1)$ be a solution of (3.17). Let us demonstrate, then, that either

$$0 < \lambda_0 < \frac{1}{2}, \qquad \frac{1}{2b} < \lambda_1 < 0,$$
$$-2 < \tau_0 < -1, \qquad -\frac{1}{b} < \tau_1 < -\frac{2}{b}, \tag{3.32.a}$$

or

$$\frac{1}{2b} < \lambda_0 < 0, \qquad 0 < \lambda_1 < \frac{1}{2},$$
$$\tau_0 < -2, \qquad -\frac{2}{b} < \tau_1. \tag{3.32.b}$$

Assume first that $0 < \lambda_0 < 1/2$, $1/(2b) < \lambda_1 < 0$. It follows from (3.17.a) that $\lambda_0 + \tau_0 + \frac{1}{2}\tau_0^2 < 0$, hence $\tau_0 + \frac{1}{2}\tau_0^2 < 0$. Since $\tau_0 \neq -1$, we have $\tau_0 \in (-2, -1)\bigcup(-1, 0)$. If $\tau_0 > -1$, in view of (3.17.a) we have

$$0 = 12\lambda_0^2 + 12\lambda_0\tau_0 + 4\tau_0^2 + \tau_0^3 >$$
$$12\lambda_0^2 + 12\lambda_0\tau_0 + 3\tau_0^2 = 3(2\lambda_0 + \tau_0)^2 \geqslant 0.$$

The contradiction proves that $-2 < \tau_0 < -1$.

It follows from (3.17.b) that $\lambda_1 + \tau_1 + \frac{b}{2}\tau_1^2 > 0$, hence $\tau_1 + \frac{b}{2}t_1^2 > 0$. Since $\tau_1 \neq -1/b$, we have $\tau_1 \in (0, -1/b) \bigcup (-1/b, -2/b)$. If $\tau_1 < -1/b$, in view of (3.17.b) we have

$$0 = 12\lambda_1^2 + 12\lambda_1\tau_1 + 4\tau_1^2 + b\tau_1^3 >$$
$$12\lambda_1^2 + 12\lambda_1\tau_1 + 3\tau_1^2 = 3(2\lambda_1 + \tau_1)^2 \geqslant 0.$$

The contradiction proves that $-1/b < \tau_1 < -2/b$.

Let us prove (3.32.b). If (3.29) holds, then (3.17.a) yields $\lambda_0 + \tau_0 + \frac{1}{2}\tau_0^2 > 0$, hence $\tau_0 + \frac{1}{2}\tau_0^2 > 0$, or, in other words, $\tau_0 \in (-\infty, -2) \bigcup (0, \infty)$. If $\tau_0 > 0$, then

$$0 = 12\lambda_0^2 + 12\lambda_0\tau_0 + 4\tau_0^2 + \tau_0^3$$
$$= 3(2\lambda_0 + \tau_0)^2 + \tau_0^2 + \tau_0^3 > 0.$$

The contradiction proves that $\tau_0 < -2$. As well, it follows from (3.17.b) that $\lambda_1 + \tau_1 + \frac{b}{2}\tau_1^2 < 0$, hence $\tau_1 + \frac{b}{2}\tau_1^2 < 0$, or, equivalently, $\tau_1 \in (-\infty, 0) \bigcup (-2/b, \infty)$. If $\tau_1 < 0$, we have

$$0 = 12\lambda_1^2 + 12\lambda_1\tau_1 + 4\tau_1^2 + b\tau_1^3$$
$$= 3(2\lambda_1 + \tau_1)^2 + \tau_1^2 + b\tau_1^3 > 0.$$

The contradiction proves that $\tau_1 > -2/b$.   **Q.E.D.**

Now we can calculate the ν–components of the solution of system (3.13):

$$\nu_0^\epsilon = \frac{\tau_0^\epsilon F(0, \sigma, -1) + \tau_1^\epsilon(1 + \tau_0^\epsilon)F(0, \sigma, 1)}{(1 + \tau_0^\epsilon)(1 + b\tau_1^\epsilon) - 1}$$

$$\nu_1^\epsilon = \frac{\tau_1^\epsilon F(0, \sigma, 1) + \tau_0^\epsilon(1 + \tau_1^\epsilon)F(0, \sigma, -1)}{(1 + \tau_0^\epsilon)(1 + b\tau_1^\epsilon) - 1} \tag{3.33}$$

Lemma 3.3 implies that mapping $\widetilde{\Phi}^2$ possesses of at least two fixed points. Denote them $M^+(0, \mu_0^+, \lambda_0^+, \nu_0^+)$ and $M^-(0, \mu_0^-, \lambda_0^-, \nu_0^-)$.

3.7 The Hyperbolic Structure of the Quotient Mapping

LEMMA 3.4. *The mapping* $\widetilde{\Phi}^2$ *is hyperbolic at* M^+ *and* M^-. *The differential* $D\widetilde{\Phi}^2(M^+)$ *has one eigenvalue less than 1 in absolute value and* $m + 2$ *eigenvalues greater than 1 in absolute value. The differential* $D\widetilde{\Phi}^2(M^-)$ *has one eigenvalue greater than 1 in absolute value and* $m+2$ *eigenvalues less than 1 in absolute value.*

Proof. Let us first consider the point M^+ and find the matrix $D^+ = D\widetilde{\Phi}^2(M^+)$ (the differential of $\widetilde{\Phi}^2$). If we differentiate (3.11) and take into account (3.12), we obtain

$$\frac{\partial \kappa_2}{\partial \kappa_0}\bigg|_{M^+} = \frac{\partial \kappa_2}{\partial \kappa_1}\bigg|_{\Phi(M^+)} \frac{\partial \kappa_1}{\partial \kappa_0}\bigg|_{M^+} = (1 + b\tau_1^+)(1 + \tau_0^+).$$

Set $\rho_0^+ = (1 + b\tau_1^+)(1 + \tau_0^+)$. In view of (3.32.a) we have $\tau_0^+ \in (-2, -1)$, $\tau_1^+ \in (-1/b, -2/b)$, hence $\rho_0^+ \in (0, 1)$. Besides that, it is easy to see that

$$\left(\frac{\partial \kappa_2}{\partial \mu_0}\right)^+ = \left(\frac{\partial \kappa_2}{\partial \lambda_0}\right)^+ = \left(\frac{\partial \kappa_2}{\partial \nu_0}\right)^+ = 0$$

at the point M^+. Here and everywhere below the superscript "+" means that the corresponding function has been calculated at the points M^+ or $\widehat{\Phi}(M^+)$, and the superscript "−" means the same for the points M^- and $\widehat{\Phi}(M^-)$.

Let

$$Q^+ = \begin{pmatrix} \left(\dfrac{\partial \lambda_2}{\partial \lambda_0}\right)^+ & \left(\dfrac{\partial \lambda_2}{\partial \mu_0}\right)^+ \\ \left(\dfrac{\partial \mu_2}{\partial \lambda_0}\right)^+ & \left(\dfrac{\partial \mu_2}{\partial \mu_0}\right)^+ \end{pmatrix}.$$

The mapping $(\lambda_0, \mu_0) \to (\lambda_2, \mu_2)$ has been defined as the composition $(\lambda_0, \mu_0) \to (\lambda_1, \mu_1) \to (\lambda_2, \mu_2)$, so $Q^+ = I_2^+ \cdot I_1^+$,

$$I_2^+ = \begin{pmatrix} \left(\dfrac{\partial \lambda_2}{\partial \lambda_1}\right)^+ & \left(\dfrac{\partial \lambda_2}{\partial \mu_1}\right)^+ \\ \left(\dfrac{\partial \mu_2}{\partial \lambda_1}\right)^+ & \left(\dfrac{\partial \mu_2}{\partial \mu_1}\right)^+ \end{pmatrix},$$

$$I_1^+ = \begin{pmatrix} \left(\dfrac{\partial \lambda_1}{\partial \lambda_0}\right)^+ & \left(\dfrac{\partial \lambda_1}{\partial \mu_0}\right)^+ \\ \left(\dfrac{\partial \mu_1}{\partial \lambda_0}\right)^+ & \left(\dfrac{\partial \mu_1}{\partial \mu_0}\right)^+ \end{pmatrix}.$$

It follows from (3.11) that $(\partial\lambda_2/\partial\nu_0)^+ = (\partial\mu_2/\partial\nu_0)^+ = 0$. Finally, $(\partial\nu_2/\partial\nu_0)^+ = 1/\rho_0^+ \cdot E_m$ where E_m is the unit $(m \times m)$–matrix. Thus, we have

$$
D^+ = \begin{pmatrix}
\rho_0^+ & 0 & 0 & 0 \\[2mm]
* & \left(\dfrac{\partial\lambda_2}{\partial\lambda_0}\right)^+ & \left(\dfrac{\partial\lambda_2}{\partial\mu_0}\right)^+ & 0 \\[4mm]
* & \left(\dfrac{\partial\mu_2}{\partial\lambda_0}\right)^+ & \left(\dfrac{\partial\mu_2}{\partial\mu_0}\right)^+ & 0 \\[4mm]
* & * & * & \dfrac{E_m}{\rho_0^+}
\end{pmatrix},
$$

where the asterisks denote quantities whose values are of no significance.

Now we are going to demonstrate that the eigenvalues of matrix Q^+ are greater than 1 in absolute value. Using (3.13) and the implicit function theorem we obtain

$$
\left(\frac{\partial\lambda_2}{\partial\lambda_1}\right)^+ = \frac{1}{(1+b\tau_1^+)^2} + \frac{1-2b\lambda_0^+}{1+\tau_1^+}\left(\frac{\partial\tau_1}{\partial\lambda_1}\right)^+ ;
$$

$$
\left(\frac{\partial\lambda_2}{\partial\mu_1}\right)^+ = \frac{1-2b\lambda_0^+}{1+b\tau_1^+}\left(\frac{\partial\tau_1}{\partial\mu_1}\right)^+ ;
$$

$$
\left(\frac{\partial\mu_2}{\partial\lambda_1}\right)^+ = \frac{\tau_1^+}{(1+b\tau_1^+)^3} + \frac{\lambda_0^+ - 3b\mu_0^+}{1+b\tau_1^+}\left(\frac{\partial\tau_1}{\partial\lambda_1}\right)^+ ;
$$

$$
\left(\frac{\partial\mu_2}{\partial\mu_1}\right)^+ = \frac{1}{(1+b\tau_1^+)^3} + \frac{\lambda_0^+ - 3b\mu_0^+}{1+b\tau_1^+}\left(\frac{\partial\tau_1}{\partial\mu_1}\right)^+ ;
$$

$$
\left(\frac{\partial\tau_1}{\partial\mu_1}\right)^+ = \frac{-1}{\frac{1}{2}\lambda_1^+ + \frac{1}{3}\tau_1^+ + \frac{b}{8}(\tau_1^+)^2} ; \qquad
\left(\frac{\partial\tau_1}{\partial\lambda_1}\right)^+ = \frac{\tau_1^+}{2}\left(\frac{\partial\tau_1}{\partial\mu_1}\right)^+ .
$$

The substitution $b = 1$ and the permutation of indices "1" and "0" give

$$
\left(\frac{\partial\lambda_1}{\partial\lambda_0}\right)^+ = \frac{1}{(1+\tau_0^+)^2} + \frac{1-2\lambda_1^+}{1+\tau_0^+}\left(\frac{\partial\tau_0}{\partial\lambda_0}\right)^+ ;
$$

$$
\left(\frac{\partial\lambda_1}{\partial\mu_0}\right)^+ = \frac{1-2\lambda_1^+}{1+\tau_0^+}\left(\frac{\partial\tau_0}{\partial\mu_0}\right)^+ ;
$$

$$
\left(\frac{\partial\mu_1}{\partial\lambda_0}\right)^+ = \frac{\tau_0^+}{(1+\tau_0^+)^3} + \frac{\lambda_1^+ - 3\mu_1^+}{1+\tau_0^+}\left(\frac{\partial\tau_0}{\partial\lambda_0}\right)^+ ;
$$

$$
\left(\frac{\partial\mu_1}{\partial\mu_0}\right)^+ = \frac{1}{(1+\tau_0^+)^3} + \frac{\lambda_1^+ - 3\mu_1^+}{1+\tau_0^+}\left(\frac{\partial\tau_0}{\partial\mu_0}\right)^+ ;
$$

$$\left(\frac{\partial \tau_0}{\partial \mu_0}\right)^+ = \frac{-1}{\frac{1}{2}\lambda_0^+ + \frac{1}{3}\tau_0^+ + \frac{1}{8}(\tau_0^+)^2}; \qquad \left(\frac{\partial \tau_0}{\partial \lambda_0}\right)^+ = \frac{\tau_0^+}{2}\left(\frac{\partial \tau_0}{\partial \mu_0}\right)^+.$$

We assert that $\det I_1 < 0$ and $\det I_2 < 0$, hence both matrices I_1 and I_2 have real nonmultiple eigenvalues. A straightforward calculation yields

$$\det I_1 = \frac{1}{(1+\tau_0^+)^5} + \frac{\lambda_1^+ - 3\mu_1^+}{(1+\tau_0^+)^3}\left(\frac{\partial \tau_0}{\partial \mu_0}\right)^+ \tag{3.34.a}$$

$$+ \frac{1-2\lambda_1^+}{(1+\tau_0^+)^4}\left(\frac{\partial \tau_0}{\partial \lambda_0}\right)^+ - \frac{1-2\lambda_1^+}{(1+\tau_0^+)^4}\tau_0^+\left(\frac{\partial \tau_0}{\partial \mu_0}\right)^+$$

$$= \frac{1}{(1+\tau_0^+)^5} + \left(\frac{\lambda_1^+ - 3\mu_1^+}{(1+\tau_0^+)^3} - \frac{1-2\lambda_1^+}{(1+\tau_0^+)^4}\frac{\tau_0^+}{2}\right)\left(\frac{\partial \tau_0}{\partial \mu_0}\right)^+;$$

$$\det I_2 = \frac{1}{(1+b\tau_1^+)^5} + \frac{\lambda_0^+ - 3b\mu_0^+}{(1+b\tau_1^+)^3}\left(\frac{\partial \tau_1}{\partial \mu_1}\right)^+ \tag{3.34.b}$$

$$+ \frac{1-2b\lambda_0^+}{(1+b\tau_1^+)^4}\left(\frac{\partial \tau_1}{\partial \lambda_1}\right)^+ - \frac{1-2b\lambda_0^+}{(1+b\tau_1^+)^4}\tau_1^+\left(\frac{\partial \tau_1}{\partial \mu_1}\right)^+$$

$$= \frac{1}{(1+b\tau_1^+)^5} + \left(\frac{\lambda_0^+ - 3b\mu_0^+}{(1+b\tau_1^+)^3} - \frac{1-2b\lambda_0^+}{(1+b\tau_1^+)^4}\frac{\tau_1^+}{2}\right)\left(\frac{\partial \tau_1}{\partial \mu_1}\right)^+.$$

In view of (3.32.a) and the relation $\mu_i^+ = \frac{1}{2}(\lambda_i^+)^2 > 0$ we have $1 + \tau_0^+ < 0$, and each term inside of the parentheses in front of the multiplier $(\partial \tau_0/\partial \mu_0)^+$ in (3.34.a) is positive. It remains to prove that for any $b < 0$ the following inequality holds:

$$\left(\frac{\partial \tau_0}{\partial \mu_0}\right)^+ = \frac{-1}{\frac{1}{2}\lambda_0^+ + \frac{1}{3}\tau_0^+ + \frac{1}{8}(\tau_0^+)^2} < 0.$$

The expression $\frac{1}{2}\lambda_0^+ + \frac{1}{3}\tau_0^+ + \frac{1}{8}(\tau_0^+)^2$ depends continuously on the parameter b. If $b = -1$, then $\frac{1}{2}\lambda_0^+ + \frac{1}{3}\tau_0^+ + \frac{1}{8}(\tau_0^+)^2 = 0.00011\ldots > 0$. To prove that this equation is positive for all $b < 0$, let us ensure that there are no $b < 0$ such that the solution τ_0^+ of (3.17) meets the equation

$$\frac{1}{2}\lambda_0^+ + \frac{1}{3}\tau_0^+ + \frac{1}{8}(\tau_0^+)^2 = 0.$$

We have

$$\frac{1}{2}(\lambda_0^+)^2 + \frac{1}{2}\lambda_0^+\tau_0^+ + \frac{1}{6}(\tau_0^+)^2 + \frac{1}{24}(\tau_0^+)^3 = 0,$$

$$\frac{1}{2}\lambda_0^+ + \frac{1}{3}\tau_0^+ + \frac{1}{8}(\tau_0^+)^2 = 0.$$

If we multiply the first equation by $-\tau_0^+/3$ and add it to the second one, we obtain

$$\frac{1}{2}(\lambda_0^+)^2 + \frac{1}{3}\lambda_0^+\tau_0^+ + \frac{1}{18}(\tau_0^+)^2 = 0,$$

$$\frac{1}{2}\lambda_0^+ + \frac{1}{3}\tau_0^+ + \frac{1}{8}(\tau_0^+)^2 = 0.$$

If we multiply the second equation by $-\lambda_0^+$ and add it to the first one, we obtain $\left(\frac{1}{18} - \frac{1}{8}\lambda_0^+\right)(\tau_0^+)^2 = 0$, hence $\lambda_0^+ = 4/9$. The corresponding value τ_0^+ meets the quadratic equation $\frac{2}{9} + \frac{1}{3}\tau_0^+ + \frac{1}{8}(\tau_0^+)^2 = 0$, hence $\tau_0^+ = -4/3$. Therefore, $\lambda_1^+ = \left(\lambda_0^+ + \tau_0^+ + \frac{1}{8}(\tau_0^+)^2\right)/(1 + \tau_0^+)^2 = 0$, which does not meet (3.32.a). Thus, $(\partial\tau_0/\partial\mu_0)^+ < 0$ for all $b < 0$. It implies that $\det I_1^+ < 0$.

To prove that $\det I_2^+ < 0$, we will repeat the preceding performances almost literally. Using (3.32.a) and the relation $\mu_i^+ = \frac{1}{2}(\lambda_i^+)^2 > 0$, one can check that $1 + b\tau_1^+ < 0$, and each term inside of the parentheses in front of the multiplier $(\partial\tau_1/\partial\mu_1)^+$ in (3.34.b) is positive for all parameters $b < 0$:

$$\left(\frac{\partial\tau_1}{\partial\mu_1}\right)^+ = \frac{-1}{\frac{1}{2}\lambda_1^+ + \frac{1}{3}\tau_1^+ + \frac{b}{8}(\tau_1^+)^2} > 0.$$

If $b = -1$, then $\frac{1}{2}\lambda_1^+ + \frac{1}{3}\tau_1^+ + \frac{b}{8}(\tau_1^+)^2 = -1.802\ldots < 0$. Assume that for some $b < 0$ the following system is compatible:

$$\frac{1}{2}(\lambda_1^+)^2 + \frac{1}{2}\lambda_1^+\tau_1^+ + \frac{1}{6}(\tau_1^+)^2 + \frac{b}{24}(\tau_1^+)^3 = 0,$$

$$\frac{1}{2}\lambda_1^+ + \frac{1}{3}\tau_1^+ + \frac{b}{8}(\tau_1^+)^2 = 0.$$

If we multiply the first equation by $-\tau_1^+/3$ and add it to the second one, we obtain

$$\frac{1}{2}(\lambda_1^+)^2 + \frac{1}{3}\lambda_1^+\tau_1^+ + \frac{1}{18}(\tau_1^+)^2 = 0,$$

$$\frac{1}{2}\lambda_1^+ + \frac{1}{3}\tau_1^+ + \frac{b}{8}(\tau_1^+)^2 = 0.$$

Thus, $\lambda_1^+ = 4/(9b)$, $\tau_1^+ = -4/(3b)$ and $\lambda_0^+ = \left(\lambda_1^+ + \tau_1^+ + \frac{b}{2}(\tau_1^+)^2\right)/(1 + b\tau_1^+)^2 = 0$, which contradicts inequalities (3.32.a). This means that $\det I_2^+ < 0$ for all $b < 0$.

Let us argue that there are no eigenvalues of I_1^+ and I_2^+ with absolute value less than or equal to 1. Consider the characteristic equations for I_j^+:

$$\xi^2 - \operatorname{tr} I_j^+ \xi + \det I_j^+ = 0, \quad j = 1, 2.$$

Let $j = 1$ first. Matrix I_1^+ having been calculated in an explicit way, it is easy to check that

$$\operatorname{tr} I_1^+$$

$$= \frac{1}{(1+\tau_0^+)^2} + \frac{1}{(1+\tau_0^+)^3} + \frac{1-2\lambda_1^+}{1+\tau_0^+}\left(\frac{\partial\tau_0}{\partial\lambda_0}\right)^+ + \frac{\lambda_1^+ - 3\mu_1^+}{1+\tau_0^+}\left(\frac{\partial\tau_0}{\partial\mu_0}\right)^+$$

$$= \frac{1}{(1+\tau_0^+)^2} + \frac{1}{(1+\tau_0^+)^3} + \left(\frac{1-2\lambda_1^+}{1+\tau_0^+}\frac{\tau_0^+}{2} + \frac{\lambda_1^+ - 3\mu_1^+}{1+\tau_0^+}\right)\left(\frac{\partial\tau_0}{\partial\mu_0}\right)^+.$$

We assert that $|\det I_1^+| > 1 + |\operatorname{tr} I_1^+|$. Since $|1 + \tau_0^+| > 1$, the absolute value of each term inside of the parentheses in front of the multiplier $(\partial\tau_0/\partial\mu_0)^+$ in (3.34) is less than the absolute value of the corresponding term in front of $(\partial\tau_0/\partial\mu_0)^+$ in (3.34.a) (both of these terms in (3.34.a) are of the same sign). We are left with proving that

$$-\frac{1}{(1+\tau_0^+)^5} > 1 + \left|\frac{1}{(1+\tau_0^+)^2} + \frac{1}{(1+\tau_0^+)^3}\right|.$$

Set $k = 1/(1+\tau_0^+)$. Since $k < -1$, this follows from the inequality $-k^5 > 1 - k^2 - k^3$ which is equivalent to the inequality $(k^3 - 1)(k^2 - 1) < 0$. Hence, matrix I_1^+ has no eigenvalues with absolute value 1. Since the matrix I_1^+ is continuous in the parameter $b < 0$, the position of eigenvalues of I_1^+ with respect to the unit circle in the $\mathbb{C}$–plane is the same for all values of $b < 0$. If $b = -1$, a straightforward calculation yields

$$I_1^+ = \begin{pmatrix} 268.57\ldots & -404.97\ldots \\ 16638.2\ldots & -26719.5\ldots \end{pmatrix},$$

and the eigenvalues are $\xi_1 = 26467.5\ldots$, $\xi_2 = -16.55\ldots$. It follows that the eigenvalues of I_1^+ are greater than 1 in absolute value for all $b < 0$.

In the case $j = 2$ the steps are completely analogous to the previous case. We have

$$\operatorname{tr} I_2 = \frac{1}{(1+b\tau_1^+)^2} + \frac{1}{(1+b\tau_1^+)^3}$$

$$+ \left(\frac{\lambda_0^+ - 3b\mu_0^+}{1+b\tau_1^+} + \frac{1-2b\lambda_0^+}{1+b\tau_1^+}\frac{\tau_1^+}{2}\right)\left(\frac{\partial\tau_1}{\partial\mu_1}\right)^+.$$

Since $|1 + b\tau_1^+| > 1$, the absolute value of the third term in the expression for $\operatorname{tr} I_2^+$ is less than the absolute value of the third term in (3.34.b). In order to prove the inequality $|\det I_2^+| > 1 + |\operatorname{tr} I_2^+|$, it remains to verify that

$$-\frac{1}{(1 + b\tau_1^+)^5} > 1 + \left|\frac{1}{(1 + b\tau_1^+)^2} + \frac{1}{(1 + b\tau_1^+)^3}\right|.$$

This last inequality is true, since $1/(1 + b\tau_1^+) < -1$. If $b = -1$, the symmetry arguments yield

$$I_2^+ = \begin{pmatrix} \left(\dfrac{\partial\lambda_1}{\partial\lambda_0}\right)^+ & -\left(\dfrac{\partial\lambda_1}{\partial\mu_0}\right)^+ \\ -\left(\dfrac{\partial\mu_1}{\partial\lambda_0}\right)^+ & \left(\dfrac{\partial\mu_1}{\partial\mu_0}\right)^+ \end{pmatrix},$$

so $\operatorname{spec} I_2^+ = \operatorname{spec} I_1^+$. Hence, for all $b < 0$, the eigenvalues of the matrix I_2^+ have absolute values greater than 1.

Since at any $b < 0$ we have $\det I_i \neq 0$ $(i = 1, 2)$ and at $b = -1$ we have $\det I_i^+ < 0$, the eigenvalues of matrices I_1^+ and I_2^+ are real and have different signs. Consider the product $I_2^+ \cdot I_1^+$. Let us note first that if some real matrix A satisfies the bound $\|A\| > C$ over the field $\mathbb{R}$, then the same bound is valid over the field $\mathbb{C}$. In this case the spectrum of the matrix A is outside the circle of radius C. Since $\|I_2^+ \cdot I_1^+ r\| > \|I_1^+ r\| > \|r\|$, the spectrum of the matrix $I_2^+ \cdot I_1^+$ is outside the unit circle.

Now we return to our original notation, getting rid of the rate (3.12). To do that let us redenote by $(\lambda_i^\epsilon)'$, $(\mu_i^\epsilon)'$, $(\nu_i^\epsilon)'$, $(\tau_i^\epsilon)'$ $(i = 0, 1$, $\epsilon = $ "+", "$-$") the solutions λ_i^ϵ, μ_i^ϵ, ν_i^ϵ, τ_i^ϵ of system (3.13). All the functions calculated for mapping (3.11) after the change (3.12) will be marked with a prime also. In this notation the coordinates of the fixed point of mapping (3.11) are the following: $M^\epsilon\left(0, A_0^{-1}(\lambda_0^\epsilon)', A_0^{-2}(\mu_0^\epsilon)', A_0^{-1}(\nu_0^\epsilon)'\right)$, where $A_0 = \alpha(\sigma) - \beta(\sigma)$, $A_1 = \alpha(\sigma) + \beta(\sigma)$, $\tau_i^\epsilon = A_0^{-1}(\tau_i^\epsilon)'$. To estimate $(\lambda_i^\epsilon)'$, $(\mu_i^\epsilon)'$, $(\tau_i^\epsilon)'$ let us set $b = A_0^{-1}A_1$ in (3.32). We obtain

$$0 < \lambda_0^+ < \frac{1}{2A_0}, \qquad \frac{1}{2A_1} < \lambda_1^+ < 0,$$

$$-\frac{2}{A_0} < \tau_0^+ < -\frac{1}{A_0}, \qquad -\frac{1}{A_1} < \tau_1^+ < -\frac{2}{A_1},$$

$$\frac{1}{2A_1} < \lambda_0^- < 0, \qquad 0 < \lambda_1^- < \frac{1}{2A_0},$$

$$\tau_0^- < -\frac{2}{A_0}, \qquad -\frac{2}{A_1} < \tau_1^-.$$

The ν–coordinates of M_i^ϵ (where $M_0^\epsilon = M^\epsilon$, $M_1^\epsilon = \Phi M_0^\epsilon$) are the following:

$$\nu_i^\epsilon = \frac{\tau_i^\epsilon F(0,\sigma,u_i) + \tau_{1-i}^\epsilon(1 + A_i\tau_i^\epsilon)F(0,\sigma,u_{1-i})}{(1 + A_i\tau_i^\epsilon)(1 + A_{1-i}\tau_{1-i}^\epsilon) - 1} \tag{3.35}$$

where $i = 0,1$, $\epsilon = $ "$+$", "$-$", and $u_0 = 1$, $u_1 = -1$.

Let us demonstrate that the matrix D^+ does not vary under the inverse transformation of variables (3.12). We have

$$\frac{\partial\kappa_2}{\partial\kappa_0} = (1 + A_1\tau_1^+)(1 + A_0\tau_0^+)$$

$$= \left(1 + \frac{A_1}{A_0}\left(\tau_1^+\right)'\right)\left(1 + \left(\tau_0^+\right)'\right) = \left(\rho_0^+\right)'$$

i.e., $\partial\kappa_2/\partial\kappa_0 = (\partial\kappa_2/\partial\kappa_0)'$. In addition,

$$\frac{\partial\lambda_1}{\partial\lambda_0} = \frac{1}{(1 + A_0\tau_0^+)^2} + \frac{1 - 2\lambda_1^+ A_0}{1 + A_0\tau_0^+}\left(\frac{\partial\tau_0}{\partial\lambda_0}\right)^+$$

$$= \frac{1}{\left(1 + (\tau_0^+)'\right)^2} + \frac{1 - 2(\lambda_1^+)'}{1 + (\tau_0^+)'}\left(\frac{\partial\tau_0}{\partial\lambda_0}\right)^+ ;$$

$$\left(\frac{\partial\tau_0}{\partial\lambda_0}\right)^+ = \frac{-\frac{1}{2}\tau_0^+}{\frac{1}{2}\lambda_0^+ + \frac{1}{3}\tau_0^+ + \frac{1}{8}A_0(\tau_0^+)^2}$$

$$= \frac{-\frac{1}{2}(\tau_0^+)'}{\frac{1}{2}(\lambda_0^+)' + \frac{1}{3}(\tau_0^+)' + \frac{1}{8}((\tau_0^+)')^2} = \left(\left(\frac{\partial\tau_0}{\partial\lambda_0}\right)^+\right)'$$

i.e., $\partial\lambda_1/\partial\lambda_0 = (\partial\lambda_1/\partial\lambda_0)'$.

It can be proved in the same way that all other terms of matrices D^+ and $\left(D^+\right)'$ coincide also. Since D^+ is a block–triangular matrix, spec D^+ does not intersect the unit circle.

Now let us consider the point M^-. Matrix $D^- = D\widetilde{\Phi}^2(M^-)$ can be represented in the following form:

$$D^- = \begin{pmatrix} \rho_0^- & 0 & 0 & 0 \\ * & \left(\dfrac{\partial\lambda_2}{\partial\lambda_0}\right)^- & \left(\dfrac{\partial\lambda_2}{\partial\mu_0}\right)^- & 0 \\ * & \left(\dfrac{\partial\mu_2}{\partial\lambda_0}\right)^- & \left(\dfrac{\partial\mu_2}{\partial\mu_0}\right)^- & 0 \\ * & * & * & \dfrac{E_m}{\rho_0^-} \end{pmatrix},$$

where $\rho_0^- \stackrel{\text{def}}{=} (1 + \tau_0^-)(1 + b\tau_0^-) > 1$. Set

$$Q^- = \begin{pmatrix} \left(\dfrac{\partial\lambda_2}{\partial\lambda_0}\right)^- & \left(\dfrac{\partial\lambda_2}{\partial\mu_0}\right)^- \\[2ex] \left(\dfrac{\partial\mu_2}{\partial\lambda_0}\right)^- & \left(\dfrac{\partial\mu_2}{\partial\mu_0}\right)^- \end{pmatrix}.$$

It follows from Section 3.5 and relation (3.15) that Q^- coincides with the matrix $(Q^+)^{-1}$ (i.e., with the inverse matrix to Q^+). Hence, $\text{spec}\, Q^- = \text{spec}\,(Q^+)^{-1}$ lies in the strict interior of the unit circle. **Q.E.D.**

3.8 Non–Degeneracy of the Fixed Point

Let us return to solutions of system (3.13) again. System (3.13.a) is linear in λ_0, λ_1, μ_0, μ_1. In the case $\tau_0 \neq -1$ and $\tau_0 \neq -1/b$ the variables λ_0, λ_1, μ_0, μ_1 can be successively expressed as smooth functions of the variables τ_0, τ_1 and then substituted in the left–hand sides of (3.13.b). Consider the functions

$$\mathcal{F}_0(\tau_0, \tau_1) = \mu_0 + \frac{1}{2}\lambda_0\tau_0 + \frac{1}{6}\tau_0^2 + \frac{1}{24}\tau_0^3,$$

$$\mathcal{F}_1(\tau_0, \tau_1) = \mu_1 + \frac{1}{2}\lambda_1\tau_1 + \frac{1}{6}\tau_1^2 + \frac{b}{24}\tau_1^3.$$

Let us show that for any $b < 0$ the mapping

$$(\tau_0, \tau_1) \longrightarrow \big(\mathcal{F}_0(\tau_0, \tau_1), \mathcal{F}_1(\tau_0, \tau_1)\big)$$

is a local diffeomorphism at points (τ_0^+, τ_1^+), (τ_0^-, τ_1^-). Indeed, the preceding construction in Section 3.8 can be interpreted in the following way. We have produced three successive changes of variables

$$(\tau_0, \tau_1) \longrightarrow (\alpha, \beta) \longrightarrow (X, Y) \longrightarrow (\xi, \eta),$$

reducing the function $\mathcal{F}_0(\tau_0, \tau_1)$ to the form (3.27.a). The function $\mathcal{F}_1(\tau_0, \tau_1)$ can be produced by the substitution of ξ/b^2 for ξ and η/b for η in (3.27.a). Denote by $L(\xi, \eta)$ the left hand–side of (3.27.a). As was demonstrated above, the function $\eta = \eta(\xi)$, defined by the implicit equation $L(\xi, \eta) = 0$, is a monotonically increasing function at $\xi < 0$. Since for any fixed $b < 0$ the mapping

$$(\xi, \eta) \to (\xi/b^2, \eta/b)$$

carries any increasing function $\eta(\xi)$ into a decreasing one, it is easy to see that the smooth curves $L(\xi, \eta) = 0$ and $L(\xi/b^2, \eta/b) = 0$ intersect each other transversally and that the point $(0,0)$ is the regular value of the mapping

$$(\xi,\eta) \quad \longrightarrow \quad \big(L(\xi,\eta),\ L(\xi/b^2,\eta/b)\big).$$

It remains to prove that each of the changes of the variables

$$(\tau_0,\tau_1) \to (\alpha,\beta); \quad (\alpha,\beta) \to (X,Y); \quad (X,Y) \to (\xi,\eta)$$

is a local diffeomorphism at points corresponding to $\tau_0^\epsilon, \tau_1^\epsilon$. There are no questions about the last one, since (3.25) is a linear mapping. The mapping $(\alpha,\beta) \to (X,Y)$, specified by (3.21), is a diffeomorphism in the case $\alpha \neq \beta$, i.e., if $\tau_1 \neq 0$. The mapping $(\tau_0,\tau_1) \to (\alpha,\beta)$ can be represented in its own turn as the composition

$$(\tau_0,\tau_1) \to (\lambda_0,\lambda_1) \to (\alpha,\beta),$$

functions α, β being linear in λ_0, λ_1. The mapping $(\tau_0,\tau_1) \to (\lambda_0,\lambda_1)$ is a diffeomorphism because $\tau_0^\epsilon \neq -1$, $\tau_1^\epsilon \neq -1/b$.

3.9 Bundles with Chattering Arcs

Since the mapping $\widetilde{\Phi}^2$ is hyperbolic at the fixed points M^ϵ, $\epsilon = $ "+", "−", the invariant manifold theorem yields that there exists some $\widetilde{\Phi}^2$–invariant C^{k_0}–curve through M^ϵ whose tangent vector at M^ϵ is an eigenvector of matrix D^ϵ with the eigenvalue ρ_0^ϵ. Since $\mu_0^\epsilon > 0$ and $\tau_0^\epsilon < 0$ at M^ϵ, condition (3.9) is satisfied on the branch of the invariant curve lying exactly in the half–plane $\kappa < 0$. We shall call the branch $\gamma_0^\epsilon(\sigma)$. Set $\gamma_1^\epsilon = \widehat{\Phi}\gamma_0^\epsilon(\sigma)$ and $\gamma^\epsilon = \gamma_0^\epsilon \bigcup \gamma_1^\epsilon$. Since the mapping $\widetilde{\Phi}^2$ smoothly depends on σ, the curve γ^ϵ can be represented as follows:

$$\lambda = \lambda_i^\epsilon(\kappa,\sigma), \quad \mu = \mu_i^\epsilon(\kappa,\sigma), \quad \nu = \nu_i^\epsilon(\kappa,\sigma),$$

where $i = 0$ if $\kappa < 0$, $i = 1$ if $\kappa > 0$; $\lambda_i^\epsilon(0,\sigma) = \lambda_i^\epsilon$, $\mu_i^\epsilon(0,\sigma) = \mu_i^\epsilon$, $\nu_i^\epsilon(0,\sigma) = \nu_i^\epsilon$; the functions $\lambda_i^\epsilon(\cdot)$, $\mu_i^\epsilon(\cdot)$, $\nu_i^\epsilon(\cdot)$ being C^{k_0} in (κ,σ). The smoothness of the curve in the parameter σ follows from the invariant manifold theorem.

Denote by Γ_σ^ϵ the inverse image of γ_σ^ϵ by transformation (3.7). The curve Γ_σ^ϵ is given by the equations

$$z_1 = 0, \qquad\qquad z_4 = \kappa,$$

$$z_2 = \mu_i^\epsilon(\kappa,\sigma)\kappa^3, \quad w = \sigma + \nu_i^\epsilon(\kappa,\sigma)\kappa, \qquad (3.36)$$

$$z_3 = \lambda_i^\epsilon(\kappa,\sigma)\kappa^2,$$

where $i = 0$ if $\kappa < 0$ and $i = 1$ if $\kappa > 0$. Let $\mathfrak{N}_\sigma^\epsilon$ be a family of solutions of system (3.5) emanating from Γ_σ^ϵ. We shall call $\Gamma^\epsilon \stackrel{\text{def}}{=} \bigcup_{\sigma \in S_0} \Gamma_\sigma^\epsilon$ the $(m+1)$–dimensional switching surface of $\mathfrak{N}_\sigma^\epsilon$. Let Σ_σ^ϵ be the manifold constituted by the trajectories of $\mathfrak{N}_\sigma^\epsilon$. Denote $\Sigma^\epsilon = \bigcup_{\sigma \in S_0} \Sigma_\sigma^\epsilon$. Let us define the mapping $p^\epsilon : \Sigma^\epsilon \to S$ as $p^\epsilon(M) = \sigma$ for any $M \in \Sigma_\sigma^\epsilon$.

DEFINITION 3.1. *Given an r–dimensional manifold Σ, its submanifold $B \subset \Sigma$ (dim $B = l < r$), and a continuous epimorphism $p : \Sigma \to B$. We call the triple (p, Σ, B) the fibre bundle of Σ with the base B and piecewise C^{k_0}–fibres $\mathcal{F} \cong \mathbb{R}^{r-l}$, iff for any $\Xi \in \Sigma$ there exist a neighborhood $U \subset \Sigma$ of Ξ and a homeomorphism $\phi : U \longrightarrow \mathbb{R}^r$ such that U can be represented as the union $U = U_+ \bigcup U_0 \bigcup U_-$ where $U_+ = \phi^{-1}\{\xi \mid \xi_1 > 0\}$, $U_0 = \phi^{-1}\{\xi \mid \xi_1 = 0\}$, $U_- = \phi^{-1}\{\xi \mid \xi_1 < 0\}$ with the following properties:*

(1) *$\phi(B \bigcap U) = \{\xi \mid \xi_1 = \ldots = \xi_{r-l} = 0\}$ and $\phi p^{-1}(M) = \{\xi \mid \xi_j = (\phi(M))_j, j = r - l + 1, \ldots, r\}$ for any $M \in B$.*

(2) *The restrictions $\phi\big|_{U_+}$, $\phi\big|_{U_-}$ are C^{k_0}–diffeomorphisms which can be prolonged as C^{k_0}–diffeomorphisms to $U_0 \backslash B$. The restriction $\phi\big|_{U_0 \backslash B}$ is also a C^{k_0}–diffeomorphism, which can be prolonged as a C^{k_0}–diffeomorphism to B. Finally, $\phi\big|_B$ is C^{k_0}–diffeomorphism.*

We see that Σ is a standard bundle with the base B and some special smooth properties of its fibres.

THEOREM 3.1. **Theorem on bundles.** *The triple $(p^\epsilon, \Sigma^\epsilon, S_0)(\epsilon =$ "+" or "$-$") is an $(m + 2)$–dimensional fibre bundle Σ^ϵ with the base S_0 and piecewise C^{k_0}–fibres $\Sigma^\epsilon_\sigma \cong \mathbb{R}^2$.*

Proof. It follows from (3.36) that if we discard the higher order terms at points of Γ^ϵ , we have

$$\frac{D(z_4, w)}{D(\kappa, \sigma)} = \begin{pmatrix} 1 & 0 \\ \nu_i^\epsilon(\sigma) & E_m \end{pmatrix}.$$

This matrix has maximal rank (equal to $m + 1$), hence the implicit function theorem implies that surface Γ^ϵ is smooth outside the plane $\kappa = 0$. Set $\Gamma^\epsilon_{\sigma 0} = \Gamma^\epsilon_\sigma \bigcap \{\kappa < 0\}$ and $\Gamma^\epsilon_{\sigma 1} = \Gamma^\epsilon_\sigma \bigcap \{\kappa > 0\}$. It follows from (3.36) that $\Gamma^\epsilon_{\sigma 0}$ and $\Gamma^\epsilon_{\sigma 1}$ are disposed in the plane $z_1 = 0$ bilaterally along the hypersurface $z_4 = 0$. Since $\lambda_i^\epsilon \neq 0$ $(i = 0, 1)$, each $\Gamma^\epsilon_{\sigma i}$ $(i = 0, 1)$ intersects the surface $z_4 = 0$ at nonzero angles. The surfaces $\Gamma^\epsilon_{\sigma 0}$ and $\Gamma^\epsilon_{\sigma 1}$ are mutually tangent at points of S_0, their common tangent plane being the span of $(\partial/\partial z_4, \partial/\partial w)$ (where $\partial/\partial z_4, \partial/\partial w$ denote the basic vectors along the z_4 and w axes). Hence, surface Γ^ϵ is C^{k_0}–imbedded into the exterior space at points $\Gamma^\epsilon \backslash S_0$ and is C^1–imbedded everywhere including the points of $S_0 \bigcap \Gamma^\epsilon$.

It is left to prove that manifold Σ^ϵ is C^{k_0}–imbedded into $\mathbb{R}^{m+4}$ at points of $\Sigma^\epsilon \backslash \Gamma^\epsilon$ and can be smoothly prolonged to $\Gamma^\epsilon \backslash S_0$ from both sides of $\Sigma^\epsilon \backslash \Gamma^\epsilon$. Let us choose the coordinates in Σ^ϵ as follows. Let $(z(t), y(t))$ be a solution of system (3.5) at $t \in [0, \kappa \tau_i(\kappa, \sigma)]$ with the initial point $(z_0(\kappa, \sigma), w_0(\kappa, \sigma)) \in \Gamma^\epsilon$. The function

$$\tau_i(\kappa, \sigma) = \tau_i\big(z_0(\kappa, \sigma), w_0(\kappa, \sigma)\big)$$

is defined by relations (3.11.b) where $i = 0$ if $\kappa < 0$ and $i = 1$ if $\kappa > 0$. Using Lemma 3.1 it can be demonstrated in the same way as for system (3.11) that if $t = \kappa\tau$, then

$$
\begin{aligned}
z_1(t) &= \kappa^3\Theta(\kappa), \\
z_2(t) &= \kappa^2\Theta(\kappa), \\
z_3(t) &= \left(\lambda_i^\epsilon + \tau + A_i\tau^2/2\right)\kappa^2 + \kappa^2\Theta(\kappa), \\
z_4(t) &= (1 + A_i\tau)\kappa + \kappa\Theta(\kappa), \\
w(t) &= \sigma + \left(\nu_i^\epsilon + \tau F(0,\sigma,u_i)\right)\kappa + \kappa\Theta(\kappa),
\end{aligned}
\tag{3.37}
$$

where $t = \kappa\tau$, $A_i = \alpha(\sigma) + u_i\beta(\sigma)$, $u_i = (-1)^{i+1}$, $i = 0$ if $\kappa < 0$ and $i = 1$ if $\kappa > 0$. It follows that, until higher order terms occur,

$$
\mathbf{D} = \frac{D(z_3, z_4, w)}{D(t, \kappa, \sigma)}
$$

$$
= \begin{pmatrix}
(1 + A_i\tau)\kappa & 2(\lambda_i^\epsilon + \tau + \frac{1}{2}A_i\tau^2)\,\kappa & \kappa\Theta(\kappa) \\
A_i & 1 + A_i\tau & \Theta(\kappa) \\
F(0,\sigma,u_i) & \nu_i^\epsilon + \tau F(0,\sigma,u_i) & E_m
\end{pmatrix}.
$$

Let us show that the $(m+2) \times (m+2)$–matrix $\mathbf{D}$ has full rank. Indeed, the determinant of the (2×2)–matrix

$$
\begin{pmatrix}
(1 + A_i\tau)\kappa & 2(\lambda_i^\epsilon + \tau + \frac{1}{2}A_i\tau^2)\kappa \\
A_i & 1 + A_i\tau
\end{pmatrix}
$$

equals $(1 - 2A_i\lambda_i^\epsilon)\kappa \neq 0$; until terms of higher order, matrix $\mathbf{D}$ has a block–triangular form. Hence, $\operatorname{rk}\mathbf{D} = m + 2$ in the case $\kappa \neq 0$. It follows that, for all sufficiently small $\kappa \neq 0$ and $\tau \in \left(0, \kappa\tau_i(\kappa,\sigma)\right)$, the surfaces Σ_σ^ϵ and Σ^ϵ are C^{k_0}–manifolds outside of Γ^ϵ, smoothly extendible to $\Gamma^\epsilon \backslash S_0$ ($\dim \Sigma_\sigma^\epsilon = 2$, $\dim \Sigma^\epsilon = m + 2$).

Remark 3.3. *Unfortunately, the mapping*

$$
(t, \kappa, \sigma) \rightarrow (z_3, z_4, w)
$$

is degenerate at points $\kappa = 0$. *The reason is that the bundles* Σ_σ^ϵ *are tangent to the switching surface* Γ^ϵ *at the singular manifold* S_0. *To calculate the basis of the tangent plane to* Σ^ϵ *at* S_0 *one must know representation (3.37) with higher order terms. Later on, this circumstance will cause some trouble in checking the projectibility of* Σ^ϵ *into the state space.*

Let us consider the mapping $\phi : \Sigma^\epsilon \rightarrow \mathbb{R}^{m+2}$ defined as follows:

$$
\phi\big(z(t), w(t)\big) = \big(\kappa, \sigma, t' = t/\tau_i(\kappa,\sigma)\big).
$$

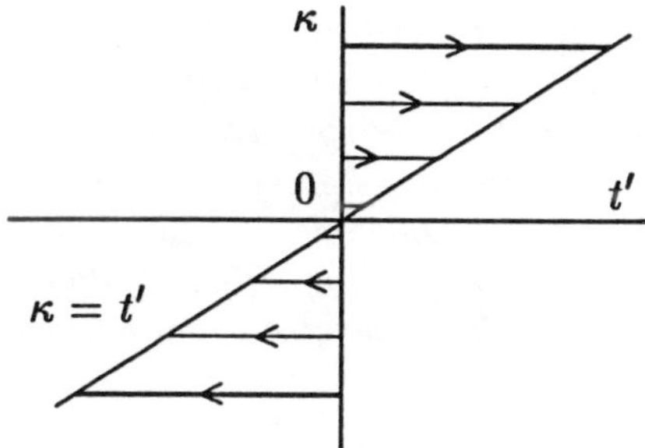

FIG. 10: THE MAP OF LOCAL COORDINATES IN THE
FIBRE OF THE CHATTERING BUNDLE

The image $\phi(\Sigma_\sigma^\epsilon)$ of the manifold Σ_σ^ϵ is the union of the straight line $t' = 0$ and the following two "triangles" on the (κ, t')–plane, first, $\kappa > t'$, $t' > 0$ and second, $\kappa < t'$, $t' < 0$ (see Fig. 10).

Consider the trajectory of system (3.5) with the initial point at $\Gamma_{\sigma i}^\epsilon$ and the end point at the first switch. Mapping ϕ carries this trajectory into the segment $\kappa = \kappa_0$, $0 \leqslant t' < \kappa_0$ if $i = 1$, and into the segment $\kappa = \kappa_0$, $-\kappa_0 < t' \leqslant 0$ if $i = 0$.

To verify that Σ^ϵ meets the conditions of Definition 3.1, one must paste together the straight half–line $\kappa = t'$, $t' > 0$ with $t' = 0$, $\kappa < 0$ and the straight half–line $\kappa = t'$, $t' < 0$ with $t' = 0$, $\kappa > 0$. **Q.E.D.**

Remark 3.4. On chattering submanifolds. *Let $\mathcal{M}$ be a smooth submanifold of the surface S_0 given by the equations $w_{k+1} = \ldots = w_m = 0$, $k = \dim \mathcal{M}$. Consider the full preimage $\Sigma_\mathcal{M} = (p^\epsilon)^{-1}\mathcal{M}$ in the fibre bundle $p^\epsilon : \Sigma^\epsilon \to S_0$. Let us demonstrate that in its own turn the manifold $\Sigma_\mathcal{M}$ is the fibre bundle with base $\mathcal{M}$ and with piecewise smooth fibres (the dimension of $\Sigma_\mathcal{M}$ equals $\dim \mathcal{M} + 2$). The switching surface of solutions of system (3.5) in $\Sigma_\mathcal{M}$ is $\Gamma_\mathcal{M} = \bigcup_{\sigma \in \mathcal{M}} \Gamma_\sigma^\epsilon$. The switching surface is defined by equations (3.36) in which one must set $\sigma \in \mathcal{M}$. Hence, $\Gamma_\mathcal{M}$ is C^{k_0}–manifold everywhere outside of $\mathcal{M}$. It follows from (3.37) that for any small $\kappa \neq 0$ the trajectories of system (3.5) intersect $\Gamma_\mathcal{M}$ at nonzero angles with nonzero velocities. Hence, $\Sigma_\mathcal{M}$ is homeomorphic to $\mathcal{M} \times \mathbb{R}^2$. The smoothness of $\Sigma_\mathcal{M}$ outside $\Gamma_\mathcal{M}$ can be argued by the same way as for the whole surface Σ^ϵ.*

Remark 3.5. The real–analytic case. *The following analytic version of Theorem 3.1 holds. Let functions f_0, f_1 and ϕ_0, ϕ_1 be real analytic. Then it can readily be verified that all the assertions of Theorem 3.1 are true. Besides that, all the bundles in question are piecewise analytic. Let us discuss what changes in the proof of Theorem 3.1 are necessary for the analytic case. All constructions and performances are preserved and even are simplified. Lemma 3.1 in this case turns out to be a trivial consequence of the Weierstrass preparation theorem [B. Malgrange, 1966]. It* *follows that the blowing up procedure (3.7) leads to system (3.11) with the analytic right hand–side. The last step is to prove that the contracting invariant manifold of the analytic diffeomorphism $\widetilde{\Phi}^2$ is an analytic manifold. Moreover, if $\widetilde{\Phi}^2$ is analytic in parameter σ then its contracting invariant manifold analytically depends on σ. This follows from the analytical form of the invariant manifold theorem. (See [J. Paleis, W. de Melou, 1982], [A. Shoshitaishvili, 1982], where the statement and the necessary references are given.)*

3.10 Lagrangian Manifolds

Suppose that system (3.1) has already been reduced to the form (3.5)–(3.6) using a smooth nondegenerate change $\phi : (\psi, x) \to (z, w)$ $(m = 2n - 4)$. Let $\mathcal{H} = H_0 + |H_1|$ be the Hamiltonian of the system. The manifold $\mathcal{M} \subset \{ z, w \mid \mathcal{H} = 0 \}$ is called *Lagrangian* (in the broad sense of the word) iff for any piecewise smooth closed curve $\gamma \subset \mathcal{M}$ we have

$$\oint_\gamma \psi \, dx = 0.$$

The standard definition of Lagrangian manifolds requires the supplementary condition $\dim \mathcal{M} = n$ that is omitted in our case.

THEOREM 3.2. **On Lagrangian manifolds.** *Let $\mathcal{M} \subset S_0$ be a Lagrangian manifold. Then the manifold $(p^\epsilon)^{-1}\mathcal{M}$ (i.e., the full preimage of $\mathcal{M}$ in the bundle Σ^ϵ) is Lagrangian.*

Recall the following classical statement of analytical dynamics.

Integral Invariant of Poincaré–Cartan.

Given the system

$$\dot{y} = I \operatorname{grad} \mathcal{H}(y) \tag{3.38}$$

$y = (\psi, x) \in \mathbb{R}^{2n}, \quad \mathcal{H} \in C^2(\mathbb{R}^{2n}), \quad I = \begin{pmatrix} 0 & -E \\ E & 0 \end{pmatrix}$, E being the unit $(n \times n)$–matrix.

Let $\Gamma_k = \{ t_k(\alpha), \psi_k(\alpha), x_k(\alpha) \mid a \leqslant \alpha \leqslant b \}$ $(k = 0, 1)$ be two piecewise smooth closed curves in $\mathbb{R}^{2n+1}$. If γ_1 can be produced by a transfer of γ_0 along the trajectories of (3.38) (i.e., for any $\alpha \in [a, b]$ the points $(t_0(\alpha), \psi_0(\alpha), x_0(\alpha))$ and $(t_1(\alpha), \psi_1(\alpha), x_1(\alpha))$ belong to the same trajectory of (3.38)), then

$$\oint_{\gamma_0} \psi \, dx - \mathcal{H} \, dt = \oint_{\gamma_1} \psi \, dx - \mathcal{H} \, dt.$$

If γ_0 lies on the zero–level surface of the Hamiltonian $\mathcal{H}$, then

$$\oint_{\mathrm{pr}\,\gamma_0} \psi\,dx \;=\; \oint_{\mathrm{pr}\,\gamma_1} \psi\,dx,$$

with $\mathrm{pr}\,\gamma_i$ $(i = 0, 1)$ being the projections of γ_i into (ψ, x)–space.

For the proof of the theorem on the integral invariant, see [V.M. Alexeev et al., 1987].

Proof of Theorem 3.2. Consider an arbitrary piecewise smooth closed curve $\gamma \subset (\Sigma_\mathcal{M} \bigcap \Gamma^\epsilon)\backslash S_0$. Let us introduce the coordinates (κ, σ) in $\Gamma^\epsilon\backslash S_0$ by means of relations (3.36).

Let $(\kappa_1, \sigma_1) = \Phi^2(\kappa, \sigma)$. The substitution of (3.36) in (3.11) brings

$$\begin{aligned}
\sigma_1 &= \sigma, \\
\kappa_1 &= (\rho_0^\epsilon + R(\kappa, \sigma))\kappa,
\end{aligned} \tag{3.39}$$

where $R(\kappa, \sigma) = \Theta(\kappa)$. Let us first consider the case $\epsilon = $ "$+$". The theorem on integral invariant of Poincaré–Cartan yields

$$\oint_\gamma \psi\,dx \;=\; \oint_{\gamma_1} \psi\,dx,$$

where $\gamma_1 = \Phi^2\gamma$. Consider the sequence of the curves $\gamma_{n+1} = \Phi^2\gamma_n$, $n = 1, 2, 3, \ldots$. Let us prove that $\gamma_n \to p^+\gamma_1$ in C^1–topology as $n \to \infty$. Indeed, let γ_1 be parameterized as follows:

$$\kappa = K_1(s), \quad \sigma = \Sigma(s), \quad s \in [0, 1].$$

Then curves γ_n can be parameterized by means of the following recurrent relations:

$$\kappa = K_n(s), \quad \sigma = \Sigma(s), \quad s \in [0, 1], \quad n = 1, 2\ldots,$$

where $K_n(s) = \left(\rho_0^+ + R(K_{n-1}(s), \Sigma(s))\right) K_{n-1}(s)$.

Hence, the projection $p^+\gamma_1$ to S_0 is given by the equations $\kappa = 0$, $\sigma = \Sigma(s)$. So

$$\left\|\gamma_n - p^+\gamma_1\right\|_{C^1} \;=\; \max_{s\in[0,1]}\left\{|K_n(s)|,\, |K_n'(s)|\right\}.$$

Recall that system (3.5) is considered in the region $|z| < \delta$ for sufficiently small $\delta > 0$. Using (3.39) we obtain

$$\left|K_{n+1}(s)\right| \;<\; \left(\rho^+ + O(\delta)\right) \cdot \left|K_n(s)\right| \tag{3.40}$$

and

$$\big(K'_{n+1}(s), \Sigma'(s)\big) = D\Phi^2 \bigg|_{\big(K_n(s),\Sigma(s)\big)} \cdot \big(K'_n(s), \Sigma'(s)\big)$$

$$= \begin{pmatrix} \rho_0^+ + R_1 & Q \\ 0 & E_n \end{pmatrix} \bigg|_{(K_n(s),\Sigma(s))} \cdot \big(K'(s), \Sigma'(s)\big).$$

Hence,

$$K'_{n+1}(s) = (\rho_0^+ + R_1)K'_n(s) + Q\Sigma'(s), \tag{3.41}$$

where $R_1 = \kappa\Theta(1), \ Q = \kappa\Theta(1)$.

Since $\rho_0^+ < -1$, inequality (3.40) implies that

$$\lim_{n\to\infty} \max_{s\in[0,1]} |K_n(s)| = 0.$$

Hence, for any $\delta_1 > 0$ there is $n_0 \in \mathbb{N}$ such that $\big|Q(K_n(s), \Sigma(s))\big| < \delta_1$ for any $n > n_0$. For any $n \geqslant n_0$ it follows from (3.41) that

$$\max_{s\in[0,1]} |K'_{n+1}(s)| < \big(\rho_0^+ + O(\delta)\big) \max_{s\in[0,1]} |K'_n(s)|$$

$$+ \delta_1 \max_{s\in[0,1]} |\Sigma'(s)|.$$

The induction arguments lead to the inequalities

$$\max_{s\in[0,1]} |K'_{n_0+m}(s)| \leqslant q^m \max_{s\in[0,1]} |K'_n(s)|$$

$$+ \delta_1\big(1 + q + \ldots + q^{m-1}\big) \max_{s\in[0,1]} |\Sigma'(s)|,$$

where $m = 1, 2, \ldots, \ q = \rho_0^+ + O(\delta) \in (0,1)$. Hence,

$$\overline{\lim_{n\to\infty}} \max_{s\in[0,1]} |K'_n(s)| < \frac{\delta_1}{1-q} \max_{s\in[0,1]} |\Sigma'(s)|.$$

Since δ_1 is an arbitrary small positive number, it follows that

$$\lim_{n\to\infty} \big\|K'_n(\cdot)\big\|_{C^0} = 0,$$

hence

$$\lim_{n\to\infty} \big\|\gamma_n - p^+\gamma\big\|_{C^1} = 0.$$

So

$$\left| \oint_{\gamma_n} \psi \, dx - \oint_{p^+\gamma_1} \psi \, dx \right| = \left| \int_0^1 (\psi_n \dot{x}_n - \psi_0 \dot{x}_0) dt \right|$$

$$\leqslant \max_{t \in [0,1]} \left\{ |\psi_n(t) - \psi_0(t)| \cdot \int_0^1 |\dot{x}(s)| ds \right.$$

$$\left. + |\psi_n(t)| \cdot \|x_n(\cdot) - x_0(\cdot)\|_{C^1} \right\} \to 0.$$

Here (ψ_n, x_n), (ψ_0, x_0) are the restrictions of the functions to γ_n and $p^+\gamma_1$ respectively. Since the integrals $\oint_{\gamma_n} \psi \, dx = \oint_\gamma \psi \, dx$ do not depend on n, we have

$$\oint_\gamma \psi \, dx = \oint_{\gamma_n} \psi \, dx = \oint_{p^+\gamma_1} \psi \, dx = 0.$$

Since $\Sigma_{\mathcal{M}}$ can be produced from $\Sigma_{\mathcal{M}} \bigcap \Sigma^\epsilon$ by mapping along the trajectories of the Hamiltonian system, we see that $\Sigma_{\mathcal{M}}$ is a Lagrangian manifold.

In order to prove that $\left(p^-\right)^{-1}\mathcal{M}$ is a Lagrangian manifold it is sufficient to replace the sequence $\gamma_n = \Phi^2 \gamma_{n-1}$ by the sequence $\gamma_n = \left(\Phi^2\right)^{-1} \gamma_{n-1}$, because $\mathfrak{N}^-$ –trajectories approach the surface S_0 at the reverse time current. $\hspace{2cm}$ **Q.E.D.**

3.11 Synthesis with Locally Optimal Chattering Arcs

Using the construction of the fibre bundle $\Sigma_{\mathcal{M}}$, we can design a sufficiently wide class of fields of extremals containing chattering trajectories, locally optimal for Problem 3.1 with appropriate boundary conditions.

Consider the trajectory $\hat{x}(\cdot)$ such that $\hat{x}(\hat{T}) \in \mathcal{M}$ for some $\hat{T} > 0$. We shall say that $\hat{x}(t)$ is *locally optimal* for Problem 3.1 iff there exists a neighborhood U of $\hat{x}(\cdot)$ such that

$$\int_0^{\hat{T}} f\big(\hat{x}(t)\big) \, dt < \int_0^T f\big(x(t)\big) \, dt$$

for any admissible trajectory $x(t) \subset U$, $x(T) \in \mathcal{M}$ for some $T > 0$. Here $f(x(t))$ means $f_0\big(x(t)\big) + u(t)f_1\big(x(t)\big)$.

Assume that $\mathcal{N}_1 \subset S_0 \bigcap \{H = 0\}$ is a connected smooth submanifold of dimension $n - 3$ such that

(1) $\left(0, \dfrac{1}{2\beta}\big((\beta - \alpha)F(0,\sigma,1) + (\beta + \alpha)F(0,\sigma,-1)\big)\right) \notin T_0\mathcal{N}_1$ (i.e., the flow of singular arcs of system (3.5) is not tangent to $\mathcal{N}_1$);

(2) $(\psi \, dx)\big|_{\mathcal{N}_1} \equiv 0$ (the transversality condition of the maximum principle).

Let us consider the manifold $\mathcal{N}_2$ having been established by the singular solutions of system (3.5) with $x(0) \in \mathcal{M}_1$ at $t \in (-\tau, \tau)$ for some sufficiently small $\tau > 0$. Let $\mathcal{N}_2^0$ be a submanifold of $\mathcal{N}_2$ corresponding to the negative half–interval $(-\tau, 0)$ (a singular trajectory starting at any point of $\mathcal{N}_2^0$ leads to $\mathcal{N}_1$ in the direct time current). Denote by $\mathfrak{N}^*$ the subset of $\mathfrak{N}^+$ –trajectories filling $\Sigma^* = (p^+)^{-1} \mathcal{N}_2$. Denote by π the projection $(\psi, x) \to x$ (or $(z, w) \to x$).

THEOREM 3.3. *Chattering optimality theorem. Assume that the Hamiltonian system (3.1) with $\psi_0 = 1$ can be reduced to (3.5)–(3.6)–form by means of a nondegenerate change $(\psi, x) \to (z, w)$. Assume that the restriction $\pi|_{\Sigma^*}$ is regular at $\Sigma^* \backslash \Gamma^*$, $\Gamma^* \stackrel{\text{def}}{=} \Sigma^* \bigcap \Gamma^+$, (i.e., $\pi|_{\Sigma^* \backslash \Gamma^*}$ is a C^1–mapping with the Jacobi matrix of maximal rank). Suppose that π can be prolonged as a regular mapping to the points of $\Gamma^* \backslash \mathcal{N}_2$. Then the π–projections of $\mathfrak{N}^*$–trajectories are locally optimal for Problem 3.1 with the target manifold $\mathcal{M} = \pi \mathcal{N}_1$, having been assumed a connected smooth manifold.*

Proof. Conditions of the theorem imply that there is a continuous function $\Psi : \mathbb{R}^n \to \mathbb{R}^n$ such that $(\Psi(x), x) \in \Sigma^*$ for any x in some open neighborhood of $\pi \mathcal{N}_1$. The smoothness conditions mean that $\Psi \in C^{k_0}$ at points of $\pi \Sigma^* \backslash \pi \Gamma^*$ and C^1–extendible to points of $\pi \Gamma^* \backslash \pi \mathcal{N}_2$. Let us demonstrate that

$$\oint_\gamma \Psi(x)\, dx = 0 \tag{3.42}$$

for any piecewise smooth closed curve $\gamma \subset \pi \Sigma^*$.

In view of Theorem 3.2 the manifold Σ^* is Lagrangian. Let us consider a curve $\gamma^* \subset \Sigma^*$ such that $\pi \gamma^* = \gamma$. By the definition of the curvilinear integral, we have

$$\oint_{\gamma^*} \psi\, dx = \oint_\gamma \Psi(x)\, dx.$$

It can readily be seen that the lifted curve γ^* does not have to be piecewise smooth (e.g., γ can intersect $\pi \mathcal{N}_2$ at a set of positive Lebesgue measure). But since codim $\pi \mathcal{N}_2 = 2$ in $\mathbb{R}^n$, in view of Thom's transversality theorem [S. Sternberg, 1964] the curve γ can be approximated in C^1–topology by a curve γ_1 that doesn't intersect $\pi \mathcal{N}_2$ and intersect $\pi \Gamma^*$ at a finite number of points. Being lifted to Σ^* by means of Ψ–mapping, the image of γ_1 is already piecewise smooth. Since Σ^* is a Lagrangian manifold, we have

$$\oint_{\gamma_1} \Psi(x)\, dx = 0.$$

This implies relation (3.42).

Assume that $x(\cdot)$ is some admissible trajectory of Problem 3.1 in the region πS^* with the initial condition x_0, $x^*(\cdot)$ being a projection of $\mathfrak{N}^*$–trajectory starting at x_0. Let T, T^* be the instants when the trajectories $x(\cdot)$ and $x^*(\cdot)$ are hitting the target manifold $\mathcal{M}$. The manifold $\mathcal{M}$ was assumed to be a connected smooth manifold such that

$$\oint_\gamma \Psi(x)\, dx = 0$$

for any curve γ, which does not need to be closed. It follows from (3.42) that

$$\oint_{x(\cdot)} \Psi(x)\, dx = \oint_{x^*(\cdot)} \Psi(x)\, dx.$$

Since $x^*(\cdot)$ meets Pontryagin's maximum principle, we have

$$\oint_{x^*(\cdot)} \Psi(x)\, dx = \int_0^{T^*} f\big(x^*(t)\big)\, dt.$$

On the other hand, for any optimal trajectory in the region $\pi\Sigma^*$ (in particular, for the trajectory $x(\cdot)$), the following inequality holds:

$$\oint_{x(\cdot)} \Psi(x)\, dx \leqslant \int_0^{T^*} f(x(t))\, dt.$$

This implies the optimality of $x^*(\cdot)$. **Q.E.D.**

Remark 3.6. Syntheses with untwisted chattering arcs. *Using Theorems 3.1 and 3.2 we can design the optimal synthesis with the chattering arcs for Problem 3.1 with more general boundary conditions if we also use the trajectories of the $\mathfrak{N}^-$ –family. In a typical situation of such synthesis there exists a singular manifold S_0 of codimension 2, whose neighborhood is partitioned into two subsets, call them U_1 and U_2, with the following properties. The region U_1 is filled with $\mathfrak{N}^+$–trajectories and the region U_2 (which includes the target manifold $\mathcal{M}$) is filled with $\mathfrak{N}^-$–trajectories. Each of the trajectories is composed of three successively adjoined segments. The first one is represented by an arc of $\mathfrak{N}^+$–trajectory reaching the singular manifold in finite time with an infinite number of switches. This segment is followed by a singular arc leading to the region U_2. The final, third, segment of an optimal trajectory is an arc of $\mathfrak{N}^-$–trajectory, which escapes from the singular manifold with an infinite number of switches and hits the target manifold. The simplest example of such synthesis is presented in Problem 2.1.*

3.12 Regular Projection of Chattering Varieties

The assumption of a regular projection of the manifold Σ^* in Theorem 3.1 is a direct analog of Jacobi's condition of the lack of focal points in the calculus of variations. As usual, this is the most complicated part to verify. The remainder of this chapter is devoted to discussing some effective sufficient conditions of the regular projecting of Σ^*. We find the basis of the tangent space $T_{(z,w)}\Gamma^*$ at $(z,w) \in S_0$ and prove that the switching surface Γ^* is a C^1–manifold (Lemma 3.5). In order to find an explicit form of the basis of $T_{(z,w)}(\Sigma^*)$ as $z \to 0$, one should first determine the values $\frac{\partial \nu}{\partial \kappa}(0,w)$, $\frac{\partial \nu}{\partial \sigma}(0,w)$ in (3.36) (see Remark 3.3). This is a rather cumbersome (and tedious) matter to perform in the general case, but, of course, for any problem of interest one could calculate $T_{(z,w)}(\Sigma^*)$ and verify whether its projection is regular. Nevertheless, under some supplementary condition on the coordinate change $(\psi, x) \to (z, w)$ (the condition is stated in Theorem 3.4 below), the limit of the projection $D\pi\big|_{(z,w)} T(\Sigma^*)$ as $z \to 0$ can be explicitly calculated already. This provides a sufficient condition of regular projection of Σ^* into x–space, which is valid, in particular, for all problems with chattering arcs in Chapter 6.

Let Σ^* be the same manifold as in Theorem 3.3. Without loss of generality we can assume that, in the vicinity of the point $(0, w_0)$, the manifold $\mathcal{N}_2$ is given by the equations $w_{n-1} = w_n = \cdots = w_{2n-4} = 0$. Denote $w' = (w_1, \ldots, w_{n-2})$, $w'' = (w_{n-1}, \ldots, w_{2n-4})$. Let $\partial/\partial w' = (\partial/\partial w_1, \ldots, \partial/\partial w_{n-2})$, $\partial/\partial w'' = (\partial/\partial w_{n-1}, \ldots, \partial/\partial w_{2n-4})$ be the basis consisting of the unit vectors $\partial/\partial w_i$ directed along the w_i–axes.

LEMMA 3.5. *Assume that the restriction* $\pi\big|_{\mathcal{N}_2}$ *is regular and that the vectors* $\phi_0(x) + \phi_1(x)$, $\phi_0(x) - \phi_1(x)$ *are not tangent to the surface* $\pi\mathcal{N}_2$ *at* $x \in \pi\mathcal{N}_2$. *Then* $\pi\Gamma^*$ *is* C^1 *everywhere and* C^{k_0} *at points outside of* $\pi\mathcal{N}_2$.

Proof. The last assertion of the lemma immediately follows from the parametric representation (3.36). Namely, the surface Γ^* consists of two C^{k_0}–surfaces, $\Gamma_i^* = \bigcup_{\sigma \in \mathcal{N}_2} \Gamma_{\sigma i}^*$, $i = 0, 1$. The surfaces Γ_0^* and Γ_1^* lie on different sides of the hypersurface $z_4 = 0$ and intersect it transversally. Let us show that Γ^* is C^1–surface. Let us calculate the limits of tangent planes $T_{(\kappa,\sigma)}\Gamma_0^*$, $\kappa < 0$, and $T_{(\kappa,\sigma)}\Gamma_1^*$, $\kappa > 0$, as $\kappa \to 0$. Recall that the values of (κ, σ) can be taken as coordinates in Γ^*. Differentiating (3.36) with respect to κ and σ at $\kappa = 0$ we obtain

$$T_{(0,\sigma)}\Gamma_i^* = \mathrm{span}\left(\frac{\partial}{\partial z_4} + \nu_i^+(0,\sigma)\frac{\partial}{\partial w}, \frac{\partial}{\partial w'}\right),$$

$\partial/\partial w'$ as the $(n-2)$–dimensional basis of $T_{(0,\sigma)}\mathcal{N}_2$.

In order to calculate the projections of $T_{(0,\sigma)}\Gamma_i^*$ let us find a more convenient form of a basis $T_{(0,\sigma)}\Gamma_i^*$. In view of (3.33) we have

$$F(0,\sigma,u_i) - A_i\nu_i^+ = \frac{\tau_{1-i}^+(1+A_i\tau_i^+)}{(1+A_i\tau_i^+)(1+A_{1-i}\tau_{1-i}^+)-1}\times$$

$$\times\big(A_{1-i}F(0,\sigma,u_i) - A_iF(0,\sigma,u_{1-i})\big) \quad (i=0,1). \tag{3.43}$$

By the definition, the velocity vector of the singular trajectories is the following one:

$$\dot{z} = 0, \quad \dot{w} = \frac{1}{2\beta}\left(A_1F(0,\sigma,-1) - A_0F(0,\sigma,1)\right).$$

Hence, both vectors $\big(F(0,\sigma,u_i) - A_i\nu_i^+\big)\cdot\partial/\partial w$ $(i=0,1)$ are directed along the singular velocity vector. Thus,

$$\big(F(0,\sigma,u_i) - A_i\nu_i^*\big)\frac{\partial}{\partial w} \in \text{span}\,\left(\frac{\partial}{\partial w'}\right),$$

hence,

$$T_{(0,\sigma)}\,\Gamma_i^* = \text{span}\,\left(A_i\frac{\partial}{\partial z_4} + A_i\nu_i^+\frac{\partial}{\partial w}, \frac{\partial}{\partial w'}\right)$$

$$= \text{span}\,\left(A_i\frac{\partial}{\partial z_4} + F(0,\sigma,u_i)\frac{\partial}{\partial w}, \frac{\partial}{\partial w'}\right).$$

This is the requisite form of the basis of $T_{(0,\sigma)}\Gamma_i^*$.

It is easy to see that the linear combination of vectors

$$A_0\frac{\partial}{\partial z_4} + F(0,\sigma,-1)\frac{\partial}{\partial w} \quad \text{and} \quad A_1\frac{\partial}{\partial z_4} + F(0,\sigma,1)\frac{\partial}{\partial w},$$

with coefficients A_1 and $-A_0$, is also collinear to the singular velocity vector and belongs to $T_{(0,\sigma)}\mathcal{N}_2 = \text{span}\,(\partial/\partial w')$. Hence, $T_{(0,\sigma)}\Gamma_1^* = T_{(0,\sigma)}\Gamma_0^*$ for any $(0,\sigma)\in\mathcal{N}_2$, which implies that Γ_i^* is C^1. Let us demonstrate now that $\pi\big|_{\Gamma^*}$ is regular.

Vector $A_i\cdot\partial/\partial z_4 + F(0,\sigma,u_i)\cdot\partial/\partial w$ is the velocity vector of nonsingular solutions with the control $u=u_i$ at the point $z=0$, $w=\sigma$. Hence, its projection to the state space equals $\phi_0(x) + u_i\phi_1(x)$, where $x=\pi(0,\sigma)$. By assumption, $\phi_0(x) + u_i\phi_1(x)$ does not belong to $T_x(\pi\mathcal{N}_2)$. Hence, the projection of $T_{(0,\sigma)}\Gamma_i^*$ being equal to the linear span of the vector $\phi_0(x) + u_i\phi_1(x)$ and $(n-2)$ basic tangent vectors to $\pi\mathcal{N}_2$ at x is the

$(n-1)$–plane, i.e., the rank of the Jacobi matrix of $\pi\big|_{\Gamma^*}$ is maximal. The statement being proved follows from the implicit function theorem.

Q.E.D.

Let us state the following fact of linear algebra. Consider two matrices, call them $\mathcal{A}$ and $\mathcal{B}$, of $(n \times k)$– and $(k \times n)$–dimensions respectively. Then the rank of the composition $\mathcal{AB}$ does not vary if we add to some column of the matrix $\mathcal{B}$ any linear combination of all other columns of $\mathcal{B}$. Indeed, let $\mathcal{A} = \begin{pmatrix} a_1 \\ \vdots \\ a_n \end{pmatrix}$, where $a_i \in (\mathbb{R}^k)^*$ means the i–th row–vector of the matrix $\mathcal{A}$, and let $\mathcal{B} = (b_1, \ldots, b_n)$, where $b_i \in \mathbb{R}^k$ means the i–th column–vector of $\mathcal{B}$. Then

$$\mathcal{AB} = \begin{pmatrix} a_1 b_1 & \ldots & a_1 b_n \\ \vdots & \ddots & \vdots \\ a_n b_1 & \ldots & a_n b_n \end{pmatrix},$$

where $a_i b_j$ means the scalar product of the row–vector a_i and the column–vector b_j. If we replace the column b_{i_0} by $b_{i_0} + \sum_{i \neq i_0} \lambda_i b_i$ ($\lambda_i \in \mathbb{R}$), then the i_0–column of the matrix $\mathcal{AB}$ is supplemented by the linear combination $\sum_{i \neq i_0} \lambda_i \begin{pmatrix} a_1 b_i \\ \vdots \\ a_n b_i \end{pmatrix}$ of the columns of matrix $\mathcal{AB}$. Of course, this does not vary the rank of $\mathcal{AB}$. Similarly, if we add to any row of the matrix $\mathcal{A}$ any linear combination of other rows of $\mathcal{A}$, then the rank of $\mathcal{AB}$ does not vary.

Assume that the conditions of Theorem 3.1 and Lemma 3.5 are fulfilled. Consider the plane, call it L, which is the span of the vector $\partial/\partial z_3$ and the plane $T_{(o,\sigma)}\Gamma^*$.

THEOREM 3.4. **On the regular projection.** *Assume that the restriction of the derivative of the projection* $\pi : (\psi, x) \to x$ *to* L *is regular. Then the restriction of* π *to* Σ^* *is regular at* $(0, w_0)$ *and hence the presumption of Theorem 3.3 on the regular projection is valid.*

COROLLARY 3.1. *Assume that* $\operatorname{codim} \pi S_0 = 2$ *and the projections of vectors* $\partial/\partial z_3$, $\partial/\partial z_4$, *and* $\partial/\partial w'$ *to* x*–space are independent at the point* $(0, w_0)$.

Then the projection of $\partial/\partial z_3$ *does not belong to the projection of*

$$T_{(0,\sigma)}\Gamma^* = span\left(\frac{\partial}{\partial z_4} + \nu_i^+ \frac{\partial}{\partial w}, \frac{\partial}{\partial w'}\right).$$

Proof of the corollary. If $\operatorname{codim} \pi S_0 = 2$, then

$$D\pi \cdot \frac{\partial}{\partial w'} = D\pi \cdot \frac{\partial}{\partial w},$$

because vectors $\partial/\partial w'$ constitute an $(n-2)$–plane, regular projecting to x–space. Hence,

$$\operatorname{span}\left(D\pi \cdot \frac{\partial}{\partial z_3}, D\pi \cdot \left(\frac{\partial}{\partial z_4} + \nu_i^* \frac{\partial}{\partial w'}\right), D\pi \frac{\partial}{\partial w}\right)$$

$$= \operatorname{span}\left(D\pi \cdot \frac{\partial}{\partial z_3}, D\pi \cdot \frac{\partial}{\partial z_4}, D\pi \frac{\partial}{\partial w}\right).$$

$$\textbf{Q.E.D.}$$

Proof of Theorem 3.4. We are to demonstrate that the rank of the composition $\mathbf{D}$ of the matrices $D_1 = \left. D\pi \right|_{(z,w)}$ and $D_2 = D(z,w)/D(t,\kappa,w')$,

$$\mathbf{D} \overset{\text{def}}{=} D_1 \cdot D_2,$$

equals n at points $(z,w) \in \Sigma^*\backslash\Gamma^*$ in a small neighborhood of $(0, w_0)$.

Consider first the matrix D_1 at the point $z = 0$,

$$D_1 = \begin{pmatrix} \dfrac{\partial x_1}{\partial z_1} & \dfrac{\partial x_1}{\partial z_2} & \dfrac{\partial x_1}{\partial z_3} & \dfrac{\partial x_1}{\partial z_4} & \dfrac{\partial x_1}{\partial w'} & \dfrac{\partial x_1}{\partial w''} \\ \vdots & \vdots & \vdots & \vdots & \vdots & \vdots \\ \dfrac{\partial x_n}{\partial z_1} & \dfrac{\partial x_n}{\partial z_2} & \dfrac{\partial x_n}{\partial z_3} & \dfrac{\partial x_n}{\partial z_4} & \dfrac{\partial x_n}{\partial w'} & \dfrac{\partial x_n}{\partial w''} \end{pmatrix},$$

D_1 being the $(n \times 2n)$ Jacobi matrix of the projection $\pi : (z,w) \to x$. Let us decompose the vector $\nu_i^* \cdot \partial/\partial w$ into its components in the basis $(\partial/\partial w', \partial/\partial w'')$:

$$\nu_i^* \frac{\partial}{\partial w} = \nu_i^{*\prime} \frac{\partial}{\partial w'} + \nu_i^{*\prime\prime} \frac{\partial}{\partial w''}.$$

By assumption, the composition of the matrix D_1 and a $(2n \times n)$–matrix whose columns consist of the coordinates of the vector $\partial/\partial z_3$ and the basis vectors of $T_{(0,\sigma)}(\Gamma^*)$, has maximal rank, i.e.,

$$\operatorname{rk} D_1 \cdot \begin{pmatrix} 0 & 0 & 0 \\ 0 & 0 & 0 \\ 1 & 0 & 0 \\ 0 & 1 & 0 \\ 0 & \nu_i^{+\prime} & E_{n-2} \\ 0 & \nu_i^{+\prime\prime} & 0 \end{pmatrix} = n. \tag{3.44}$$

LEMMA 3.6. *There exists a coordinate system* $x_1, \ldots, x_n$ *such that the matrix* D_1 *has the form*

$$
D_1 = \begin{pmatrix} * & * & 1 & * & 0 & 0 \\ * & * & 0 & \dfrac{\partial x_2}{\partial z_4} & 0 & \dfrac{\partial x_2}{\partial w''} \\ * & * & 0 & * & \dfrac{\partial x'}{\partial w'} & * \end{pmatrix} \tag{3.45}
$$

where $x' = (x_3, \ldots, x_n)$, $\partial x_1/\partial z_3 = 1$, $\partial x_2/\partial z_4 + \partial x_2/\partial w \cdot \nu_i^{+''} \neq 0$, *and* $\mathrm{rk}\,(\partial x'/\partial w') = n - 2$.

Proof of Lemma 3.6. It is readily seen that there exist coordinates in the x–plane such that $\partial x_1/\partial z_3 \neq 0$, $\partial x_1/\partial w'' = 0$. Indeed, $\partial x_1/\partial z_i$ is the x_1–coordinate of vector $\partial/\partial z_i$, expressed in (ψ, x)–coordinates:

$$
\frac{\partial}{\partial z_i} = \frac{\partial x_1}{\partial z_i}\frac{\partial}{\partial x_1} + \cdots + \frac{\partial x_n}{\partial z_i}\frac{\partial}{\partial x_n} + \frac{\partial \psi_1}{\partial z_i}\frac{\partial}{\partial \psi_1} + \cdots + \frac{\partial \psi_n}{\partial z_i}\frac{\partial}{\partial \psi_n}
$$

(the same can be said on $\partial x_1/\partial w_j$). Let us change the coordinates w', w'' to achieve the projection of $\partial/\partial z_3$ to be transversal to the projections of $\partial/\partial w''$. Choose as $\partial/\partial x_1$ a vector which is directed along $D\pi(\partial/\partial z_3)$ and as $\partial/\partial x_2, \ldots, \partial/\partial x_n$ vectors whose span contains the projections $D\pi \cdot \partial/\partial w''$. Hence, we can assume that

$$
\frac{\partial x_1}{\partial z_1} = a, \ \frac{\partial x_1}{\partial z_2} = b, \ \frac{\partial x_1}{\partial z_3} = 1, \ \frac{\partial x_1}{\partial z_4} = c, \ \frac{\partial x_1}{\partial w'} = d, \ \frac{\partial x_1}{\partial w''} = 0
$$

for some a, b, c, d. Besides that, in this basis we have $\partial x_2/\partial z_3 = \cdots = \partial x_n/\partial z_3 = 0$. It follows that

$$
D_1 = \begin{pmatrix} a & b & 1 & c & d & 0 \\ * & * & 0 & * & * & * \\ * & * & 0 & * & * & * \end{pmatrix}.
$$

Now (3.45) is a direct consequence of (3.44).   **Q.E.D.**

Let us return to the case $z \neq 0$. Consider the matrix

$$
D_2 = \begin{pmatrix} \dfrac{\partial z}{\partial t} & \dfrac{\partial z}{\partial \kappa} & \dfrac{\partial z}{\partial w'} \\ \dfrac{\partial w}{\partial t} & \dfrac{\partial w}{\partial \kappa} & \dfrac{\partial w}{\partial w'} \end{pmatrix}.
$$

It follows from (3.36) that, up to higher order terms, at points $(z, w) \in \Sigma^* \backslash \Gamma^*$ we have

$$D_2 = \begin{pmatrix} \kappa^4\Theta(1) & \kappa^3\Theta(1) & \kappa^4\Theta(1) \\ \kappa^3\Theta(1) & \kappa^2\Theta(1) & \kappa^3\Theta(1) \\ (1 + A_i\tau)\kappa + \kappa^2\Theta(1) & 2(\lambda_i^+ + \tau + \frac{1}{2}A_i\tau^2)\kappa + \kappa^2\Theta(1) & \kappa^2\Theta(1) \\ A_i + \kappa\Theta(1) & 1 + A_i\tau + \kappa\Theta(1) & \kappa\Theta(1) \\ F(0, \sigma, u_i) + \kappa\Theta(1) & \nu_i^+ + \tau F(0, \sigma, u_i) + \kappa\Theta(1) & \dfrac{\partial w}{\partial w'} \end{pmatrix}.$$

By means of a natural transformation of the columns of D_2 (we multiply the first column by $-\tau$ and add it to the second one, then the second column, with the coefficient $-A_i$, is added to the first one), we reduce D_2 to the matrix

$$D_3 = \begin{pmatrix} \kappa^3\Theta(1) & \kappa^3\Theta(1) & \kappa^3\Theta(1) \\ \kappa^2\Theta(1) & \kappa^2\Theta(1) & \kappa^2\Theta(1) \\ (1 - 2A_i\lambda_i^+)\kappa + \kappa^2\Theta(1) & (2\lambda_i^+ + \tau)\kappa + \kappa^2\Theta(1) & \kappa^2\Theta(1) \\ \kappa\Theta(1) & 1 + \kappa\Theta(1) & \kappa\Theta(1) \\ F(0, \sigma, u_i) - A_i\nu_i^+ + \kappa\Theta(1) & \nu_i^+ + \kappa\Theta(1) & \dfrac{\partial w}{\partial w'} \end{pmatrix}.$$

It follows from (3.43) that $F(0, \sigma, u_i) - A_i\nu_i^+ \in T_\sigma \mathcal{N}_2$, i.e., vector $F(0, \sigma, u_i) - A_i\nu_i^+$ equals some linear combination of vector–columns of the matrix $\frac{\partial w}{\partial w'}(0, \sigma)$. Adding a proper combination of the last $(n - 2)$ columns of the matrix D_3 to the first one and then eliminating the term $\kappa\Theta(1)$ in the first column by means of the second one, etc., we reduce D_3 to the matrix

$$D_4 = \begin{pmatrix} \kappa^3\Theta(1) & \kappa^3\Theta(1) & \kappa^3\Theta(1) \\ \kappa^2\Theta(1) & \kappa^2\Theta(1) & \kappa^2\Theta(1) \\ (1 - 2A_i\lambda_i^+)\kappa + \kappa^2\Theta(1) & \kappa^2\Theta(1) & \kappa^2\Theta(1) \\ \kappa^2\Theta(1) & 1 + \kappa\Theta(1) & \kappa^2\Theta(1) \\ \kappa\Theta(1) & \nu_i^+ + \kappa\Theta(1) & \dfrac{\partial w}{\partial w'} + \kappa\Theta(1) \end{pmatrix}.$$

The last $(2n - 4)$ rows of the matrix D_4 (i.e., the row $\left(\kappa\Theta(1), \nu_i^+ + \kappa\Theta(1), \partial w/\partial w' + \kappa\Theta(1)\right)$ in vector notation) can be reduced to the form

$$\begin{pmatrix} 0 & 0 & E_{n-2} + \kappa\Theta(1) \\ \kappa\Theta(1) & {\nu_i^+}'' + \kappa\Theta(1) & \kappa\Theta(1) \end{pmatrix},$$

where E_{n-2} is the unit $(n-2) \times (n-2)$–matrix associating with vectors tangent to $\mathcal{N}_2$. What is left is to multiply the matrices D_1 and D_4 and discard the higher order terms.

Thus, each of two smooth parts of Σ^*, call them Σ_i^*, $i = 0, 1$, has a regular projection into x–space at $\kappa \neq 0$. In view of (3.37), the velocity vector $\partial(z, w)/\partial t$ of system (3.5) and $n-1$ vectors of a basis of the tangent plane Γ^* constitute a basis of $T\Sigma_i^*$ at $(z, w) \in \Gamma_i^* \backslash \{x = 0\}$. Both planes have regular projections into x–space. Therefore, the trajectories of system (3.1) intersect $\pi(\Gamma^* \backslash \mathcal{N}_2)$ transversally. The surface $\pi\Gamma^*$ partitions x–space into two open regions. In view of (3.29) the term $(1 - 2A_i\lambda_i^+)\kappa$ of matrix D_4 has the same sign for both $i = 0$ and $i = 1$. It follows that, at points of $\pi(\Gamma^* \backslash \mathcal{N}_2)$, vectors of the state velocity are directed to the same half–space for both $u = 1$ and $u = -1$. Hence, the trajectories of Pontryagin's maximum principle intersect $\pi\Gamma^*$ having to pass from one of these open regions into the other. Therefore, the trajectories do not intersect each other in $\pi(\Sigma^* \backslash \mathcal{N}_2)$. **Q.E.D.**

Chapter 4

THE UBIQUITY OF FULLER'S PHENOMENON

This chapter contains a study of the *ubiquity* of Fuller's phenomenon. The term "ubiquity" is somewhat broad and can be interpreted in different ways, depending on the choice of the class of problems in question, the choice of a group acting on this class which determines the topology, and so on.

We deal with optimization problems that are affinely generated by a scalar control $u \in [-1, 1]$. Their Pontryagin systems are described by a pair of functions $H_0(\psi, x)$, $H_1(\psi, x)$. Pontryagin's function equals $H_0 + u H_1$ where $|u| \leqslant 1$. Since the optimal control u follows the sign of H_1, we have the non–smooth Hamiltonian $H = H_0 + |H_1|$. The right hand side of the corresponding Hamiltonian system is discontinuous along the smooth hypersurface $\Sigma = \{\psi, x \mid H_1 = 0\}$. The jump of the gradient of H is orthogonal to Σ, and hence the jump of the Hamiltonian vector field is tangent to Σ. Thus, we can consider the infinite–dimensional manifold of the Hamiltonian vector fields with tangent discontinuity along a smooth hypersurface. Denote this manifold by $\mathfrak{H}$ and equip it with C^∞–Whitney topology.

According to the tradition going back to H. Poincaré, the study of the behavior of trajectories begins from "generic" points of $\mathfrak{H}$, i.e., we suppose that "almost all" perturbations of the corresponding system preserve the topological structure of the set of trajectories. All other systems represent various kinds of singularities. In this context the term *singularity* relates to a class of systems with an equivalent structure of trajectories. We say that the codimension of the singularity $\mathfrak{h}$ equals k if this class of systems is locally described by k independent equations and some collection of inequalities in $\mathfrak{H}$. As a rule, a small perturbation of the system violates the structure of optimal trajectories, but any k–parametric family of perturbations (in the case it is "generic", i.e., transversal to $\mathfrak{h}$) contains a system with the same structure. Alternatively, suppose that the dimension of the state space is greater than k and that equations of the singularity are given in terms of the functions H_0, H_1 (as in our case). Then these equations define some manifold $\mathfrak{K}$ of codimension k lying in the state space. In the general case the manifold $\mathfrak{K}$ is stable for small perturbations of H_0, H_1. In this sense the fact of existence of Fuller's points appears generic.

The research of singularities of Hamiltonian systems with tangent discontinuity has been partially realized by I . Kupka in his remarkable works [1985, 1988, 1990]. In particular, he found some explicit sufficient conditions for Fuller's phenomenon containing eight equations, so the codimension of Fuller's singularity was bounded by 8.

Here we outline Kupka's results and make some improvements using our theory. We prove that the codimension of Fuller's singularity is less than or equal to 7.

4.1 Kupka's Results

Given a $2m$–dimensional symplectic space $\mathcal{M}$ and a regular hypersurface Σ in a small neighborhood U of a point $z_0 \in \Sigma$. Let Σ partition U into two open connected subsets U_+ and U_-. Let X_+, X_- be two smooth canonical vector fields with the Hamiltonians H_+ and H_- defined on $U_+ \bigcup \Sigma$ and $U_- \bigcup \Sigma$ respectively, such that $H_+\big|_\Sigma = H_-\big|_\Sigma$.

I. Kupka researched the singularity of the pair (X_+, X_-) at the point $z_0 \in \Sigma$ such that the vector $X_+(z_0) - X_-(z_0)$ belongs to the tangent space $T_{z_0}\Sigma$. Let us define a trajectory of the Hamiltonian vector field X as an absolutely continuous curve $z(t)$, $t \in [t_0, t_1]$, such that

$$\dot{z}(t) = X_\sigma\big(z(t)\big)$$

if $z(t)$ belongs to U_σ ($\sigma = $ "+" or "−"), and

$$\dot{z}(t) \in \bigcup_{0 \leqslant \lambda \leqslant 1} \Big[\lambda X_+\big(z(t)\big) + (1 - \lambda)X_-\big(z(t)\big)\Big]$$

if $z(t)$ belongs to Σ.

DEFINITION 4.1. **Fuller pair.** *A pair of curves* $\Gamma_+, \Gamma_- \subset \Sigma$ *is called a Fuller pair at the point* $z_0 = \Gamma_+ \bigcap \Gamma_-$ *if the following statements hold. For any point* $m \in \Gamma_+$ *there exists a trajectory* $z(t)$ *starting at* m *and ending at* z_0 *whose switching instants constitute an increasing sequence* $\{t(n) \mid n \in \mathbb{N}\}$ *such that*

(a) $z\big(t(n)\big)$ *belongs to* Γ_+ *if* n *is even and belongs to* Γ_- *if* n *is odd,*

(b) the sequence $t(n)$ *converges to* $T(z_0) < \infty$,

(c) there exists a constant $k > 1$ depending only on the pair (Γ_+, Γ_-) such that

$$\lim_{n \to \infty} \left(t(2n+1) - t(2n)\right)/k^{2n} = \gamma_1(z_0),$$
$$\lim_{n \to \infty} \left(t(2n) - t(2n-1)\right)/k^{2n} = \gamma_2(z_0),$$

γ_1 and γ_2 being some positive constants.

Now we can state the main result of I. Kupka concerning the ubiquity of Fuller's phenomenon. Given two smooth functions H_+ and H_-, define the mapping $\alpha(H_+, H_-) : R^{2n} \to R^6$ whose components are

$$\{\operatorname{ad}^2 F(G), \operatorname{ad} F(G)\}, \quad \{\operatorname{ad}^2 G(F), \operatorname{ad} G(F)\},$$
$$\tfrac{1}{6} \operatorname{ad}^3 F\big(\operatorname{ad} G(F)\big), \quad \tfrac{1}{4} \operatorname{ad}^2 F\big(\operatorname{ad}^2 G(F)\big),$$
$$\tfrac{1}{6} \operatorname{ad} F\big(\operatorname{ad}^3 G(F)\big), \quad \tfrac{1}{24} \operatorname{ad}^4 G(F),$$

where $F = (H_+ + H_-)/2$, $G = (H_+ - H_-)/2$. By $\operatorname{ad} f(g) = \{f, g\}$ we mean the Poisson bracket of functions f and g.

Theorem K [I. Kupka, 1990]. *There exists a semi–algebraic set $\mathfrak{F}$ in $\mathbb{R}^6$ with nonempty interior, such that, if (H_+, H_-) is any pair of functions on a symplectic manifold $\mathcal{M}$ satisfying the following assumptions:*

(i) *the differentials $dH_+(z_0)$, $dH_-(z_0)$, $d\{H_+, H_-\}(z_0)$ are independent,*

(ii) *all Poisson brackets of length less than or equal to four of H_+ and H_- are zero at z_0,*

(iii) *$\alpha(H_+, H_-)(z_0) \in \mathfrak{F}$,*

then the Hamiltonian vector field (X_+, X_-) associated with (H_+, H_-) has a Fuller pair at z_0.

COROLLARY 4.1. *The codimension of Fuller's singularity is eight.*

Proof of the corollary. There exist eight different Poisson brackets of length less than or equal to four of functions H_+ and H_-:

$$H_+, \quad H_-, \quad \{H_+, H_-\},$$
$$\{H_+, \{H_+, H_-\}\}, \quad \{H_-, \{H_+, H_-\}\},$$
$$\big\{H_+, \{H_+, \{H_+, H_-\}\}\big\}, \quad \big\{H_-, \{H_-, \{H_+, H_-\}\}\big\},$$
$$\big\{H_+, \{H_-, \{H_+, H_-\}\}\big\} \equiv \big\{H_-, \{H_+, \{H_+, H_-\}\}\big\}.$$

It can readily be proved that these functions are functionally independent in the general position.

Since conditions (i)–(iii) impose a finite number of restrictions in the form of equality on the Hamiltonians H_+, H_-, the Fuller point z_0 as a singular point of the Hamiltonian vector field cannot be eliminated by an arbitrary small field perturbation in dimension $n \geqslant 8$.   **Q.E.D.**

Let us outline the proof of Theorem K. It is based on the reduction of the Hamiltonian system, which meets conditions (i) and (iii), to a certain semi–canonical form (Proposition K below). This reduction is connected with the degree of degeneration of the Hamiltonian vector field at the point z_0 rather than with the existence of the singular manifolds.

Given two smooth functions H_+ and H_-, defined in an open neighborhood of 0 in $\mathbb{R}^{2n}$ with the coordinates $x_1, \ldots, x_n, p_1, \ldots, p_n$, and the standard symplectic structure $\sum_{i=1}^{2n} dp_i \wedge dx_i$.

Proposition K [I. Kupka, 1990]. *Assume that*

(1) *all Poisson brackets of length less than or equal to 4 of H_+ and H_- are zero at 0,*

(2) *the differentials $dH_+(0)$, $dH_-(0)$, $d\{H_+, H_-\}(0)$ are independent. Then there exists a symplectic system of coordinates $x_1, \ldots, x_n$, $p_1, \ldots, p_n$ in a neighborhood of 0 such that if we set $F = (H_+ + H_-)/2$, $G = (H_+ - H_-)/2$ then*

(3) $G = p_1$,

(4) $F = p_2 - \frac{1}{2}ax_3^2 + x_1(p_3 + x_1 F_1 + x_2 F_2)$, *where F_1, F_2 are smooth functions of (x,p),*

(5) *in the gradation where the weights of the variables are*
$w(x_1) = w(x_2) = 1$, $w(x_3) = 2$, $w(x_k) = 3$ *if* $k \geqslant 4$,
$w(p_1) = w(p_2) = 4$, $w(p_k) = 3$ *if* $k \geqslant 3$, *until there are terms of order greater than or equal to three, we have*

$$F_1 = \frac{1}{2}bx_3 + c_3 x_1^2,$$
$$F_2 = c_0 x_2^2 + c_1 x_1 x_2 + c_2 x_1^2;$$

(6) *the coefficients a, b, c_i $(0 \leqslant i \leqslant 3)$ have the following values:*

$$a = \{\mathrm{ad}^2 F(G), \mathrm{ad}\, F(G)\}(0),$$

$$b = \{\mathrm{ad}^2 G(F), \mathrm{ad}\, G(F)\}(0),$$

$$c_0 = \frac{1}{6}\mathrm{ad}^4 F(G)(0), \qquad c_1 = \frac{1}{4}\mathrm{ad}^2 F\big(\mathrm{ad}^2 G(F)\big)(0),$$

$$c_2 = \frac{1}{6}\mathrm{ad}\, F\big(\mathrm{ad}^3 G(F)\big)(0), \quad c_3 = \frac{1}{24}\mathrm{ad}^4 G(F)(0).$$

A change in variables, reducing the system to the semi–canonical form, is non–unique. Furthermore, Kupka searched for a generating function of the change in an implicit form as a solution of Cauchy's problem for a certain hyperbolic equation. This circumstance complicates the study of chattering arcs geometry in optimal problems.

The semi–canonical form in Proposition K can be considered to be a small perturbation of the equations of Pontryagin's maximum principle for the following problem. Let us add the equation $\dot{z} = 1$ to Fuller's system $\dot{x} = y$, $\dot{y} = u$, $u \in [-1, 1]$. The control system

$$\dot{x} = y, \quad \dot{y} = u, \quad \dot{z} = 1,$$
$$u \in [-1, 1],$$

or, in Kupka's notation,

$$\dot{x}_3 = x_1, \quad \dot{x}_1 = u, \quad \dot{x}_2 = 1,$$
$$u \in [-1, 1], \tag{4.1}$$

is homogeneous under the $\mathfrak{G}$–action

$$\mathfrak{g}_\lambda : (x_3, x_2, x_1) \longrightarrow (\lambda^2 x_3, \lambda x_2, \lambda x_1),$$

and the weight of x_2 equals 1. Denote by ξ the tuple (x_3, x_2, x_1, u). Let us consider the homogeneous functional

$$\mathcal{J}\big(\xi(\cdot), T\big) = \int_0^T \Big(\frac{a}{2} x_3^2 - \frac{b}{2} x_3 x_1^2 - c_0 x_1 x_2^3$$
$$- c_1 x_1^2 x_2^2 - c_2 x_1^3 x_2 - c_3 x_1^4 \Big) dt$$

Now if the tuple $\xi(t) = \big(x_3(t), x_2(t), x_1(t), u(t)\big)$ is a solution of system (4.1), then for every $\lambda > 0$ the tuple

$$\xi_\lambda(t) = \big(\lambda^2 x_3(t/\lambda), \lambda x_2(t/\lambda), \lambda x_1(t/\lambda), u(t/\lambda)\big)$$

is also a solution of (4.1) and $\mathcal{J}\big(\xi_\lambda(\cdot), \lambda T\big) = \lambda^5 \mathcal{J}(\xi(\cdot), T)$. The problem of minimization of $\mathcal{J}(\cdot)$ on solutions of (4.1) will be called *Kupka's problem*. In Fuller's case we have $a = 1$, $b = 0$, $c_i = 0$ $(i = 1, \ldots, 4)$. If we wrote out the equations of Pontryagin's maximum principle for Kupka's problem, we would arrive at the Hamiltonian system of functions F and G as in Proposition K with the same coefficients a, b, c_i and zero terms of higher order. In exactly the same way as for Fuller's problem it can be shown that if a, b, c_i are small enough, then there exists at least one one–parameter family of chattering arcs at $x_1 = x_2 = x_3 = 0$. The Fuller pair (Γ_+, Γ_-) is represented by two $\mathfrak{G}$–orbits whose parameters could be calculated as roots of some algebraic equation. It is left to prove that if

we add any term of higher order to the right–hand side of the Hamiltonian
system for Kupka's problem, then the perturbed system will also possess a
chattering arc.

Thus, we deal with the system

$$\frac{dy}{dt} = \mathcal{H}(y(t)), \tag{4.2}$$

where $y = (p, x) \in \mathbb{R}^{2n}$, $\quad \mathcal{H} = I \operatorname{grad}\big(F(p, x)+uG(p, x)\big)$, $I = \begin{pmatrix} O & -E \\ E & O \end{pmatrix}$,
$u = \operatorname{sgn} G(p, x)$. Everywhere else in this section the functions F and G
are considered already to have been reduced to the semi–canonical form as
in Proposition K.

Let us denote by σ the Poincaré mapping of the switching surface $\Sigma =
\big\{ y \mid G(y) = 0 \big\}$ along solutions of system (4.2). The Fuller pair Γ_+ and
Γ_- would be represented by two contracting σ^2–invariant curves such that
$\sigma(\Gamma_+) \subset \Gamma_-$ and $\sigma(\Gamma_-) \subset \Gamma_+$. Since the differential $D\sigma$ is degenerate
at $y = 0$, it is convenient to resolve the singularity of system (4.2) at
$y = 0$ by means of a graduated blowing up procedure. Let us consider the
action of the one–parameter group $(t \cdot y) = (t^{w_1}y_1, \ldots, t^{w_{2n}}y_{2n})$, where
$(w_1, \ldots, w_{2n})$ are the weights of the variables as in Proposition K. The
associated norm $\|y\|_w$ can be defined as follows:

$$\|y\|_w = \left[\sum_{i=1}^{2n} y_i^{2u_i} \right]^{1/(2u)},$$

where $u_i = u/w_i$, u being the least common multiple of the integers
$w_1, \ldots, w_{2n}$. Let ξ be the coordinates on the unit sphere $\mathcal{S}_1$ in the space
$\mathbb{R}^{2n}$ with the norm $\|\cdot\|_w$: $\xi = y/\|y\|_w$.

Define the mapping $\pi : \mathcal{S}_1 \times \overline{\mathbb{R}}_+ \longrightarrow \mathbb{R}^{2n}$ to be

$$\pi(\xi, \tau) = \big(\tau^{w_1}\xi_1, \ldots, \tau^{w_{2n}}\xi_{2n}\big).$$

It is clear that π is a one–to–one correspondence in the interior of $\mathcal{S}_1 \times \mathbb{R}_+$.
It collapses the boundary $B = \mathcal{S}_1 \times 0$ onto 0. Furthermore, $\pi^{-1}(t \cdot y) =
(\xi, t\tau)$, where $\pi(\xi, \tau) = y \neq 0$. It is easy to see that the lifting of the
map σ, $\hat{\sigma}(\xi, \tau) = \pi^{-1}\big(\sigma(\pi(\xi, \tau))\big)$, can be prolonged smoothly from the
interior $\mathcal{S}_1 \times \mathbb{R}_+$ to the boundary B.

It can be shown that the mapping $\hat{\sigma}$ has no fixed points on the
boundary $B = \pi^{-1}(0)$, so we look for $\hat{\sigma}^2$–fixed points lying on B and
a $\hat{\sigma}^2$–invariant contracting manifold in its neighborhood. The π–image of
such an invariant manifold would give the Fuller pair (Γ_+, Γ_-) of system
(4.2) at $0 \in \mathbb{R}^{2n}$. In the case of Kupka's problem it can be demonstrated
by a straightforward calculation that the mapping $\hat{\sigma}^2$ has a fixed point
$\zeta \in B$ with a single eigenvalue λ lying in the interval $(0, 1)$ while the

other eigenvalues have absolute values greater than 1. The eigenvector of λ is transversal to $T_\zeta B$. The disturbance of system (4.2) by terms of higher order (the quasihomogeneous vector field in Kupka's terminology) has the same property. In particular, the restrictions $\widehat{\sigma}^2\big|_B$ and $D\widehat{\sigma}^2\big|_{TB}$ coincide both for the perturbed problem and for Kupka's problem itself. Now the statement of Theorem K follows from the invariant manifold theorem.

Let us denote by $\mathfrak{S}$ the set of Fuller's points y_0 of system (4.2). The last result of Kupka concerns the projection of $\mathfrak{S}$ into the n–dimensional space of the original optimal problem. Let $\mathfrak{M}$ be a smooth n–manifold and $\mathcal{U}$ be a closed subset of $\mathbb{R}^u$. Denote by $\mathfrak{DS}$ the set of all pairs (V, C) of a smooth vector field V ($V : \mathfrak{M} \times \mathcal{U} \longrightarrow T\mathfrak{M}$) and a smooth function $C : \mathfrak{M} \times \mathcal{U} \longrightarrow \mathbb{R}$. We endow the set $\mathfrak{DS}$ with C^∞–topology. Let us consider system (4.2) with the Hamiltonian $H(p, x, u) = pV(x, u) - \lambda C(x, u)$, where λ is a parameter.

Corollary K [I. Kupka, 1988]. *Assume that the dimension of $\mathfrak{M}$ is at least 4. Then there exists a nonempty submanifold $\mathfrak{S}(V, C) \subset \mathfrak{M}$ of codimension 4 such that all points of $\mathfrak{S}(V, C)$ are Fuller's points for (V, C).*

To prove the statement Kupka produces the Hamiltonian

$$H_0 = H(p, x, u) = F_0 + u\, G_0$$

which is associated with the following control system in $\mathbb{R}^4$:

$$
\begin{aligned}
\dot{x}_1 &= u, & \dot{x}_3 &= x_1 + Bx_1^2, \\
\dot{x}_2 &= 1, & \dot{x}_4 &= D_1 x_1^2 + D_2 x_1 x_2, \\
& & u &\in [-1, 1],
\end{aligned}
$$

and with the cost functional

$$
\begin{aligned}
C = {} & \frac{a}{2}x_3^2 - \frac{b}{2}x_1^2 x_3 - c_0 x_1 x_2^3 - c_1 x_1^2 x_2^2 - c_2 x_1^3 x_2 - c_3 x_1^4 \\
& - A_1 x_1^2 x_4 - E_1 x_1^3 x_3 - A_2 x_1 x_2 x_4 - E_2 x_1 x_2 x_3,
\end{aligned}
$$

B, D_1, D_2, a, b, c_i ($i = 0, \dots, 3$), A_1, A_2, E_1, E_2 being constants. Assume that

(i) $(a, b, c_0, c_1, c_2, c_3)$ is an interior point of the set $\mathfrak{F}$ (where $\mathfrak{F}$ has been defined by Theorem K),

(ii) both determinants Δ_1 and Δ_2 are nonzero,

$$
\Delta_1 = \begin{vmatrix} A_1 & D_1 \\ A_2 & D_2 \end{vmatrix}, \qquad
\Delta_2 = \begin{vmatrix} 24c_3 & 6c_2 & E_1 \\ b + 4c_2 & 4c_1 & aB \\ a + 2c_1 & 6c_0 & 6E_2 \end{vmatrix}.
$$

Condition (i) implies that system $\mathfrak{S}_0$ associated with the Hamiltonian H_0 has a Fuller point at $m_0 = 0$. Using (ii) it can be shown that the differentials of all eight Poisson brackets of functions F_0 and G_0 of length less than or equal to four are independent at 0. It follows from Theorem K that any system $\mathfrak{S}$ lying near $\mathfrak{S}_0$ in C^∞–topology has a Fuller point near the point m_0. It implies the statement of Corollary K.

Let us show the distinctions between Kupka's results and ours. Both methods are founded on the reduction of the system to a certain semi–canonical form and on the resolution of its singularity at the Fuller point by means of a blowing up procedure. The differences can be explained by the different purposes. Kupka's object was the proof of the ubiquity of Fuller's phenomenon. This presupposes the establishment of a finite number of restrictions on the Hamiltonian at a single Fuller point. Hence, the information about the neighboring points can be reduced to a minimum. Really, Kupka considers six–dimensional Pontryagin's system for Kupka's problem imbedded in a space of higher dimension as a direct factor in a direct space product. This is easy to see that the single Fuller point of the six–dimensional system gives the whole $(2n - 6)$–dimensional manifold $\mathcal{S}$ such that for any $m_0 \in \mathcal{S}$ there exists a Fuller pair $(\Gamma_+, \Gamma_-)(m)$. Recall that every variable of the associated six–dimensional Hamiltonian system has nonzero weight under the Fuller's group action. There exist two different ways to disturb the system. Kupka uses the most general one, prescribing nonzero weights to the supplemental variables. Such disturbances retain a single Fuller point on the switching surface. In our approach, the disturbances preserve all Fuller points on the switching surface of the optimal synthesis. Of course, it presupposes essentially more restrictions on the Hamiltonian (even an infinite number of them), but more restrictions provide more results. We have already mentioned that Kupka's results on the ubiquity are not advanced enough to prove the optimality of the chattering extremals. The advantage of our technique includes the possibility of the determination of the chattering arcs manifolds (at least their tangent spaces at Fuller points) and the explicit construction of the extremals fields with chattering arcs. It allows us to apply our results to numerous optimal problems in applied mathematics, engineering, mechanics.

To conclude the survey of Kupka's results let us mention that his approach was connected with an extension of the symmetric Fuller problem, where $u \in [-1, 1]$. We have considered a disturbation of a more general nonsymmetric problem of A.T. Fuller–C. Marchal (Problem 3.2) depending on the real parameter $b < 0$, $u \in [b, 1]$.

In the next section we slightly improve Kupka's result and decrease the estimation of the codimension of Fuller's phenomenon from eight to seven.

4.2 Codimension of the Set of Fuller points

We now apply the techniques developed in Chapter 3 to the investigation of the ubiquity of Fuller's phenomenon. We look for a certain convenient form of system (4.2) in the vicinity of an extremal of second local order. As was done above, we denote by z_1 the function G and by z_2 the function $\{F, G\}$. Since for any function $L(p, x)$ the derivative of L along solutions of system (4.2) is the Poisson bracket $\{F + uG, L\}$, we have $\dot{z}_1 = z_2$. As well, $\dot{z}_2 = \{F + uG, z_2\} = \{F, \{F, G\}\} + u\{G, \{F, G\}\}$. Let us set $z_3 = \{F, \{F, G\}\}$, $w_1 = \{G, \{F, G\}\}$. If the intrinsic order of system (4.2) is greater than 1, then $\{G, \{F, G\}\} \equiv 0$ as the function of (p, x). This case was considered in Chapter 3. Here we will suppose only that there exists a singular solution $r(t) = (p(t), x(t))$ of second local order such that $w_1(r(t)) = 0$ and hence $dw_1(r(t))/dt = 0$. We are to establish a canonical form of the Hamiltonian system in a small neighborhood of $r(t)$. We have

$$\dot{z}_3 = \{\mathrm{ad}^3 F, G\} + u\Big\{G, \{F, \{F, G\}\}\Big\},$$
$$\dot{w}_1 = \Big\{F, \{G, \{F, G\}\}\Big\} + u\Big\{G, \{G, \{F, G\}\}\Big\}.$$

Jacobi's identity implies
$$\Big\{G, \{F, \{F, G\}\}\Big\} \equiv \Big\{F, \{G, \{F, G\}\}\Big\}.$$

Let us put

$$z_4 = \Big\{F, \{F, \{F, G\}\}\Big\},$$
$$w_2 = \Big\{G, \{F, \{F, G\}\}\Big\},$$
$$w_3 = \Big\{G, \{G, \{F, G\}\}\Big\}.$$

Then
$$\dot{z}_3 = z_4 + u\,w_2, \qquad \dot{w}_1 = w_2 + u\,w_1.$$

Finally,

$$\dot{z}_4 = \alpha(p, x) + u\,\beta(p, x),$$
$$\dot{w}_2 = \gamma(p, x) + u\,\delta(p, x),$$
$$\dot{w}_3 = \epsilon(p, x) + u\,\zeta(p, x),$$

where

$$\alpha = \mathrm{ad}^4 F(G), \qquad\qquad \beta = \{G, \{\mathrm{ad}^3 F(G)\}\},$$
$$\gamma = \Big\{F, \{G, \{F, \{F, G\}\}\}\Big\}, \qquad \delta = \Big\{G, \{G, \{F, \{F, G\}\}\}\Big\},$$
$$\epsilon = \Big\{F, \{G, \{G, \{F, G\}\}\}\Big\}, \qquad \zeta = \Big\{G, \{G, \{G, \{F, G\}\}\}\Big\}.$$

Set $r = (z_1, z_2, z_3, z_4, w_1, w_2, w_3, v) \in \mathbb{R}^{2n}$ $(v \in \mathbb{R}^{2n-7})$. Let us consider the system

$$
\begin{aligned}
\dot{z}_1 &= z_2, & \dot{w}_1 &= w_2 + u\,w_3, \\
\dot{z}_2 &= z_3 + u\,w_1, & \dot{w}_2 &= \gamma + u\,\delta, \\
\dot{z}_3 &= z_4 + u\,w_2, & \dot{w}_3 &= \epsilon + u\,\zeta, \\
\dot{z}_4 &= \alpha + u\,\beta, & \dot{v} &= V(r, u),
\end{aligned}
\tag{4.3}
$$

where $u = \operatorname{sgn} z_1$; $\alpha, \beta, \gamma, \delta, \epsilon, \zeta$ are C^∞–functions of r, and V is a C^∞ vector–function of r, u. Let us consider a point $r_0 = (0, v_0) \in \mathbb{R}^{2n}$ and the mapping $\mathcal{A} : \mathbb{R}^{2n} \longrightarrow \mathbb{R}^6$, $\mathcal{A}(r) = (\alpha, \beta, \gamma, \delta, \epsilon, \zeta)(r)$.

THEOREM 4.1. **Theorem on ubiquity.** *There exists a set $\mathfrak{F} \in \mathbb{R}^6$ with a nonempty interior such that if $\mathcal{A}(r_0) \in \mathfrak{F}$, then system (4.3) has at least one Fuller pair (Γ_+, Γ_-) at r_0.*

Proof. We will repeat the main stages of the proof of Theorems 3.1–3.3. Fortunately, the essential part of the procedure has already been completed. As usual, we denote by $\Sigma = \{ r \mid z_1 = 0 \}$ the switching surface in $\mathbb{R}^{2n}$ and consider the following change of variables in $\Sigma \backslash \{r_0\}$:

$$
(z_1, z_2, z_3, z_4, w_1, w_2, w_3, v) \longrightarrow (\tau, \xi_1, \xi_2, \xi_3, \xi_4, \xi_5, \nu),
$$
$$
z_2 = \xi_1 \tau^3, \quad z_3 = \xi_2 \tau^2, \quad z_4 = \tau,
$$
$$
w_1 = \xi_3 \tau^2, \quad w_2 = \xi_4 \tau, \quad w_3 = \xi_5 \tau,
$$
$$
v = v_0 + \nu\tau.
$$

We would like to write out the Poincaré mapping $\Sigma \to \Sigma$ for system (4.3) in terms of (τ, ξ, ν). Let $r(0) \in \Sigma \backslash \{r_0\}$ be a starting point for the trajectory $r(t)$ of system (4.3) such that $u(t) \equiv 1$ for $t \in (0, t_1)$ and $u(t) \equiv -1$ for $t \in (t_1, t_1 + t_2)$. Assume that $r(t)$ intersects the switching surface Σ at the instants $t = 0, t_1, t_1 + t_2$ transversally (i.e., $z_2(0) \neq 0$, $z_2(t_1) \neq 0$, $z_2(t_1 + t_2) \neq 0$). In this case the coordinates of the point $\tilde{r} = r(t_1 + t_2)$ and functions t_1, t_2 are smooth in $r(0)$. Integration of the equation $\dot{w}_3 = \epsilon(r) + u\,\zeta(r)$ on the interval $(0, t_1 + t_2)$ gives

$$
w_3(t_1 + t_2) = w_3(0) + \int_0^{t_1} (\epsilon + \zeta)\, dt + \int_{t_1}^{t_1 + t_2} (\epsilon - \zeta)\, dt.
$$

Let us explore functions t_1, t_2 with $t_1 = \sigma_1 \tau$, $t_2 = \sigma_2 \tau$, where $\sigma_1 \neq 0$, $\sigma_2 \neq 0$. In view of the mean value theorem we have

$$
\epsilon\big(r(t)\big) - \epsilon\big(r(0)\big) = \frac{\partial \epsilon}{\partial r}(\bar{r})\,\big(r(t) - r(0)\big)
$$
$$
= \frac{\partial \epsilon}{\partial r}(\bar{r}) \frac{\partial r}{\partial t}(\bar{t})\, t = \Theta(t, r(0))\, t,
$$

Θ being a C^1–function defined in some open neighborhood of $r_0 = (0, v_0)$. Hence,

$$\widetilde{w}_3 = \widetilde{\xi}_5\widetilde{\tau} = \Big(\xi_5(0) + (\epsilon(r_0) + \zeta(r_0))\sigma_1 + (\epsilon(r_0) - \zeta(r_0))\sigma_2\Big)\tau + \Theta(1)\tau^2.$$

Here, and everywhere below in this chapter, $\Theta(1)$ denotes an arbitrary C^1–function (one's own in every concrete case) of variables τ, ξ, ν, σ_1, σ_2. The tilde over the letter means that the function has been calculated at the point $r(t_1 + t_2)$ of the second switching. It can be shown in a similar way that

$$\widetilde{z}_4 = \quad \widetilde{\tau} = \Big(1 + (\alpha + \beta)\sigma_1 + (\alpha - \beta)\sigma_2\Big)\tau + \Theta(1)\tau^2,$$

$$\widetilde{w}_3 = \widetilde{\xi}_5\widetilde{\tau} = \Big(\xi_5 + (\epsilon + \zeta)\sigma_1 + (\epsilon - \zeta)\sigma_2\Big)\tau + \Theta(1)\tau^2,$$

$$\widetilde{w}_2 = \widetilde{\xi}_4\widetilde{\tau} = \Big(\xi_4 + (\gamma + \delta)\sigma_1 + (\gamma - \delta)\sigma_2\Big)\tau + \Theta(1)\tau^2,$$

$$\widetilde{w}_1 = \widetilde{\xi}_3\widetilde{\tau}^2 = \Big(\xi_3 + (\xi_4 + \xi_5)\sigma_1 + (\gamma + \delta + \epsilon + \zeta)\frac{\sigma_1^2}{2}$$

$$+ \Big(\xi_4 - \xi_5 + (\gamma + \delta - \epsilon - \zeta)\sigma_1\Big)\sigma_2$$

$$+ (\gamma - \delta - \epsilon + \zeta)\frac{\sigma_2^2}{2}\Big)\tau^2 + \Theta(1)\tau^3, \qquad (4.4.\text{a})$$

$$\widetilde{z}_3 = \widetilde{\xi}_2\widetilde{\tau}^2 = \Big(\xi_2 + (1 + \xi_4)\sigma_1 + (\alpha + \beta + \gamma + \delta)\frac{\sigma_1^2}{2}$$

$$+ \Big(1 - \xi_4 + (\alpha - \beta - \gamma + \delta)\sigma_1\Big)\sigma_2$$

$$+ (\alpha - \beta - \gamma + \delta)\frac{\sigma_2^2}{2}\Big)\tau^2 + \Theta(1)\tau^3,$$

$$\widetilde{z}_2 = \widetilde{\xi}_1\widetilde{\tau}^3 = \Big(\xi_1 + (\xi_2 + \xi_3)\sigma_1 + (1 + 2\xi_4 + \xi_5)\frac{\sigma_1^2}{2}$$

$$+ (\alpha + \beta + 2(\gamma + \delta) + \epsilon + \zeta)\frac{\sigma_1^3}{6}$$

$$+ \Big(\xi_2 - \xi_3 + (1 - \xi_5)\sigma_1 + (\alpha + \beta - \epsilon - \zeta)\frac{\sigma_1^2}{2}\Big)\sigma_2$$

$$+ \Big(1 - 2\xi_4 + \xi_5 + (\alpha + \beta - 2(\gamma + \delta) + \epsilon + \zeta)\sigma_1\Big)\frac{\sigma_2^2}{2}$$

$$+ (\alpha - \beta - 2(\gamma - \delta) + \epsilon - \zeta)\frac{\sigma_2^3}{6}\Big)\tau^3 + \Theta(1)\tau^4,$$

$$v_0 + \widetilde{\widetilde{\nu}}\widetilde{\widetilde{\tau}} = v_0 + \big(\nu + V(0, v_0, 1)\sigma_1 + V(0, v_0, -1)\sigma_2\big)\tau + \Theta(1)\tau^2.$$

The switching instants $t_1 = \sigma_1\tau$ and $t_2 = \sigma_2\tau$ are specified by the roots of the equations

$$\xi_1 + (\xi_2 + \xi_3)\frac{\sigma_1^2}{2} + (1 + 2\xi_4 + \xi_5)\frac{\sigma_1^2}{6}$$
$$+ (\alpha + \beta + 2(\gamma + \delta) + \epsilon + \zeta)\frac{\sigma_1^3}{24} + \Theta(1)\tau = 0 \tag{4.4.b}$$

and

$$\xi_1 + (\xi_2 + \xi_3)\sigma_1 + (1 + 2\xi_4 + \xi_5)\frac{\sigma_1^2}{2}$$
$$+ (\alpha + \beta + 2(\gamma + \delta) + \epsilon + \zeta)\frac{\sigma_1^3}{6} \tag{4.4.c}$$
$$+ \left(\xi_2 - \xi_3 + (1 - \xi_5)\sigma_1 + (\alpha + \beta - \epsilon - \zeta)\frac{\sigma_1^2}{2}\right)\frac{\sigma_2}{2}$$
$$+ \left(1 - 2\xi_4 + \xi_5 + (\alpha + \beta - 2(\gamma + \delta) + \epsilon + \zeta)\sigma_1\right)\frac{\sigma_2^2}{6}$$
$$+ (\alpha - \beta - 2(\gamma - \delta) + \epsilon - \zeta)\frac{\sigma_2^3}{24} + \Theta(1)\tau = 0.$$

The letters α, β, γ, δ, ϵ, ζ in (4.4) mean the values of the corresponding functions at $r_0 = (0, v_0)$. We seek the fixed point (τ^*, ξ^*, ν^*) of mapping (4.4) on the surface $\tau = 0$. Let us set $\widetilde{\tau} = \rho\tau$, $\widetilde{\xi} = \xi = \xi^*$, $\widetilde{\nu} = \nu = \nu^*$ in (4.4), cancel out the τ–monomial factors, and then put $\tau = 0$. Since system (4.4.a) is linear in ξ^*, ν^* in this case, the variables ξ^* and ν^* can be expressed through σ_1, σ_2 and then substituted into (4.4.b), (4.4.c). Let the equations $\Sigma_1(\sigma_1, \sigma_2) = 0$, $\Sigma_2(\sigma_1, \sigma_2) = 0$ be the result of such a substitution. It is easy to see that $\xi^*, \nu^*, \Sigma_1, \Sigma_2$ are smooth in σ_1, σ_2, and the parameters $\mathfrak{p} = (\alpha, \beta, \gamma, \delta, \epsilon, \zeta)$ if $\rho \overset{\text{def}}{=} 1 + (\alpha + \beta)\sigma_1 + (\alpha - \beta)\sigma_2 \neq 1$. The following observation eliminates the necessity of writing out the explicit form of functions Σ_1, Σ_2.

Let functions $\mathfrak{p}$ in (4.3) all be constant and $V \equiv V(0, v_0, u)$. Let us demonstrate that there exists an open set $\mathfrak{F}_0$ in the six–dimensional space of parameters $\mathfrak{p}$ such that the equations $\Sigma_1 = 0$, $\Sigma_2 = 0$ have a unique solution (σ_1, σ_2) depending on $\mathfrak{p}$ with the following properties:

(1) $\rho = 1 + (\alpha + \beta)\sigma_1 + (\alpha - \beta)\sigma_2 \in (0, 1)$;

(2) $\sigma_1 > 0$, $\sigma_2 > 0$;

(3) the solution $r(t)$ of system (4.3), whose starting point $r(0)$ is defined by ξ^*, ν^* and by an arbitrary small $\tau > 0$, meets the conditions: $z_1(t) > 0$ for $t \in (0, \tau\sigma_1)$ and $z_2(t) < 0$ for $t \in (\tau\sigma_1, \tau(\sigma_1 + \sigma_2))$.

Let us put $\gamma = \delta = \epsilon = \zeta = 0$, $\alpha - \beta = 1$, $\alpha + \beta = b < 0$ (b is an arbitrary real parameter) and put $w_1(0) = w_2(0) = w_3(0) = 0$. Then the uniqueness theorem for ordinary differential equations yields that $w_1(t) = w_2(t) = w_3(t) = 0$ identically. We arrive at Fuller's problem, since the restriction of system (4.3) on the surface $w_1 = w_2 = w_3 = 0$ coincides with system (3.5) in Chapter 3. This is the circumstance that saves us from the recurrence of many complicated calculations later on. As follows from Lemma 3.2, there exists at least one solution (σ_1^*, σ_2^*) of the equations $\Sigma_1 = 0$, $\Sigma_2 = 0$, meeting the conditions (1)–(3). Moreover, the pair (σ_1^*, σ_2^*) is an isolated solution and

$$\det \left. \frac{D(\Sigma_1, \Sigma_2)}{D(\sigma_1, \sigma_2)} \right|_{(\sigma_1^*, \sigma_2^*)} \neq 0$$

(see the remark on the nondegeneracy of the fixed point in Section 3.8). The implicit function theorem gives that for all $\mathfrak{p}$ in some open neighborhood of the point $\mathfrak{p}_0 = \left((1+b)/2, (b-1)/2, 0, 0, 0, 0 \right)$ there exists a solution $(\sigma_1^*, \sigma_2^*)(\mathfrak{p})$ meeting conditions (1)–(3). Hence, the set $\mathfrak{F}$ has a nonempty interior.

It is left to prove that $(0, \xi^*, \nu^*)$ is the hyperbolic fixed point for the transformation $(\tau, \xi, \nu) \to (\widetilde{\tau}, \widetilde{\xi}, \widetilde{\nu})$ specified by (4.4). We have

$$D = \left. \frac{D(\widetilde{\tau}, \widetilde{\xi}_1, \widetilde{\xi}_2, \widetilde{\nu}, \widetilde{\xi}_3, \widetilde{\xi}_4, \widetilde{\xi}_5)}{D(\tau, \xi_1, \xi_2, \nu, \xi_3, \xi_4, \xi_5)} \right|_{(0, \xi^*, \nu^*)} = \begin{pmatrix} \mathfrak{a} & \mathfrak{b} \\ \mathfrak{c} & \mathfrak{d} \end{pmatrix},$$

where

$$\mathfrak{a} = \frac{D(\widetilde{\tau}, \widetilde{\xi}_1, \widetilde{\xi}_2, \widetilde{\nu})}{D(\tau, \xi_1, \xi_2, \nu)}, \quad \mathfrak{b} = \frac{D(\widetilde{\tau}, \widetilde{\xi}_1, \widetilde{\xi}_2, \widetilde{\nu})}{D(\xi_3, \xi_4, \xi_5)},$$

$$\mathfrak{c} = \frac{D(\widetilde{\xi}_3, \widetilde{\xi}_4, \widetilde{\xi}_5)}{D(\tau, \xi_1, \xi_2, \nu)}, \quad \mathfrak{d} = \frac{D(\widetilde{\xi}_3, \widetilde{\xi}_4, \widetilde{\xi}_5)}{D(\xi_3, \xi_4, \xi_5)}$$

are $(2n-4) \times (2n-4)$, $(2n-4) \times 3$, $3 \times (2n-4)$, and 3×3 matrices respectively. We search $\operatorname{spec} D$ at the fixed point $(0, \xi^*, \nu^*)(\mathfrak{p}_0)$. By the definition, at this point, the matrix $\mathfrak{a}$ coincides with the matrix D^+, calculated in Lemma 3.4. So $\operatorname{spec} \mathfrak{a}$ consists of a single eigenvalue less than 1 in absolute value (equal to ρ_0^+) and $2n-5$ eigenvalues greater than 1 in absolute value. The structure of the matrix $\mathfrak{b}$ is of no significance. Let us note only that the first row of D is $\operatorname{grad} \tau \big|_{(0, \xi^*, \nu^*)} = (\rho_0^+, 0, 0, 0, 0, 0, 0)$. Taking into account that $\gamma = \delta = \epsilon = \zeta = 0$, $\xi_3^* = \xi_4^* = \xi_5^* = 0$, and differentiating (4.4.a), we get $\mathfrak{c} = \begin{pmatrix} * \\ * & \mathbb{O} \\ * \end{pmatrix}$, where the asterisks denote the partial derivatives $\partial \widetilde{\xi}_3 / \partial \tau$, $\partial \widetilde{\xi}_4 / \partial \tau$, $\partial \widetilde{\xi}_5 / \partial \tau$, and $\mathbb{O}$ denotes the zero $3 \times (2n-5)$–matrix. Therefore, $\operatorname{spec} D$ is the union of $\operatorname{spec} \mathfrak{a}$ and $\operatorname{spec} \mathfrak{d}$. A straightforward calculation gives

$$
\mathfrak{d} = \begin{pmatrix} \rho^{-2} & (\sigma_1 + \sigma_2)\rho^{-2} & (\sigma_1 - \sigma_2)\rho^{-2} \\ 0 & \rho^{-1} & 0 \\ 0 & 0 & \rho^{-1} \end{pmatrix}.
$$

It follows that three eigenvalues of $\mathfrak{d}$ have absolute values greater than 1, so D is hyperbolic at the "Fuller" point $(0, \xi^*, \nu^*)$ and remains hyperbolic at all fixed points corresponding to the parameters $\mathfrak{p}$ in some open neighborhood $\mathfrak{F}_1 \subset \mathfrak{F}_0$ of $\mathfrak{p}_0$. The eigenvector of D with the eigenvalue $\rho \in (-1,1)$ is continuous in $\mathfrak{p}$, so it is transversal to the surface $\tau = 0$ for all $\mathfrak{p}$ in some open $\mathfrak{F} \subset \mathfrak{F}_1$. The Fuller pair (Γ_+, Γ_-) of system (4.3) at $r_0 = (0, v_0)$, such that $\mathcal{A}(\mathfrak{p}(r_0)) \in \mathfrak{F}$, can be generated by the usual way on the basis of the invariant manifold theorem. That ends the proof of Theorem 4.1.

Q.E.D.

Remark 4.1. *The set $\mathfrak{F}$ in Theorem 4.1 contains some open neighborhood of the point $\mathfrak{p}_0 = \big((1+b)/2, (b-1)/2, 0, 0, 0, 0\big)$ $(b < 0)$. This gives an explicit description of an open set of Hamiltonian systems which possess Fuller pairs (Γ_+, Γ_-).*

Remark 4.2. *A simplest sufficient condition for the reducibility of system (4.2) to the (4.3)–form is the functional independence of the Poisson brackets $z_1, z_2, z_3, z_4, w_1, w_2, w_3$ at r_0. But this condition can be slightly extended to include some degenerate cases. For example, we can suppose that the function $w_2 = \big\{ G, \{F, \{F, G\}\} \big\}$ is identically zero (or both $w_2 = \big\{ G, \{F, \{F, G\}\} \big\}$ and $w_3 = \big\{ G, \{G, \{F, G\}\} \big\}$ are identically zero, or any other collection of functions w_i is identically zero). Then system (4.2) can be reduced to the form (4.3) where functions γ and δ are identically zero (or $\gamma \equiv \delta \equiv \epsilon \equiv \zeta \equiv 0$ relatively, etc.). Hence, the surface $\mathfrak{P} = \{z, w \mid w_2 = 0\}$ (or $\mathfrak{P} = \{z, w \mid w_2 = w_3 = 0\}$, etc.) is an integral variety of (4.3) and the whole Fuller pair (Γ_+, Γ_-), generated as an invariant manifold in Theorem 3.1, lies in the surface $\mathfrak{P}$.*

Let $\mathfrak{H}$ be a subset of the direct product $C^\infty(\mathbb{R}^{2n}) \times C^\infty(\mathbb{R}^{2n})$ with the following properties. For any $H = F + uG \in \mathfrak{H}$ there exist an open subset $\mathcal{U} \subset \mathbb{R}^{2n}$ and a smooth manifold $\mathcal{S} \subset \mathcal{U}$ of codimension 7 such that for any $m \in \mathcal{S}$ there exists a Fuller pair $(\Gamma_+, \Gamma_-)(m)$ of system (4.3) with the Hamiltonian H.

COROLLARY 4.2. *The set $\mathfrak{H}$ has a nonempty interior in the slight Witney's topology of the direct product. Hence, the codimension of the Fuller's singularity does not exceed 7.*

Proof. Let us consider the Hamiltonian $H_0 = F_0 + uG_0$, $u = \operatorname{sgn} G_0$, associated with the control system in $\mathbb{R}^{2n}$,

$$\dot{x}_1 = u, \quad \dot{x}_2 = 1, \quad \dot{x}_3 = x_1, \quad \dot{x}_4 = \epsilon_1 x_1^2,$$
$$\dot{r}_1 = 0, \quad \dot{r}_2 = 0, \ldots, \dot{r}_{2n-4} = 0,$$
$$u \in [-1, 1],$$

and the functional

$$\int_0^T x_3^2 + \epsilon_2 x_1^2 x_2^2 + \epsilon_3 x_1^3 x_4 \, dt,$$

i.e., $F_0 = \psi_3 x_1 + \epsilon_1 \psi_4 x_1^2 + \psi_2 - x_3^2 - \epsilon_2 x_1^2 x_2^2 - \epsilon_3 x_1^3 x_4$, $G_0 = z_1 = \psi_1$.
A straightforward calculation yields

$$z_2 = \{F_0, G_0\} = -\psi_3 - 2\epsilon_1 \psi_4 x_1 + 2\epsilon_2 x_1 x_2^2 + 3\epsilon_3 x_1^2 x_4,$$

$$z_3 = \{F_0, \{F_0, G_0\}\} = \epsilon_1 \epsilon_3 x_1^4 + 4\epsilon_2 x_1 x_2 - 2x_3,$$

$$w_1 = \{G_0, \{F_0, G_0\}\} = -2\epsilon_1 \psi_4 + 2\epsilon_2 x_2^2 + 6\epsilon_3 x_1 x_4,$$

$$z_4 = \left\{F_0, \{F_0, \{F_0, G_0\}\}\right\} = (-2 + 4\epsilon_2)x_1,$$

$$w_2 = \left\{G_0, \{F_0, \{F_0, G_0\}\}\right\} = 4\epsilon_1 \epsilon_3 x_1^3 + 4\epsilon_2 x_2,$$

$$w_3 = \left\{G_0, \{G_0, \{F_0, G_0\}\}\right\} = 6\epsilon_3 x_4,$$

where

$$\alpha = \operatorname{ad}^4 F_0(G_0) = 0,$$

$$\beta = \operatorname{ad} G_0 \left(\operatorname{ad}^3 F_0 (G_0)\right) = -2 + 4\epsilon_3,$$

$$\gamma = \operatorname{ad} F_0 \left(\operatorname{ad} G_0 \left(\operatorname{ad}^2 F_0 (G_0)\right)\right) = 4\epsilon_2,$$

$$\delta = \operatorname{ad}^2 G_0 \left(\operatorname{ad}^2 F_0 (G_0)\right) = 12\epsilon_1 \epsilon_3 x_1^2,$$

$$\epsilon = \operatorname{ad} F_0 \left(\operatorname{ad}^2 G_0 \left(\operatorname{ad} F_0 (G_0)\right)\right) = 6\epsilon_1 \epsilon_3 x_1^2,$$

$$\zeta = \operatorname{ad}^3 G_0 \left(\operatorname{ad} F_0 (G_0)\right) = 0.$$

For all sufficiently small nonzero parameters ϵ_1, ϵ_2, ϵ_3, the equations $z_i = 0$ $(i = 1, \ldots, 4)$ and $w_j = 0$ $(j = 1, 2, 3)$ imply $x_i = 0$ $(i = 1, \ldots, 4)$ and $\psi_j = 0$ $(j = 1, 3, 4)$. It follows from Theorem 4.1 that all $(\psi, x) \in \mathcal{L} \overset{\text{def}}{=} \{(\psi, x) \mid x_i = 0, \ i = 1, \ldots, 4, \ \psi_j = 0, \ j = 1, 3, 4\}$ are Fuller points of system (4.3) with the Hamiltonian H. Since the

Hamiltonian H has been chosen in such a way that each of the Poisson brackets z_i $(i = 1, \ldots, 4)$ and w_j $(j = 1, 2, 3)$, contains the only one first order term, it can readily be seen that the differentials of z_i, w_j are independent at all $(\psi, x) \in \mathcal{L}$. So $\mathcal{L}$ is a smooth manifold in $\mathbb{R}^{2n}$ and codim $\mathcal{L} = 7$. Hence, all the Hamiltonians in some neighborhood of H_0 in $C^\infty(\mathbb{R}^{2n}) \times C^\infty(\mathbb{R}^{2n})$ belong to $\mathfrak{H}$.

Q.E.D.

4.3 Structural Stability of the Optimal Synthesis in the Two–Dimensional Fuller Problem

We begin with some general result closely related to the previous theme. Consider system (3.5) in Chapter 3 under the same assumptions on its right–hand side except for condition (3.6). Assume that the functions $f_i(z, w, u)$ depend on some small parameter ϵ, $f_i(z, w, u) = \epsilon\phi_i(z, w, u)$, $i = 1, 2, 3, 4$. Let us replace condition (3.6) by the following one:

$$\varlimsup_{\kappa \to +0} \left| \phi_i\big(\mathfrak{g}_\kappa(z), w, u\big) \right| \kappa^{-(4-i)} < C \tag{4.5}$$

for all (z, w, u) in some open neighborhood of the set

$$\bigcup_{u \in [-1, 1]} (0, w_0, u)$$

(i.e., we decrease the order of the perturbation terms f_i with respect to $\mathfrak{g}_\kappa$–action by 1).

It can readily be proved that if ϵ is small enough, then all the results on the existence of the bundles with chattering arcs, on the existence of Lagrangian submanifolds with chattering arcs and their optimality, and on the nice projectibility of these manifolds are still valid.

PROPOSITION 4.1. *For any sufficiently small $\epsilon \geqslant 0$ the statements of Theorems 3.1–3.4 hold for system (3.5) whose right–hand side meets (4.5).*

Proof. In view of (4.5), the mapping $\widetilde{\Phi}^2$ in Section 3.4 is continuous with respect to ϵ and is defined by equations (3.11), supplemented with functions of $O(\epsilon)$ form. Set $\widetilde{\Phi}^2 = F(\epsilon)$. Since $F(0)$ is a diffeomorphism in the vicinity of its two hyperbolic fixed points (Lemma 3.4), for all sufficiently small $\epsilon \geqslant 0$ there are two fixed points of the mapping $F(\epsilon)$ with the same hyperbolic structure in their neighborhood. The rest of the proof remains the same.

Q.E.D.

As an application of Proposition 4.1 let us consider the following

PROBLEM 4.1. *Minimize*

$$\int_0^T \left(x^2(1 + \epsilon f(x,y)) + \epsilon y^4 g(x,y) \right) dt$$

subject to

$$\dot{x} = y + \epsilon\Big(\phi_1(x,y) + u\,\phi_2(x,y)\Big),$$

$$\dot{y} = u + \epsilon\Big(\phi_3(x,y) + u\,\phi_4(x,y)\Big),$$

$$u \in [A_0, A_1]$$

with boundary conditions

$$x(0) = x_0, \quad y(0) = y_0, \quad x(T) = 0, \quad y(T) = 0.$$

Here f, g, ϕ_i $(i = 1, 2, 3, 4)$ are C^2–functions, $g \geqslant 0$, $\phi_1(0,0) = \phi_2(0,0) = 0$, A_0 and A_1 are some constants, $A_0 < 0$, $A_1 > 0$. Let the initial point belong to the domain $x_0^2 + y_0^2 < \delta^2$. Problem 4.1 is a perturbation of Fuller's problem in the general form. We prove the optimality of a chattering mode for this perturbation.

PROPOSITION 4.2. *For any sufficiently small $\epsilon \geqslant 0$ and $\delta > 0$ the optimal switching curve to Problem 4.1 has the form*

$$\Gamma = \begin{cases} x = \lambda_1(y)\,y^2, & y > 0, \\[2mm] x = \lambda_0(y)\,y^2, & y < 0, \end{cases}$$

with $\lambda_0(\cdot)$, $\lambda_1(\cdot)$ being C^1–functions such that $\lambda_1(0) \in (1/(2A_0), 0)$, $\lambda_0(0) \in (0, 1/(2A_1))$. The optimal control $\widehat{u}(t)$ equals A_0 on the left–hand side of the curve Γ and equals A_1 on its right–hand side. The optimal trajectories of Problem 4.1 attain the origin in finite time with an infinite number of switches of control.

Proof. Let us consider Pontryagin's maximum principle for Problem 4.1 with the Hamiltonian

$$H = \psi_1\big(y + \epsilon(\phi_1 + u\phi_2)\big) + \psi_2\big(u + \epsilon(\phi_3 + u\phi_4)\big) - \big(x^2(1 + \epsilon f) + \epsilon y^4 g\big).$$

We have

$$
\begin{aligned}
\dot{\psi}_2 &= -\psi_1 + \epsilon\Big(x^2 f_y' + 4y^3 g + y^4 g_y' - \psi_1(\phi_{1y}' + u\phi_{2y}') \\
&\qquad\qquad - \psi_2(\phi_{3y}' + u\phi_{4y}')\Big), \\
\dot{\psi}_1 &= 2x + \epsilon\Big(2xf + x^2 f_x' + y^4 g_x' - \psi_1(\phi_{1x}' + u\phi_{2x}') \\
&\qquad\qquad - \psi_2(\phi_{3x}' + u\phi_{4x}')\Big), \\
\dot{x} &= y + \epsilon(\phi_1 + u\phi_2), \\
\dot{y} &= u + \epsilon(\phi_3 + u\phi_4),
\end{aligned}
\tag{4.6}
$$

where

$$
u = \begin{cases}
A_0, & \text{if} \quad \psi_2(1 + \epsilon\phi_4) + \epsilon\psi_1\phi_2 > 0, \\
A_1, & \text{if} \quad \psi_2(1 + \epsilon\phi_4) + \epsilon\psi_1\phi_2 < 0.
\end{cases}
$$

Let us set

$$
\begin{aligned}
z_1 &= \psi_2 + \epsilon(\psi_2\phi_4 + \psi_1\phi_2), & z_3 &= -2x, \\
z_2 &= -\psi_1, & z_4 &= -2y.
\end{aligned}
$$

If ϵ is small enough, then the functions (z_1, z_2, z_3, z_4) are functionally independent. We are to ensure that system (4.6) meets relations (4.5) with respect to the mapping

$$\mathfrak{g}_\kappa(z_1, z_2, z_3, z_4) = (\kappa^4 z_1, \kappa^3 z_2, \kappa^2 z_3, \kappa z_4), \quad \kappa > 0.$$

First of all let us show that the variable ψ_2 is of the order at least four, i.e., $\overline{\lim}_{\kappa \to +0} \psi_2\big(\mathfrak{g}_\kappa(z)\big)\kappa^{-4} <$ const. Indeed, $\psi_2 = (z_1 - \epsilon\psi_1\phi_2)/(1 + \epsilon\phi_4)$, and it suffices to demonstrate that the function ϕ_2 has not less than the first order, i.e.,

$$\overline{\lim_{\kappa \to +0}} \; \frac{\phi_2(\kappa^2 x, \kappa y)}{\kappa} < \infty$$

for any (x, y) in the vicinity of the origin. By the assumption, we have $\phi_2(0, 0) = 0$. In view of the mean value theorem, we have

$$\big|\phi_2(x, y)\big| < \big|\phi_{2x}'(\Theta_1 x, \Theta_2 y)\big| \cdot |x| + \big|\phi_{2y}'(\Theta_1 x, \Theta_2 y)\big| \cdot |y|,$$

functions Θ_1 and Θ_2 being bounded to the segment $[0, 1]$. It follows that

$$\left| \phi_2(\kappa^2 x, \kappa y) \right| < O(1) \cdot \kappa,$$

which has been just required. Now the conditions on the right–hand side of (4.6) can be immediately verified by checking the weights of the variables in the expressions $\dot{z}_1$, $\dot{z}_2$, etc. The statement being proved follows from Theorems 3.1 and 3.3, and Proposition 4.1.   **Q.E.D.**

Let us determine the codimension of Problem 4.1 among all the two–dimensional problems with affine single input and an integral performance index,

$$\int_0^T f(x, y)\, dt \to \min,$$

$$\dot{x} = \phi_1 + u\phi_2, \quad \dot{y} = \phi_3 + u\phi_4,$$

$$u \in [A_0, A_1],$$

$$x(0) = x_0, \quad y(0) = y_0, \quad x(T) = 0, \quad y(T) = 0.$$

To obtain Problem 4.1 one must impose a finite number of restrictions in the form of equality on the functions f, ϕ_1, ϕ_2, ϕ_3, ϕ_4, namely,

$$f(0,0) = f'_x(0,0) = f'_y(0,0) = f''_{xy}(0,0) = f''_{yy}(0,0) = 0,$$

$$f'''_{xyy}(0,0) = f'''_{yyy}(0,0) = 0,$$

$$\phi_1(0,0) = \phi_2(0,0) = 0.$$

Besides that, some restrictions in the form of inequality are to be satisfied:

$$f''_{xx}(0,0) > 0, \quad f^{IV}_{yyyy}(0,0) > 0.$$

Thus, the chattering control is a phenomenon of a finite codimension d for the two–dimensional problems ($d = 9$).

In conclusion, let us describe the results of V. Telesnin [1986], closely connected with Problem 4.1. Telesnin designed the optimal synthesis for the problem

$$\int_0^T x^2(t)\, dt \longrightarrow \min,$$

$$\dot{x} = y,$$
$$\dot{y} = -x + u, \qquad u \in [-1, 1],$$

$$x(0) = x_0, \quad y(0) = y_0, \quad x(T) = y(T) = 0,$$

(x_0, y_0) being arbitrary initial conditions. In some small neighborhood of the origin this problem can be considered to be a particular case of Problem 4.1, and the statement of Proposition 4.2 holds. The technique used by V.R. Telesnin differs from ours. He has considered the perturbed problem

$$\int_0^T (x^2 + \epsilon y^2)\, dt \longrightarrow \min$$

with the same differential constraints. For all $\epsilon > 0$ there is no chattering mode for such perturbation and the switching curve, call it Γ_ϵ, can be synthesized by traditional means of backward integrating the adjoint system. The switching curve of the original problem can be obtained as a limit (in C^0-topology) of Γ_ϵ as $\epsilon \to +0$.

Chapter 5

HIGHER ORDER SINGULAR EXTREMALS

This chapter deals with the geometry of chattering arcs near the manifold of singular extremals of order greater than or equal to 3. It appears that in this case there exists a number of two–dimensional piecewise smooth manifolds which represent α– and ω–limit sets of all chattering trajectories in orbit space. The main result of the chapter is a solution of the three–dimensional Fuller problem. Besides that, we state a number of verisimilar conjectures concerning the problems of higher order. The theoretical and numerical foundations of these conjectures are presented.

5.1 Conjectures Concerning Higher Order Singular Modes

It was demonstrated in Chapter 3 that solutions to Fuller's problem determine the asymptotic behavior of chattering arcs for general Hamiltonian systems in the vicinity of the manifold of second order singular arcs. It would appear reasonable that solutions of the k–dimensional analog of Fuller's problem (Problem 5.1 below) determine the asymptotic behavior of chattering arcs near the manifold of singular trajectories of order k. It motivates the interest in a complete description of optimal solutions to the following problem:

PROBLEM 5.1. ***The Multidimensional Fuller Problem.***
Minimize

$$\int_0^\infty x^2\, dt$$

subject to

$$x^{(k)} = u \in [\beta, 1], \quad \beta < 0$$

with initial conditions

$$x^{(m)}(0) = x_m, \quad m = 0, \ldots, k-1.$$

The simplest case of Problem 5.1, where $k = 3$ and $\beta = -1$, was treated in [A.T. Fuller, P.E. Grensted, 1965], [C. Dorling, E. Ryan, 1981], and [C. Marchal, 1973]. In these papers, the solutions leading to the origin without switches were found as well as a family of automodelling chattering solutions. We shall say that a family of solutions is *automodelling* (self–similar) if it is invariant under the action of some group of symmetries. The behavior of all other optimal solutions and the structure of the switching surface for $k = 3$, $\beta = -1$ were treated in [V. Borisov, 1988]. All these results are presented and discussed in this chapter.

Before speaking on the results and conjectures concerning the multidimensional Fuller problem, we consider the trivial case of the "one–dimensional" Fuller problem, when $k = 1$, $\beta = -1$. Pontryagin's maximum principle gives the following second–order system:

$$\dot{\psi} = x, \quad \dot{x} = \operatorname{sgn} \psi.$$

The behavior of trajectories near the origin resembles a flattened saddle and is depicted in Fig. 11.

The half–parabolas $\mathcal{OA}$ and $\mathcal{OC}$ in Fig. 11 constitute the stable separatrix while $\mathcal{OB}$ and $\mathcal{OD}$ constitute the unstable one.

Now let $k = 2$, $\beta = -1$. Then Problem 5.1 coincides with Problem 2.1. Its Hamiltonian system can be written as follows:

$$\dot{z}_1 = z_2, \quad \dot{z}_2 = z_3, \quad \dot{z}_3 = z_4, \quad \dot{z}_4 = -\operatorname{sgn} z_1. \tag{5.1}$$

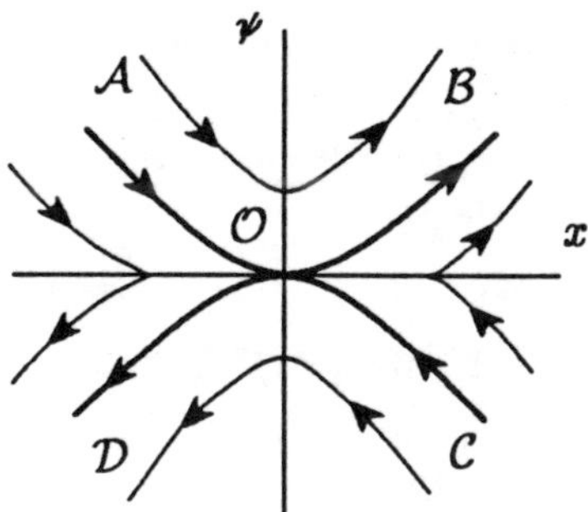

FIG. 11: FLATTENED SADDLE

Let us agree to depict two–dimensional integral varieties of system (5.1) as one–dimensional curves. Then the picture in Fig. 11 admits the following interpretation. System (5.1) has exactly two two–dimensional integral varieties $\mathfrak{N}^+$ and $\mathfrak{N}^-$ in which the trajectories approach the origin as time is increasing and decreasing respectively. These varieties are tangent to each other and correspond to the stable and unstable separatrices in Fig. 11. It follows from Lemmas 2.5–2.6 that all trajectories on the zero–level surface of the Hamiltonian tend to $\mathfrak{N}^+$ as $t \to -\infty$ and to $\mathfrak{N}^-$ as $t \to \infty$ in the "angular" metric generated by $\widetilde{\mathfrak{G}}$–orbits. The inner structure of the chattering arcs inside of $\mathfrak{N}^+$ and $\mathfrak{N}^-$ is lost in Fig. 11, although it can be exposed separately in its own turn (see Figs. 6 and 8, on pages 26, 33).

Consider the Hamiltonian system for Problem 5.1 with $k = 3$, $\beta = -1$:

$$\dot{z}_1 = z_2, \quad \dot{z}_2 = z_3, \ \ldots, \ \dot{z}_5 = z_6, \quad \dot{z}_6 = \operatorname{sgn} z_1 \tag{5.2}$$

System (5.2) is treated in Section 5.4. It is proved that there exist two three–dimensional integral varieties $\mathfrak{N}^+$ and $\mathfrak{N}^-$, filled by the trajectories of system (5.2) through the origin. The manifold $\mathfrak{N}^-$ corresponds to the stable separatrix of the saddle point, and $\mathfrak{N}^+$ corresponds to the unstable one. The term *separatrix* falls within the realm of the two–dimensional theory. It is not adequate in the multidimensional case since "separatrix" does not separate anything. We use this term because of the lack of a better one. The inner structure of $\mathfrak{N}^+$ is the following. There exist two trajectories with $u \equiv 1$ and $u \equiv -1$ leading to the origin without switches. In addition to that, there exists a two–dimensional manifold filled by automodelling chattering solutions. Their behavior is analogous to that of the chattering arcs in Fig. 6. All other solutions in $\mathfrak{N}^+$ are not automodelling. They attain the origin in finite time with an infinite number of switches. However, if $t \to -\infty$, then non–automodelling chattering solutions in $\mathfrak{N}^+$ have a finite number of switches. The switching surface in $\mathfrak{N}^+$ is piecewise smooth. A similar structure with the inversion of the time current is inherent in the manifold $\mathfrak{N}^-$ also.

The results of this chapter lead us to the following conjectures on the structure of solutions of the Hamiltonian system for the multidimensional Fuller problem. Consider the system

$$\dot{z}_1 = z_2, \quad \dot{z}_2 = z_3, \dots, \dot{z}_{2k-1} = z_{2k},$$

$$\dot{z}_{2k} = \begin{cases} 1 & \text{if} \quad (-1)^k z_1 < 0, \\ \beta & \text{if} \quad (-1)^k z_1 > 0. \end{cases} \tag{5.3}$$

System (5.3) is homogeneous under the action of the following group of symmetries

$$\mathfrak{g}_\lambda : z \mapsto \mathfrak{g}_\lambda(z), \quad \lambda > 0,$$

$$\left(\mathfrak{g}_\lambda(z)\right)_i = \lambda^{2k+1-i} z_i, \quad i = 1, \dots, 2k.$$

One can verify that if $z(t)$ is a solution of system (5.3), then for any $\lambda > 0$ the trajectory $\mathfrak{g}_\lambda\left(z(t/\lambda)\right)$ is a solution also.

CONJECTURE 5.1. **The saddle structure of Fuller's system.** *There exist two k–dimensional integral varieties $\mathfrak{N}^+$ and $\mathfrak{N}^-$ of system (5.3) with chattering trajectories such that the trajectories in $\mathfrak{N}^+$ approach the origin as time is increasing while the trajectories in $\mathfrak{N}^-$ approach it as time is decreasing. All other trajectories of system (5.3) tend to $\mathfrak{N}^-$ as $t \to \infty$ and to $\mathfrak{N}^+$ as $t \to -\infty$ in the "angular" metric generated by the orbits of the group $\mathfrak{g}_\lambda$.*

CONJECTURE 5.2. **The number of cycles in the orbit space.** *If $k = 2m$, then there are m one–parameter families of the automodelling chattering trajectories in $\mathfrak{N}^+$. If $k = 2m + 1$, then in $\mathfrak{N}^+$ there are m one–parameter families of the automodelling chattering trajectories and two trajectories leading to the origin with constant controls $u = 1$ and $u = \beta$. The same structure with the inversion of the time current is inherent in $\mathfrak{N}^-$.*

Each one–parameter family of the automodelling trajectories corresponds to a cycle in the orbit space that motivates the title of the conjecture.

CONJECTURE 5.3. **Smoothness of the switching surface.** *Manifolds $\mathfrak{N}^+$ and $\mathfrak{N}^-$ are imbedded into $\mathbb{R}^{2k}$ as piecewise analytic manifolds whose intersection with the switching surface $z_1 = 0$ is also a piecewise analytic manifold homeomorphic to $\mathbb{R}^{k-1}$.*

The preceding conjectures concern the k–dimensional Fuller problems. Now let us consider the perturbation of the k–dimensional Fuller problems under the condition specified below. Assume that the system of Pontryagin's maximum principle in Problem 5.1 can be reduced to the following form:

$$\dot{z}_i = z_{i+1} + f_i(z, w, u), \quad i = 1, \ldots, 2k-1,$$

$$\dot{z}_{2k} = \alpha(w) + u\,\beta(w) + f_{2k}(z, w, u),$$

$$\dot{w} = F(z, w, u),$$

$$u = \operatorname{sgn} z_1,$$

where

$$w \in \mathbb{R}^m, \quad z = (z_1, \ldots, z_{2k}) \in \mathbb{R}^{2k}.$$

Assume that

$$\varlimsup_{\lambda \to +0} \lambda^{-(2k+1-i)} \left\| f_i\big(\mathfrak{g}_\lambda(z), w, u\big) \right\|_{C^0(U)} < \infty,$$

with U having been some open neighborhood of the point $(0, w_0)$ for some fixed $w_0 \in \mathbb{R}^m$. Let us define an open neighborhood W of the point $(0, w_0)$ on the plane $z = 0$ by the inequalities

$$(-1)^k \beta(w) < 0, \quad |\alpha(w)| < |\beta(w)|.$$

It is easy to see that W is filled by singular solutions of order k.

CONJECTURE 5.4. **Bundles with chattering arcs.** *There exists an open neighborhood of the region W in $\mathbb{R}^{m+4}$ containing two piecewise smooth fibre bundles $p^+ : \mathfrak{M}^+ \to W$, $p^- : \mathfrak{M}^- \to W$ whose common base is the region W and whose k–dimensional fibres $\mathfrak{M}_\sigma^+$ and $\mathfrak{M}_\sigma^-$ are integral varieties of system (5.3). The asymptotic behavior of trajectories in $\mathfrak{M}_\sigma^+$ and $\mathfrak{M}_\sigma^-$ is orbitally equivalent to those in $\mathfrak{N}^+$ and $\mathfrak{N}^-$ where the latter two manifolds are as in conjectures 5.1, 5.2, 5.3.*

Conjecture 5.5. ***Structure of Lagrangian manifolds.*** *If $\mathcal{M} \subset W$ is a Lagrangian submanifold of W, then the full preimages $(p^+)^{-1}\mathcal{M}$ and $(p^-)^{-1}\mathcal{M}$ are Lagrangian submanifolds of $\mathbb{R}^{2k+m}$.*

The last conjecture implies that the chattering arcs are optimal for an appropriate optimization problem provided that the projection of the Lagrangian manifold is regular. This chapter is devoted to partial substantiation of these conjectures.

5.2 Problems with Linear Constraints

Let us consider the following

Problem 5.2. ***Problem with linear constraints.***
Minimize

$$\int_0^\infty f(x)\,dt$$

subject to

$$\dot{x} = \mathbf{A}x + \mathbf{b}u, \quad u \in [\beta, 1], \tag{5.4}$$

with an initial point

$$x(0) = x_0.$$

Here $x \in \mathbb{R}^n$, $\mathbf{A}$ is a constant $(n \times n)$–matrix, $\mathbf{b} \in \mathbb{R}^n$ is a constant n–vector, $f : \mathbb{R}^n \to \mathbb{R}$ is a C^2–function, $f \geqslant 0$, $f(0) = 0$, $\beta < 0$. The admissible solutions of (5.4) are absolutely continuous functions, the admissible controls belong to $L_\infty[0, \infty)$.

Assume that the following condition of the *complete local controllability* is fulfilled:

$$\mathrm{rk}\,(\mathbf{b}, \mathbf{A}\mathbf{b}, \dots, \mathbf{A}^{n-1}\mathbf{b}) = n. \tag{5.5}$$

Assumption 5.1. *For any x_0 there exists an admissible solution of system (5.4) hitting the origin in finite time.*

It will be demonstrated (see Remark 5.2 below) that Assumption 5.1 is always valid for a class of problems which are invariant under the action of some symmetry group and meet condition (5.5).

Lemma 5.1. *For any $x_0 \in \mathbb{R}^n$ there exists a solution to Problem 5.2.*

Proof. Denote by $\mathcal{X}$ a set of absolutely continuous functions $x(\cdot)$: $\mathbb{R}_+ \to \mathbb{R}^n$ such that $x(0) = x_0$. Let $\mathcal{K}$ be the operator $L_\infty[0,\infty) \to \mathcal{X}$ defined as follows:

$$(\mathcal{K}u)(t) = e^{\mathbf{A}t}x_0 + \int_0^\infty \rho_t(\tau)u(\tau)\,d\tau,$$

where

$$\rho_t(\tau) = \begin{cases} e^{\mathbf{A}(t-\tau)}\mathbf{b}, & \text{if } 0 \leqslant \tau \leqslant t, \\ 0, & \text{if } \tau > t. \end{cases}$$

It is easy to see that $\mathcal{K}$ assigns to the function $u(\cdot) \in L_\infty[0,\infty)$ the corresponding solution $x(t)$ of system (5.4). It follows from the Banach–Alaoglu theorem [N. Dunford, J. Schwartz, 1958; p. 424] that the unit ball of $L_\infty[0,\infty)$ is weak* compact (because $L_\infty[0,\infty) = L_1^*[0,\infty)$). Assume that the control sequence $\{u_n(\cdot)\}_{n=1}^\infty$ minimizes the functional in Problem 5.2. Since we can always choose a convergent subsequence, we can assume that there is a limit $u_0(\cdot)$ of the sequence $\{u_n(\cdot)\}_{n=1}^\infty$ in the weak* topology of $L_\infty[0,\infty)$. By the definition of $L_\infty[0,\infty)$, for any $\rho(\cdot) \in L_1[0,\infty)$ we have

$$\lim_{n\to\infty} \int_0^\infty \rho(\tau)u_n(\tau)\,d\tau = \int_0^\infty \rho(\tau)u_0(\tau)\,d\tau.$$

Since the function $\rho_t(\cdot)$ belongs to $L_1[0,\infty)$, the sequence $x_n(t) = (\mathcal{K}u_n)(t)$ $(n = 1, 2, \ldots)$ is pointwise convergent on $[0,\infty)$ and its limit is a function $x_0(t) = (\mathcal{K}u_0)(t)$. The form of the operator $\mathcal{K}$ includes that the sequence $\{x_n(\cdot)\}_{n=1}^\infty$ is uniformly Lipschitz, so the convergence is uniform on any compact subset of $[0,\infty)$.

In view of Assumption 5.1, the minimum of the functional in Problem 5.2 is positive and finite. It follows from the Fatou theorem [N. Dunford, J. Schwartz, 1958; p. 152] that

$$\lim_{n\to\infty} \int_0^\infty f(x_n(t))\,dt \;\geqslant\; \int_0^\infty f(x_0(t))\,dt.$$

Hence, $x_0(t)$ is a solution to Problem 5.2. $\hfill$ **Q.E.D.**

ASSUMPTION 5.2. *The function f does not depend on $x_{k+1}, \ldots, x_n$ and is strictly convex in $x_1, \ldots, x_k$ $(1 \leqslant k \leqslant n)$.*

Set $\mathcal{L} = \{x \mid x_1 = \ldots = x_k = 0\}$.

ASSUMPTION 5.3. *For any* $x \in \mathcal{L}\backslash\{0\}$ *we have*

$$\min_{u \in [\beta, 1]} \max_{1 \leq j \leq k} \left| (\mathbf{A}x + \mathbf{b}u)_j \right| > 0. \tag{5.6}$$

It follows that the set of admissible velocities of system (5.4) (the *maneuverability domain*) strictly supports the surface $\mathcal{L}$ at all points outside the origin.

Remark 5.1. *Let us clarify the sense of inequality (5.6). This is a sufficient condition for any admissible trajectory $x(t)$ of system (5.4) with the initial point $x(0) \in \mathcal{L}$ intersecting the surface $\mathcal{L}$ at a discrete set of points. This means that there exists a number $\epsilon = \epsilon\big((x(\cdot)\big) > 0$ such that for any $t \in (-\epsilon, \epsilon)$ the condition $x(t) \in \mathcal{L}$ implies $t = 0$. We see that (5.6) is a rather rough sufficient condition for this property of solutions of system (5.4). A more refined sufficient condition is the following.*

Assume that the strict inequality (5.6) is fulfilled everywhere in $\mathcal{L}$ except for the points of some submanifold $\Sigma_1 \subset \mathcal{L}$ of codimension 1. This implies that for any $x_1 \in \Sigma_1$ there exists an admissible vector $Ax_1 + bu_1$ which is tangent to $\mathcal{L}$. Assume that all tangent vectors with this property are not tangent to the submanifold Σ_1 everywhere in Σ_1 except for the points of some submanifold $\Sigma_2 \subset \Sigma_1$ of codimension 1 (we see that Σ_2 is a submanifold of $\mathcal{L}$ of codimension 2) and so on. In this way we arrive at a decreasing chain of imbedded submanifolds

$$\mathcal{L} \supset \Sigma_1 \supset \Sigma_2 \supset \Sigma_3 \supset \ldots \supset \Sigma_s$$

such that for any $x \in \Sigma_k$ $(k = 1, \ldots, s)$ we have $\mathbf{A}x + \mathbf{b}u_k \in T_x\Sigma_{k-1}$ (for some $u_k \in [-1, 1]$) and for any $x \in \Sigma_k\backslash\Sigma_{k+1}$ we have $\mathbf{A}x + \mathbf{b}u \notin T_x\Sigma_k$ for all $u \in [-1, 1]$. It can readily be seen that if the last manifold Σ_s in this chain is empty, then all admissible solutions do intersect $\mathcal{L}$ at a discrete set of points.

Denote by ω the *Bellman function* of Problem 5.2, i.e., the minimum of the functional as a function of the initial point x_0.

LEMMA 5.2. *For any $x_0 \in \mathbb{R}^n$ there exists a unique solution to Problem 5.2. The Bellman function ω is strictly convex.*

Proof. Assume that $x^{(i)}(t)$ $(i = 1, 2)$ are two distinct optimal solutions to Problem 5.2 under controls $u^{(i)}(t)$ respectively. Consider the trajectory

$$x^{(\lambda)}(t) = \lambda x^{(1)}(t) + (1 - \lambda)x^{(2)}(t)$$

under the control

$$u^{(\lambda)}(t) = \lambda u^{(1)}(t) + (1 - \lambda)u^{(2)}(t).$$

Let us define the map **pr** as the operator of an orthogonal projection of $\mathbb{R}^n$ onto the plane $x_{k+1} = \ldots = x_n = 0$. Put $c_0 = \max\{2, 1 - \beta\}$. If $\mathbf{pr}\, x^{(1)}(t) = \mathbf{pr}\, x^{(2)}(t)$ and $x^{(i)}(t) \not\equiv 0$ while $t \in (t_0, t_1)$, then the trajectory $(x^{(1)}(t) - x^{(2)}(t))/c_0$ goes along $\mathcal{L}\backslash\{0\}$ and satisfies (5.4) with the admissible control $u(t) = (u^{(1)}(t) - u^{(2)}(t))/c_0$. This contradicts Assumption 5.3. Hence $\mathbf{pr}\, x^{(1)}(t) \neq \mathbf{pr}\, x^{(2)}(t)$ at some set of positive Lebesgue measure. The strict convexity of the function f in the variables $x_1, x_2, \ldots, x_k$ implies

$$\int_0^\infty f\big(x^{(\lambda)}(t)\big)\, dt < \lambda \int_0^\infty f\big(x^{(1)}(t)\big)\, dt + (1 - \lambda) \int_0^\infty f\big(x^{(2)}(t)\big)\, dt.$$

If the initial points $x^{(1)}(0)$ and $x^{(2)}(0)$ coincide, then the inequality implies that $x^{(1)}(\cdot)$ and $x^{(2)}(\cdot)$ are not optimal. If $x^{(1)}(0) \neq x^{(2)}(0)$, then it implies that $\omega(x)$ is strictly convex. **Q.E.D.**

COROLLARY 5.1. *The Bellman function ω is Lipschitz.*

Proof. Function ω is strictly convex and finite, hence it is Lipschitz [R. Rockafellar, 1970; p. 86]. **Q.E.D.**

COROLLARY 5.2. *For any $T > 0$ the mapping*

$$\mathbb{R}^n \longrightarrow C[0, T], \quad x_0 \mapsto x(\cdot, x_0),$$

is continuous.

Here and everywhere below in this section $x(\cdot, x_0)$ denotes the solution to Problem 5.2 with the initial point x_0, and $u(\cdot, x_0)$ denotes the optimal control on $x(\cdot, x_0)$.

Proof of Corollary 5.2. Assume that the sequence $\{x_n\}_{n=1}^{\infty}$ converges to x_0 as $n \to \infty$. Denote by $u_0(\cdot)$ the weak* limit of the sequence $u(\cdot, x_n)$ (if there is no limit, we can take an appropriate convergent subsequence). Put $x_0(t) = (\mathcal{K}u_0)(t)$ with the given $x(0) = x_0$. It can be proved by the same way as in Lemma 5.1 that for any fixed $t \in (0, \infty)$ we have

$$\lim_{n \to \infty} x(t, x_n) = x_0(t),$$

and the convergence is uniform on any compact subset of $[0, \infty)$. It follows from the Fatou theorem that

$$\int_0^{\infty} f(x_0(t))\, dt \leqslant \varliminf_{n \to \infty} \int_0^{\infty} f(x(t, x_n))\, dt = \varliminf_{n \to \infty} \omega(x_n).$$

On account of Corollary 5.1 we have $\varliminf_{n \to \infty} \omega(x_n) = \omega(x_0)$, hence

$$\int_0^{\infty} f(x_0(t))\, dt \leqslant \omega(x_0),$$

and $x_0(t) = x(t, x_0)$.   **Q.E.D.**

5.3 Problems with Symmetries

DEFINITION 5.1. *We shall say that Problem 5.2 admits the action of the one–parameter group $\mathfrak{G} \cong \mathbb{R}_+$ (where $\mathbb{R}_+$ is the multiplicative group of positive reals) iff there exists a continuous mapping*

$$\Lambda : \mathbb{R}_+ \longrightarrow Diff\,(\mathbb{R}^n),$$

$$\lambda \mapsto \mathfrak{g}_\lambda(\cdot)$$

($Diff\,(\mathbb{R}^n)$ being the group of diffeomorphisms of $\mathbb{R}^n$) with the following properties.

(1) $\mathfrak{g}_{\lambda\mu}(x) = \mathfrak{g}_\lambda(\mathfrak{g}_\mu(x))$ *for any $\lambda, \mu \in \mathbb{R}_+$ and any $x \in \mathbb{R}^n$.*
(2) $\Lambda^{-1}(\mathbf{id}) = 1$, *where $\mathbf{id}$ is the identity mapping on $\mathbb{R}^n$.*
(3) $\mathfrak{g}_\lambda(0) = 0$ *for any $\lambda \in \mathbb{R}_+$.*
(4) $D\mathfrak{g}_\lambda\big|_x (\mathbf{A}x + \mathbf{b}u) = \lambda^\alpha(\mathbf{A}\mathfrak{g}_\lambda(x) + \mathbf{b}u)$
and $f(\mathfrak{g}_\lambda(x)) = \lambda^\gamma f(x)$ for any $x \in \mathbb{R}^n$ and $u \in [\beta, 1]$, ($D\mathfrak{g}_\lambda\big|_x$ being the differential of $\mathfrak{g}_\lambda$ at the point x).
(5) *If $\mathfrak{g}_\lambda(x) = x$ and $x \neq 0$, then $\lambda = 1$, i.e., for any $x \neq 0$ the stationary subgroup of x is trivial.*
(6) *For any $x \in \mathbb{R}^n$ there exists $\lim_{\lambda \to +0} \mathfrak{g}_\lambda(x) = 0$.*

Let us note that the action of the group $\mathfrak{G}$ is characterized by two real numbers α, γ which are usually assumed to be positive below. Everywhere below Problem 5.2 is assumed to admit the action of the symmetry group $\mathfrak{G}$.

The following statement establishes the automodelling properties of solutions to problems with symmetries. It appears that contracting the trajectories along $\mathfrak{G}$–orbits together with the proper time–scaling do not lead the solution out of the class of optimal solutions.

LEMMA 5.3.

(1) $x(t, \mathfrak{g}_\lambda(x_0)) = \mathfrak{g}_\lambda(x(\lambda^{-\alpha}t, x_0))$,
   $\quad u(t, \mathfrak{g}_\lambda(x_0)) = u(\lambda^{-\alpha}t, x_0)$.
(2) $\omega(\mathfrak{g}_\lambda(x_0)) = \lambda^{\alpha+\gamma}\omega(x_0)$.

Proof. Assume that $x(t)$ is an admissible trajectory of system (5.4) with the control $u(t)$. A straightforward differentiation gives

$$\frac{d}{dt}\mathfrak{g}_\lambda\big(x(\lambda^{-\alpha}t)\big) = D\,\mathfrak{g}_\lambda\Big|_{x(\lambda^{-\alpha}t)} \cdot \frac{d}{dt}x(\lambda^{-\alpha}t)$$

$$= D\mathfrak{g}_\lambda\Big|_{x(\lambda^{-\alpha}t)} \cdot \Big(\mathbf{A}x(\lambda^{-\alpha}t) + \mathbf{b}u(\lambda^{-\alpha}t)\Big)\lambda^{-\alpha}$$

$$= \mathbf{A}\mathfrak{g}_\lambda\Big(x(\lambda^{-\alpha}t)\Big) + \mathbf{b}u(\lambda^{-\alpha}t),$$

i.e., the trajectory $\mathfrak{g}_\lambda\big(x(\lambda^{-\alpha}t)\big)$ is a solution of system (5.4) with the control $u = u(\lambda^{-\alpha}t)$. Further on,

$$\int_0^\infty f\Big(\mathfrak{g}_\lambda\big(x(\lambda^{-\alpha}t)\big)\Big)\,dt = \lambda^\gamma \int_0^\infty f\big(x(\lambda^{-\alpha}t)\big)\,dt$$

$$= \lambda^{\alpha+\gamma}\int_0^\infty f\big(x(t)\big)\,dt.$$

It can readily be seen that for any $\lambda > 0$ the mapping $x(t) \to \mathfrak{g}_\lambda\big(x(\lambda^{-\alpha}t)\big)$ is a one–to–one correspondence. Therefore, it carries an optimal solution to Problem 5.2 with the initial point x_0 into an optimal one with the initial point $\mathfrak{g}_\lambda(x_0)$. $\qquad$ **Q.E.D.**

Remark 5.2. *Now we can argue why the statement of Assumption 5.1 is always valid for the problem which admits the action of the group $\mathfrak{G}$. In view of the complete local controllability (5.5) for any sufficiently small x_0 there exists an admissible solution $x(t)$ that attains the origin in a finite time T. Then for any $\lambda > 0$ the trajectory $\mathfrak{g}_\lambda\big(x(t/\lambda)\big)$ is also a solution of system (5.4) and attains the origin in the finite time λT. Since g_λ is continuous in λ and g_1 is the identity mapping, it is easy to see that for any $\lambda > 0$ the diffeomorphism g_λ is an epimorphism. So the local controllability of system (5.4) in a small neighborhood of the origin provides controllability everywhere in $\mathbb{R}^n$.*

LEMMA 5.4. *For any* $x_0 \in \mathbb{R}^n$ *there exists* $T_0 = T_0(x_0) > 0$ *such that the optimal solution* $x(t, x_0)$ *to Problem 5.2 is also optimal to the following problem with fixed time.*

PROBLEM 5.3. *Minimize*

$$\int_0^{T_0} f(x)\, dt$$

subject to

$$\dot{x} = \mathbf{A}x + \mathbf{b}u, \quad u \in [\beta, 1]$$

with boundary conditions

$$x(0) = x_0, \quad x(T_0) = 0.$$

Proof. We are to demonstrate that for any $x_0 \in \mathbb{R}^n$ the optimal solution $x(\cdot, x_0)$ to Problem 5.2 hits the origin in finite time. Let us show first that

$$\lim_{t \to \infty} x(t, x_0) = 0.$$

Assume the contrary: there exist $x_0 \in \mathbb{R}^n$, a number $\epsilon_0 > 0$, and an increasing sequence $\{t_n\}_{n=1}^{\infty}$, $\lim_{n \to \infty} t_n = \infty$, such that for any $n \in \mathbb{N}$ the points $y_n = x(t_n, x_0)$ lie outside the ball B_{ϵ_0} of radius ϵ_0 about the origin. Consider the function

$$\delta(\epsilon) = \inf \left\{ \omega(x) : x \in \partial B_{\epsilon} \right\}.$$

The Bellman function ω is continuous, nonnegative, strictly convex and hence nonzero everywhere outside the origin, so $\delta(\epsilon) > 0$. Since $\int_0^{\infty} f(x(t, x_0))\, dt < \infty$, for any $\delta > 0$ there exists $T(\delta) > 0$ such that $\int_{T(\delta)}^{\infty} f(x(t, x_0))\, dt < \delta$. Let us take $\delta < \delta(\epsilon_0)$ and choose the index n with $t_n > T(\delta)$. It has been assumed that $\omega(y_n) \geqslant \delta(\epsilon_0)$. Hence,

$$\int_{t_n}^{\infty} f(x(t, x_0))\, dt = \int_0^{\infty} f(x(t, y_n))\, dt = \omega(y_n) \geqslant \delta(\epsilon_0).$$

On the other hand,

$$\int_{t_n}^{\infty} f(x(t, x_0))\, dt \leqslant \int_{T(\delta)}^{\infty} f(x(t, x_0))\, dt < \delta,$$

which contradicts the inequality $\delta < \delta(\epsilon_0)$.

The set $\omega_C = \{ x \mid \omega(x) = C \}$ is called the *C–level surface* of the Bellman function ω. Let t_C be a solution of the equation $\omega(x(t_C, x)) =$

$C/2$. Since $\omega\big(x(t,x)\big)$ decreases with the growth of t, the function $t_C(x)$ is well–defined for any $x \in \omega_C$, $C \neq 0$. Put

$$\tau(C) = \sup_{x \in \omega_C} \{t_C(x)\}.$$

Let us demonstrate that $\tau(C) < \infty$. Assume the contrary: there exists a sequence $\{x_n\}_{n=1}^{\infty}$, $x_n \in \omega_C$, such that $\omega\big(x(t_n,x_n)\big) = C/2$, where $t_n \stackrel{\text{def}}{=} t_C(x_n) \to \infty$ as $n \to \infty$. We can regard the sequence $\{x_n\}_{n=1}^{\infty}$ to be converging to some $x_0 \in \mathbb{R}^n$ (otherwise, we can take some convergent subsequence). Then the optimal solution $x(t,x_0)$ is a pointwise limit of the optimal solutions $x(t,x_n)$ (Corollary 5.2). Hence, the trajectory $x(t,x_0)$ occurs outside the set $\{x \,|\, \omega(x) < C/2\}$ at the instants t_n. This set contains some open neighborhood of the origin that contradicts the condition $\lim_{t\to\infty} x(t,x_0) = 0$. Thus, $\tau(C) < \infty$.

Let us prove the identity $\tau(C/2) = \sigma^{\alpha}\tau(C)$, where $\sigma = 2^{-1/(\alpha+\gamma)}$. In view of Lemma 5.3, the level–surface ω_C meets the relation

$$\omega_{C_1} = \mathfrak{g}_\nu(\omega_{C_2}), \tag{5.7}$$

with C_1, C_2 having been arbitrary nonzero constants; $\nu = (C_1/C_2)^{1/(\alpha+\gamma)}$. Indeed, if $\omega(x) = C_2 \neq 0$, then

$$\omega\big(\mathfrak{g}_\nu(x)\big) = \nu^{\alpha+\gamma}\omega(x) = \frac{C_1}{C_2}\omega(x) = \frac{C_1}{C_2}C_2 = C_1$$

(Lemma 5.3). In the case $C_2 = 2C_1 = C$, this means $\omega_{C/2} = \mathfrak{g}_\sigma\omega_C$.

Assume that for some $x_0 \in \omega_C$ we have $x(t_0,x_0) \in \omega_{C/2}$. If we set $\tau = \sigma^{-\alpha}t_0$, then

$$\omega\Big(x\big(\tau,\mathfrak{g}_\sigma(x_0)\big)\Big) = \omega\Big(\mathfrak{g}_\sigma\big(x(\sigma^{-\alpha}\tau,x_0)\big)\Big)$$

$$= \sigma^{\alpha+\gamma}\omega\big(x(t_0,x_0)\big) = \sigma^{\alpha+\gamma}\frac{C}{2} = \frac{1}{2}\frac{C}{2} = \frac{C}{4},$$

i.e., $x\big(\tau,\mathfrak{g}_\sigma(x_0)\big) \in \omega_{C/4}$. This fact and the invertibility of the mapping $\mathfrak{g}_\sigma : \omega_C \longrightarrow \omega_{C/2}$ prove that $\tau(C/2) = \sigma^{\alpha}\tau(C)$.

Denote by $T(x_0)$ the instant when the optimal solution $x(t,x_0)$ hits the origin. If $\omega(x_0) = C$, then it follows from the definition of the function τ that

$$T(x_0) \leqslant \tau(C) + \tau\left(\frac{C}{2}\right) + \ldots + \tau\left(\frac{C}{2^n}\right) + \ldots$$

$$= \tau(C)\big(1 + \sigma^{\alpha} + \ldots + \sigma^{n\alpha} + \ldots\big) = \frac{\tau(C)}{1 - \sigma^{\alpha}} < \infty. \tag{5.8}$$

Q.E.D.

Corollary 5.3. *The function* $T(x)$ *is continuous.*

Proof. Let us first verify the relation

$$T\big(\mathfrak{g}_\lambda(x)\big) = \lambda^\alpha T(x). \tag{5.9}$$

Indeed, on the one hand, using Lemma 5.3 we obtain

$$x\big(t, \mathfrak{g}_\lambda(x)\big)\Big|_{t=\lambda^\alpha T(x)} = \mathfrak{g}_\lambda\big(x(\lambda^{-\alpha}t, x)\big)\Big|_{t=\lambda^\alpha T(x)}$$

$$= \mathfrak{g}_\lambda\big(x(T(x), x)\big) = \mathfrak{g}_\lambda(0) = 0.$$

Hence, $T\big(\mathfrak{g}_\lambda(x)\big) \leqslant \lambda^\alpha T(x)$. On the other hand,

$$T(x) = T\Big(\mathfrak{g}_{1/\lambda}\big(\mathfrak{g}_\lambda(x)\big)\Big) = \big(\tfrac{1}{\lambda}\big)^\alpha T\big(\mathfrak{g}_\lambda(x)\big),$$

which implies $\lambda^\alpha T(x) \leqslant T\big(\mathfrak{g}_\lambda(x)\big)$ and (5.9) is valid.

The Bellman function ω being continuous (Corollary 5.1), it follows that $\lim_{x\to 0}\omega(x) = 0$. If $\omega(x) = C$, then $T(x) \leqslant \sup_{y\in\omega_C} T(y)$. In view of (5.7) we have

$$\sup_{y\in\omega_C} T(y) = \sup_{z\in\omega_1} T\big(\mathfrak{g}_k(z)\big) = k^\alpha \sup_{z\in\omega_1} T(z),$$

where $k = C^{1/(\alpha+\gamma)}$, this fact together with (5.8) implies

$$\varlimsup_{x\to 0} T(x) \leqslant \frac{\tau(1)}{1-\sigma^\alpha} \cdot \lim_{x\to 0} \omega(x)^{\alpha/(\alpha+\gamma)} = 0.$$

Since $T(x) \geqslant 0$, there exists $\lim_{x\to 0} T(x) = 0$, i.e., the function $T(x)$ is continuous at the origin.

Consider an arbitrary sequence $\{x_n\}_{n=1}^\infty$ such that $\lim_{n\to\infty} x_n = x_0$ and set $T^+ = \varlimsup_{n\to\infty} T(x_n)$. It follows from (5.8) that $T^+ < \infty$. Let us demonstrate that $T^+ \leqslant T(x_0)$. Assume the contrary: $T(x_0) = T^+ - \epsilon < T^+$ for some $\epsilon > 0$. Then there exists a subsequence $\{x_{n_k}\}_{k=1}^\infty$ such that $T(x_{n_k}) > T^+ - \epsilon/2$. On account of Corollary 5.2 we have $\lim_{k\to\infty} x(T^+ - \epsilon, x_{n_k}) = x(T^+ - \epsilon, x_0) = 0$. Set $y_k = x(T^+ - \epsilon, x_{n_k})$. Then $\lim_{k\to\infty} y_k = 0$, though $T(y_k) \geqslant \epsilon/2$. This contradicts the continuity of the function $T(x)$ at the origin.

Let $T^- = \varliminf_{n\to\infty} T(x_n)$. Let us demonstrate that $T(x_0) \leqslant T^-$. Assume the contrary: $T(x_0) = T^- + \epsilon/2$ for some $\epsilon > 0$. Then there exists a subsequence $\{x_{n_k}\}_{k=1}^\infty$ such that $T(x_{n_k}) < T^- + \epsilon/2$. On account of Corollary 5.2 we have $\lim_{k\to\infty} x(T^- + \epsilon/2, x_{n_k}) = 0$. Hence, $x(T^- + \epsilon/2, x_0) = 0$ though at the same time $T(x_0) > T^- + \epsilon/2$. Thus, $\varlimsup_{n\to\infty} T(x_n) = \varliminf_{n\to\infty} T(x_n) = T(x_0)$.   **Q.E.D.**

Let us write the equations of Pontryagin's maximum principle for Problem 5.2:

$$\dot{\psi} = -\mathbf{A}^*\psi + f_x'\big(x(t,x_0)\big), \tag{5.10.a}$$

$$\max_{u\in[\beta,1]} \langle u\psi(t),\mathbf{b}\rangle = \langle u(t,x)\psi(t),\mathbf{b}\rangle \quad \text{(a.e.)}. \tag{5.10.b}$$

Denote by $\Psi(x_0)$ the mapping which joins the point x_0 with the initial point $\psi(0)$ of a solution of the adjoint system (5.10). The following lemma implies that the function $\Psi(x)$ is well–defined.

LEMMA 5.5.

(1) *For any* x_0 *there exists a unique solution* $\psi(t)$ *of equations (5.10).*
(2) *The Bellman function* ω *in Problem 5.2 is* C^1 *and* $-\Psi(x_0)$
  $= \mathbf{grad}\ \omega(x_0)$ *for any* $x_0 \in \mathbb{R}^n$.

Proof. Let $x_0 \neq 0$ and take $T_0 > T(x_0)$. In view of Lemma 5.4 the optimal solution $x(t,x_0)$ to Problem 5.2 is also optimal to Problem 5.3 with the fixed terminal time T_0. Pontryagin's maximum principle implies that there exists an absolutely continuous vector–function $\psi(t)$ and a number $\lambda \geqslant 0$, simultaneously nonzero, such that both the equation

$$\dot{\psi} = -\mathbf{A}^*\psi + \lambda f_x'\big(x(t,x)\big) \tag{5.11}$$

and the maximum condition (5.10.b) are valid. We assert that $\lambda \neq 0$. Let us consider the solution of system (5.11) at $t \in \big[T(x_0),T_0\big]$. Since the optimal control $u(t,x_0)$ is identically zero on this interval, we have $\langle\psi(t),\mathbf{b}\rangle = 0$. In the case $\lambda = 0$ it follows from (5.11) that $\psi(t) = e^{-\mathbf{A}^*t}\psi_0$, and successive differentiation yields

$$\frac{d}{dt}\langle\psi(t),\mathbf{b}\rangle = -\langle\psi(t),\mathbf{A}\mathbf{b}\rangle = 0,$$

$$\ldots$$

$$\frac{d^k}{dt^k}\langle\psi(t),\mathbf{b}\rangle = \frac{d}{dt}\langle(-1)^{k-1}\psi(t),\mathbf{A}^{k-1}\mathbf{b}\rangle$$
$$= (-1)^k\langle\psi(t),\mathbf{A}^k\mathbf{b}\rangle = 0,$$

$$k = 1,2,\ldots,n.$$

The condition of local controllability (5.5) implies $\psi(t) \equiv 0$ at $t \in \big[T(x_0),T_0\big]$. In the case $\lambda = 0$ system (5.11) is linear in ψ, so $\psi(t) \equiv 0$ for all t. This proves that $\lambda \neq 0$ and everywhere below λ is assumed to be equal to 1.

Assume that for some $x(t, x_0)$ there exist two different solutions $\psi_1(t) \neq \psi_2(t)$ of system (5.11) with $\lambda = 1$. Then the function $\Delta\psi \overset{\text{def}}{=} \psi_1(t) - \psi_2(t)$ satisfies (5.11) with $\lambda = 0$. The case $\lambda = 0$ has just been considered. It has been proved that $\Delta\psi(t) = 0$ at $t \in [T(x_0), t_0]$. So $\psi_1(t)$ and $\psi_2(t)$ have the same boundary conditions at $t = T(x_0)$ and the uniqueness theorem for system (5.11) yields $\Delta\psi(t) \equiv 0$. This proves the first statement of Lemma 5.5 and, in particular, the uniqueness of the function $\Psi(x_0)$.

Now let us consider Problem 5.3 in which the initial condition $x(0) = x_0$ is replaced by the condition $x(0) \in l$ with l having been some hyperplane supporting the level–surface of the Bellman function at the point x_0. Since function ω is strictly convex, the trajectory $x(t, x_0)$ is also a solution to the problem with new boundary conditions for any choice of l. It follows from Pontryagin's maximum principle that there exists a solution $\psi^{(l)}(t)$ of system (5.10) meeting the supplementary transversality condition $\psi^{(l)}(0) \perp l$. As was proved above, equation (5.10) has a unique solution, namely $\Psi(x(t, x_0))$. So for any l we have $\psi^{(l)}(0) = \Psi(x_0)$ and hence $\Psi(x_0) \perp l$. In the case $\Psi(x_0) \neq 0$ this implies that the function ω is differentiable at x_0 and $\Psi(x_0) = \lambda(x_0)\,\text{grad}\,\omega(x_0)$, $\lambda(x_0)$ being a coefficient of proportionality. In the case $\Psi(x_0) = 0$ the inclusion $0 \in \partial\omega(x_0)$ is valid because the subdifferential of a convex function is upper semicontinuous [R. Rockafellar, 1970; p. 234]. The function $\omega(x)$ is strictly convex, so the inclusion $0 \in \partial\omega(x_0)$ is a necessary and sufficient condition for the minimum. Therefore, the relation $\Psi(x_0) = 0$ implies $x_0 = 0$. It is well–known that if a convex function has a single supporting hyperplane at any point of some open region Q, then this function is C^1 in Q [R. Rockafellar, 1970; p. 246]. Hence, $\omega \in C^1(\mathbb{R}^n \backslash \{0\})$. It is known that the C^1–Bellman function satisfies the Bellman equation

$$-\langle \text{grad}\,\omega(x),\, \mathbf{A}x + \mathbf{b}\hat{u} \rangle + f(x) = 0,$$

$$\hat{u} \overset{\text{def}}{=} u(0, x).$$

To argue this, it is sufficient to differentiate the identity

$$\omega(x_0) = \omega(x(t, x_0)) + \int_0^t f(x(s, x_0))\, ds$$

with respect to t at $t = 0$. Since the function $\Psi(x) = \lambda(x)\text{grad}\,\omega(x)$ satisfies the condition $H(\Psi(x), x, \hat{u}) = 0$, it follows that $\lambda(x) = -1$ in the case $f(x) \neq 0$, so $\Psi(x) = -\text{grad}\,\omega(x)$. If $x \in \mathcal{L}\backslash\{0\}$ (i.e., $f(x) = 0$), the identity $\lambda(x) = -1$ is valid in view of the continuity of function $\Psi(x)$ along solutions $x(t, x_0)$.

We are left with proving that function $\omega(x)$ is smooth at the origin. Recall that the trajectory $x(t, x_0)$ hits the origin in a finite time $T(x_0)$.

Hence, $\Psi(0) = 0$. The value of function $\Psi(x_0)$ can be obtained by integration of system (5.10) on the interval $(0, T(x_0))$. Since function $T(x_0)$ is continuous (Corollary 5.3) and system (5.10) is linear in ψ, we have $\lim_{x \to 0} \Psi(x) = 0$. Thus, $\Psi \in C(\mathbb{R}^n)$ and $\omega \in C^1(\mathbb{R}^n)$. **Q.E.D.**

Denote by Π the optimal switching surface of Problem 5.2, defined as the set where $\langle \Psi(x), \mathbf{b} \rangle = 0$.

LEMMA 5.6. *Assume that Problem 5.2 admits the action of the symmetry group $\mathfrak{G}$. Then the switching surface Π is a continuous manifold homeomorphic to $\mathbb{R}^{n-1}$.*

Proof. Let us consider the set of points on the level–surface ω_C $(C \neq 0)$ where $\mathrm{grad}\, \omega(x) \perp \mathbf{b}$. Denote it by Π_C. Since $\omega \in C^1$, the inclusion $x \in \Pi_C$ induces that $\mathbf{b} \in T_x \omega_C$ (i.e., $\mathbf{b}$ is a tangent vector to ω_C). On the other hand, if $\mathbf{b} \in T_x \omega_C$ at some $x \in \omega_C$, then $\mathrm{grad}\, \omega(x) \perp \mathbf{b}$, i.e., $x \in \Pi_C$. Recall that function ω is strictly convex. Let $\mathbf{p}$ be the projection of $\mathbb{R}^n$ onto the hyperplane $\langle x, \mathbf{b} \rangle = 0$ along the vector $\mathbf{b}$. The image of Π_C under the map $\mathbf{p}$ is obviously the boundary of the image of ω_C. Since $\mathbf{p}(\omega_C)$ is a compact convex subset of the hyperplane $\langle x, \mathbf{b} \rangle = 0$ with nonempty interior, its boundary is homeomorphic to S^{n-2}. Being restricted to Π_C, $\mathbf{p}$ is a one–to–one continuous correspondence. Hence, $\Pi_C \cong S^{n-2}$. The homogeneity of Problem 5.2 implies $\Pi = \bigcup_{\lambda > 0} \mathfrak{g}_\lambda (\Pi_C)$, hence $\Pi \cong \mathbb{R}^{n-1}$. The surface Π partitions $\mathbb{R}^n$ into two half–spaces Π^+ and Π^-, where either $\langle \psi(x), \mathbf{b} \rangle > 0$, $\hat{u} = 1$ (in Π^+), or $\langle \psi(x), \mathbf{b} \rangle < 0$, $\hat{u} = \beta$ (in Π^-). **Q.E.D.**

The following example supplies a family of problems homogeneous under the action of a symmetry group $\mathfrak{G}$.

PROBLEM 5.4. *A family of homogeneous problems.*
Minimize

$$\int_0^\infty \sum_{i=1}^n c_i x_i^{\sigma n/(n+1-i)} \, dt$$

subject to

$$\dot{x}_1 = x_2, \ldots, \dot{x}_{n-1} = x_n, \dot{x}_n = u \in [\beta, 1],$$

with initial conditions

$$x_i(0) = x_i^{(0)}$$

$$(c_i \geqslant 0, \quad i = 1,\ldots,n, \quad \sigma \geqslant 2).$$

The $\mathfrak{G}$–action can be defined as follows:

$$\mathfrak{g}_\lambda \,:\, \mathbb{R}^n \longrightarrow \mathbb{R}^n,$$
$$\mathfrak{g}_\lambda(x) = (\lambda^n x_1,\, \lambda^{n-1} x_2,\ldots,\lambda^{n-k} x_k,\ldots,\lambda x_n),\quad \lambda \geqslant 0.$$

Constants α and γ, mentioned in Definition 5.1, equal 1 and σn.

5.4 Bi–Constant Ratio Solutions of Fuller's Problems

In this section we treat the automodelling (self–similar) solutions of the k–dimensional Fuller problem (Problem 5.1). First of all, we will investigate the dependence of the number of the one–parameter families of automodelling chattering solutions of equations (5.12)–(5.13) on the parameter $\beta < 0$, and, secondly, we will calculate these numbers for $k \leqslant 10$.

Consider the system

$$\dot{x}_1 = u,\ \dot{x}_2 = x_1,\ \ldots,\ \dot{x}_n = x_{n-1},\ \ldots \qquad (5.12)$$

where $x = (x_1, x_2, \ldots, x_n, \ldots) \in \mathbb{R}^d$ (d is a sufficiently large integer) and u is an arbitrary measurable function with two values only: $u(t) \in \{\beta, 1\}$, $\beta < 0$.

Our final purpose is to specify some class of automodelling solutions of system (5.12) subjected to the following supplementary limitation (maximum condition):

$$u(t) = \begin{cases} 1, & \text{if } (-1)^k x_{2k}(t) < 0, \\ \beta, & \text{if } (-1)^k x_{2k}(t) > 0. \end{cases} \qquad (5.13)$$

To define this class let us consider the following one–parameter group $\mathfrak{G}$ of diffeomorphisms of $\mathbb{R}^n$,

$$\mathfrak{G} = \{\mathfrak{g}_\lambda\},\quad \lambda \in \mathbb{R}_+,$$
$$\mathfrak{g}_\lambda(x) \overset{\text{def}}{=} (\lambda x_1,\, \lambda^2 x_2,\, \ldots,\, \lambda^n x_n,\, \ldots).$$

It can be immediately verified that if the pair $\big(x(t),\, u(t)\big)$ is a solution of (5.12), then for any $\lambda > 0$ the pair $\big(\mathfrak{g}_\lambda\big(x(t/\lambda)\big),\, u\big(t/\lambda\big)\big)$ is a solution of (5.12) also.

DEFINITION 5.2. *After E. Ryan [C. Dorling, E. Ryan, 1981], we shall say that $\big(x(t), u(t)\big)$ is a b.c.r. (bi–constant ratio) solution of system (5.12) if there exist $t_1 > t_0 > 0$ and $\mu > 0$ such that*

$$u(t) = \begin{cases} 1 & \text{if} \ \ 0 < t < t_0, \\ \beta & \text{if} \ \ t_0 < t < t_1, \end{cases}$$

and

$$x(t + t_1) = \mathfrak{g}_\mu x(t) \quad \text{for all } t \geqslant 0.$$

Let us clarify the etymology of the term. Assume that $t_0, t_1, t_2, \ldots$ are successive switching instants on a **b.c.r.** solution. It is possible to prove that the ratios $(t_{2k+2} - t_{2k+1})/(t_{2k+1} - t_{2k})$ are independent of k. The same is true for the ratios $(t_{2k+3} - t_{2k+2})/(t_{2k+2} - t_{2k+1})$. Within the action of the homogeneity group, the solution is completely characterized by these two constants. This determines the abbreviation **b.c.r.** When the constants coincide (as it does in the case $\beta = -1$), the solution is called a *constant ratio* solution.

We will return to restriction (5.13) later on; now we try to specify the initial points x_0 of **b.c.r.** solutions of system (5.12) which do not need to meet (5.13). If we take into account the homogeneity of system (5.12) under $\mathfrak{G}$–action, it remains to determine a single point of the **b.c.r.** solution on each $\mathfrak{G}$–orbit. So we can set $t_0 = 1$, $\tau = t_1 - 1$ and consider a **b.c.r.** solution $\big(x(t), u(t)\big)$ corresponding to two real parameters $\tau > 0$ and $\mu > 0$. To obtain the stable **b.c.r.** solutions we consider the values $\mu \in (0, 1)$.

In matrix notation, system (5.12) is equivalent to

$$\dot{x} = M x + u\,\mathbf{m},$$

where

$$M = \begin{pmatrix} 0 & 0 & 0 & \ldots \\ 1 & 0 & 0 & \ldots \\ 0 & 1 & 0 & \ldots \\ \vdots & \vdots & \vdots & \ddots \end{pmatrix} \quad \text{and} \quad \mathbf{m} = \begin{pmatrix} 1 \\ 0 \\ 0 \\ \vdots \end{pmatrix}.$$

If the control $u(t)$ is constant, then the solution $x(t)$ of (5.12) starting at x_0 can be explicitly calculated as follows:

$$x(t) = e^{Mt} \left(x_0 + u \int_0^t e^{-Ms}\mathbf{m}\,ds \right),$$

where

$$
e^{Mt} = \begin{pmatrix}
1 & 0 & 0 & \ldots & 0 & \ldots \\
t & 1 & 0 & \ldots & 0 & \ldots \\
\dfrac{t^2}{2} & t & 1 & \ldots & 0 & \ldots \\
\vdots & \vdots & \vdots & \ddots & \vdots & \ldots \\
\dfrac{t^{n-1}}{(n-1)!} & \dfrac{t^{n-2}}{(n-2)!} & \dfrac{t^{n-3}}{(n-3)!} & \ldots & 1 & \ldots \\
\vdots & \vdots & \vdots & \ddots & \vdots & \ddots
\end{pmatrix}
$$

$$
\int_0^t e^{-Ms}\,\mathbf{m}\,ds = \begin{pmatrix}
t \\
-t^2/2 \\
\vdots \\
(-1)^{n+1}\,t^n/n! \\
\vdots
\end{pmatrix}.
$$

Everywhere below in this section the expression $\big[v\big]_n$ means the value of the n–th coordinate of a vector $v \in \mathbb{R}^d$. A straightforward calculation yields

$$
\left[e^{Mt} \int_0^t e^{-Ms}\,\mathbf{m}\,ds \right]_n = t \cdot \frac{t^{n-1}}{(n-1)!} - \frac{t^2}{2} \cdot \frac{t^{n-2}}{(n-2)!} + \ldots
$$

$$
+ (-1)^{n+1} \cdot \frac{t^n}{n!} \cdot 1 = \frac{t^n}{n!}\big(1 - (1-1)^n\big) = \frac{t^n}{n!},
$$

and, finally,

$$
x(t) = e^{Mt} x_0 + uw(t)
$$

$$
\big[w(t)\big]_n = \frac{t^n}{n!}.
$$

Set

$$
K_\mu = \begin{pmatrix}
\mu & 0 & 0 & \ldots \\
0 & \mu^2 & 0 & \ldots \\
0 & 0 & \mu^3 & \ldots \\
\vdots & \vdots & \vdots & \ddots
\end{pmatrix}.
$$

The definition of a **b.c.r.** solution with the parameters $t_0 = 1$, $t_1 - t_0 = \tau$ and μ involves the following equation for the initial point x_0,

$$
K_\mu x_0 = e^{M\tau}\big(e^M x_0 + w(1)\big) + \beta w(\tau)
$$

$$
= e^{M(\tau+1)} x_0 + e^{M\tau} w(1) + \beta w(\tau).
$$

Using Newton's binomial formula one can check that

$$\left[e^{M\tau}\,w(1)\right]_n = \frac{\tau^{n-1}}{(n-1)!}\cdot 1 + \frac{\tau^{n-2}}{(n-2)!}\cdot\frac{1}{2!} + \ldots + 1\cdot\frac{1}{n!}$$

$$= \frac{(\tau+1)^n - \tau^n}{n!}.$$

Thus, for $x = x_0$, the following equation holds:

$$\left(K_\mu - e^{M(\tau+1)}\right)\cdot x = p(\tau) \tag{5.14}$$

where

$$\left[p(\tau)\right]_n = \frac{(\tau+1)^n + (\beta-1)\tau^n}{n!}.$$

To write out the operator $\left(K_\mu - e^{M(\tau+1)}\right)^{-1}$, let us note first that

$$MK_\mu = \begin{pmatrix} 0 & 0 & 0 & \ldots \\ \mu & 0 & 0 & \ldots \\ 0 & \mu^2 & 0 & \ldots \\ \vdots & \vdots & \vdots & \ddots \end{pmatrix}, \quad K_\mu M = \begin{pmatrix} 0 & 0 & 0 & \ldots \\ \mu^2 & 0 & 0 & \ldots \\ 0 & \mu^3 & 0 & \ldots \\ \vdots & \vdots & \vdots & \ddots \end{pmatrix},$$

i.e.,

$$K_\mu M = \mu\,MK_\mu.$$

As a result, we obtain

$$K_\mu\,e^{M\sigma} = K_\mu\left(E + M\sigma + M^2\frac{\sigma^2}{2!} + \ldots + M^n\frac{\sigma^n}{n!} + \ldots\right)$$

$$= \left(E + \mu M\sigma + \mu^2 M^2\frac{\sigma^2}{2!} + \ldots + \mu^n M^n\frac{\sigma^n}{n!} + \ldots\right)K_\mu$$

$$= e^{\mu M\sigma}\cdot K_\mu.$$

Since $(AB)^{-1} = B^{-1}A^{-1}$, the inverse operator to the operator $K_\mu - e^{M(\tau+1)}$ can be expanded in the series as follows:

$$\left(K_\mu - e^{M(\tau+1)}\right)^{-1} = \left[\left(E - K_\mu e^{-M(\tau+1)}\right)\cdot\left(-e^{M(\tau+1)}\right)\right]^{-1}$$

$$= -e^{-M(\tau+1)}\cdot\left(E + K_\mu e^{-M(\tau+1)} + \ldots + \left(K_\mu e^{-M(\tau+1)}\right)^n + \ldots\right)$$

$$= -e^{-M(\tau+1)}\cdot\left(E + e^{-M(\tau+1)\mu}\cdot K_\mu + \ldots\right.$$

$$\cdots + e^{-M(\tau+1)(\mu+\mu^2+\ldots+\mu^n)} \cdot K_{\mu^n} + \cdots\Big)$$

$$= -\sum_{\alpha=0}^{\infty} e^{-M(\tau+1)\frac{1-\mu^{\alpha+1}}{1-\mu}} \cdot K_{\mu^\alpha}.$$

It follows from (5.14) that

$$x = -\sum_{\alpha=0}^{\infty} e^{-M(\tau+1)\frac{1-\mu^{\alpha+1}}{1-\mu}} \cdot K_{\mu^\alpha} p(\tau). \tag{5.15}$$

We need below the explicit expression of coordinates of the vector $x(1)$, which is the first switching point on the **b.c.r.** solution. We have

$$x(1) = e^M x + w(1)$$

$$= w(1) - \sum_{\alpha=0}^{\infty} e^{M\left(1-(\tau+1)\frac{1-\mu^{\alpha+1}}{1-\mu}\right)} \cdot K_{\mu^\alpha} p(\tau). \tag{5.16}$$

Using Newton's binomial formula one can check that

$$\left[e^{M\sigma} \cdot K_{\mu^\alpha} p(\tau)\right]_n = \sum_{l=1}^{n} \frac{\sigma^{n-l}}{(n-l)!} \cdot \mu^{\alpha l} \frac{(\tau+1)^l + (\beta-1)\tau^l}{l!}$$

$$= \frac{(\sigma + \mu^\alpha(\tau+1))^n - \sigma^n}{n!} + (\beta-1)\frac{(\sigma+\mu^\alpha\tau)^n - \sigma^n}{n!}$$

$$= \frac{1}{n!}\left\{(\sigma + \mu^\alpha(\tau+1))^n - \beta\sigma^n + (\beta-1)(\sigma+\mu^\alpha\tau)^n\right\}.$$

The n–th coordinate of vector (5.15) can be represented in the form

$$[x]_n = -\frac{1}{n!}\sum_{\alpha=0}^{\infty}\left\{\left((\tau+1)\frac{\mu^{\alpha+1}-1}{1-\mu} + \mu^\alpha(\tau+1)\right)^n\right.$$

$$-\beta\left((\tau+1)\frac{\mu^{\alpha+1}-1}{1-\mu}\right)^n + (\beta-1)\left((\tau+1)\frac{\mu^{\alpha+1}-1}{1-\mu} + \mu^\alpha\tau\right)^n\right\}$$

$$= -\frac{1}{n!(1-\mu)^n}\sum_{\alpha=0}^{\infty}\left\{(\tau+1)^n(\mu^\alpha-1)^n - \beta(\tau+1)^n(\mu^{\alpha+1}-1)^n\right.$$

$$+ (\beta-1)\left(\mu^\alpha(\mu+\tau) - (\tau+1)\right)^n\right\}.$$

Let us note that the first and the second summands in the series cannot be directly combined with an appropriate multiplier because the term $\beta(1+$

$\tau)^n\left(\mu^{\alpha+1} - 1\right)^n$ does not approach zero as $\alpha \to \infty$. Consider therefore the N–th partial sum of the series:

$$\sum_{\alpha=0}^{N}\left\{(\mu^\alpha - 1)^n - \beta(\mu^{\alpha+1} - 1)^n\right\}$$

$$= \sum_{\alpha=0}^{N} (\mu^\alpha - 1)^n - \sum_{\alpha=1}^{N} \beta(\mu^\alpha - 1)^n - \beta(\mu^{N+1} - 1)^n$$

$$= (1 - \beta) \sum_{\alpha=0}^{N} (\mu^\alpha - 1)^n - \beta(\mu^{N+1} - 1)^n.$$

Discarding the infinitesimally small terms we arrive at the expression

$$[x]_n = \frac{-1}{n!(1 - \mu)^n}\left((-1)^{n+1}\beta(1 + \tau)^n\right.$$

$$\left.+\sum_{\alpha=0}^{\infty}\left\{(1 - \beta)(1 + \tau)^n(\mu^\alpha - 1)^n + (\beta - 1)\left(\mu^\alpha(\mu + \tau) - (1 + \tau)\right)^n\right\}\right).$$

Let $\nu = (\mu + \tau)/(1 + \tau)$ $(0 < \nu < 1$ in view of the inequalities $0 < \mu < 1,\ 0 < \tau)$. If we remove the parenthesis in the n–th power of the corresponding summand and change the order of summation (on account of the uniform convergence of the series), we obtain

$$[x]_n = \frac{(-1)^{n+1}(1 - \beta)}{n!(1 - \nu)^n} \cdot \left(\frac{\beta}{\beta - 1} + \sum_{l=1}^{n}(-1)^l C_n^l \frac{1 - \nu^l}{1 - \mu^l}\right). \tag{5.17}$$

This is the explicit expression for $[x]_n$ in terms of τ and μ as desired. The expression for $[x(1)]_n$ can be obtained from (5.16) by repeating the preceding calculation:

$$\left[e^M x + w(1)\right]_n$$

$$= \frac{1}{n!}\left[1 - \sum_{\alpha=0}^{\infty}\left\{\left(1 + (1 + \tau)\frac{\mu^{\alpha+1} - 1}{1 - \mu} + \mu^\alpha(1 + \tau)\right)^n\right.\right.$$

$$- \beta\left(1 + (1 + \tau)\frac{\mu^{\alpha+1} - 1}{1 - \mu}\right)^n$$

$$\left.\left.+ (\beta - 1)\left(1 + (1 + \tau)\frac{\mu^{\alpha+1} - 1}{1 - \mu} + \mu^\alpha\tau\right)^n\right\}\right]$$

$$= \frac{1}{n!}\left[1 - \frac{1}{(1-\mu)^n}\sum_{\alpha=0}^{\infty}\left\{\Big(\mu^{\alpha}(1+\tau) - (\mu+\tau)\Big)^n\right.\right.$$

$$- \beta\Big(\mu^{\alpha+1}(1+\tau) - (\mu+\tau)\Big)^n$$

$$\left.\left.+ (\beta-1)\Big(\mu^{\alpha}(\mu+\tau) - (\mu+\tau)\Big)^n\right\}\right]$$

$$= \frac{1}{n!}\left[1 - \frac{1}{(1-\mu)^n}\Big((-1)^{n+1}\beta(\mu+\tau)^n + (1-\mu)^n\right.$$

$$+ \sum_{\alpha=1}^{\infty}\left\{\Big(\mu^{\alpha}(1+\tau) - (\mu+\tau)\Big)^n(1-\beta)\right.$$

$$\left.\left.+ (\beta-1)\Big(\mu^{\alpha}(\mu+\tau) - (\mu+\tau)\Big)^n\right\}\Big)\right]$$

$$= \frac{1}{n!}\left(\frac{(-1)^n\beta(\mu+\tau)^n}{(1-\mu)^n} + \frac{\beta-1}{(1-\mu)^n}\times\right.$$

$$\left.\times \sum_{\alpha=1}^{\infty}\left\{\Big(\mu^{\alpha}(1+\tau) - (\mu+\tau)\Big)^n - (\mu^{\alpha}-1)^n(\mu+\tau)^n\right\}\right)$$

$$= (-1)^n\frac{(\beta-1)(\mu+\tau)^n}{n!(1-\mu)^n}\left(\frac{\beta}{\beta-1}\right.$$

$$\left.+ \sum_{\alpha=1}^{\infty}\left\{\Big(1-\frac{\mu^{\alpha}}{\nu}\Big)^n - (1-\mu^{\alpha})^n\right\}\right)$$

$$= \frac{(-1)^n(\beta-1)(\mu+\tau)^n}{n!(1-\mu)^n}\left(\frac{\beta}{\beta-1} + \sum_{l=1}^{n}(-1)^l\frac{C_n^l(\nu^{-l}-1)}{1-\mu^l}\mu^l\right).$$

Hence, the explicit expression of the vector $e^M x + w(1)$ is the following one:

$$\left[e^M x + w(1)\right]_n \tag{5.18}$$

$$= \frac{(-1)^{n+1}(1-\beta)\nu^n}{n!(1-\nu)^n}\cdot\left(\frac{1}{\beta-1} + \sum_{l=1}^{n}(-1)^l\frac{C_n^l(\nu^{-l}\mu^l-1)}{1-\mu^l}\right).$$

Equations (5.17) and (5.18) are mutually connected. Assume that the collection (β,τ,μ) determines some **b.c.r.** solution of system (5.12), that is,

$$x(t+1+\tau) = \mathfrak{g}_{\mu}\, x(t), \quad t \geqslant 0.$$

Using the homogeneity property of system (5.12) it is possible to generate another **b.c.r.** solution dual to the **b.c.r.** solution $x(t)$. Consider first the trajectory

$$x^{(1)}(t) \overset{\text{def}}{=} \frac{1}{\beta}\, \mathfrak{g}_{-\beta}\, x\left(-\frac{t}{\beta}\right),$$

i.e., $x_1^{(1)} = -x_1(-t/\beta)$, $x_2^{(1)} = \beta x_2(-t/\beta)$, $x_3^{(1)} = -\beta^2 x_3(-t/\beta)$, etc. We see that $x^{(1)}(t)$ meets system (5.12), where

$$\dot{x}_1^{(1)}(t) = \frac{1}{\beta}\cdot \dot{x}_1(s)\bigg|_{s=-\frac{t}{\beta}} \in \left\{\frac{1}{\beta}, 1\right\}.$$

Let us note that $\dot{x}_1^{(1)}(t) = 1/\beta$ at $t \in (0, -\beta)$ and $\dot{x}_1^{(1)}(t) = 1$ at $t \in \big(-\beta, -\beta(1+\tau)\big)$. The coefficient of contraction is common for both $x(t)$ and $x^{(1)}(t)$,

$$\frac{x_1^{(1)}\big(-\beta(1+\tau)\big)}{x_1^{(1)}(0)} = \frac{-x_1(1+\tau)}{-x_1(0)} = \mu.$$

We would like to change the order of the sign alternations of the control to obtain $u = 1$ on the first interval and $u = 1/\beta$ on the second one. So consider the trajectory

$$x^{(2)}(t) \overset{\text{def}}{=} x^{(1)}(t-\beta) = \frac{1}{\beta}\mathfrak{g}_{-\beta}\, x\left(\frac{t-\beta}{-\beta}\right).$$

It can readily be checked that $\dot{x}_1^{(2)}(t) = 1$ on the interval $t \in (0, -\beta\tau)$ and $\dot{x}_1^{(2)}(t) = 1/\beta$ on $\big(-\beta\tau, -\beta(\tau+\mu)\big)$. It remains to normalize the trajectory $x^{(2)}(t)$ as follows:

$$\widetilde{x}(t) \overset{\text{def}}{=} \mathfrak{g}_{-1/(\beta\tau)}\, x^{(2)}(-\beta\tau t) = \frac{1}{\beta}\mathfrak{g}_{1/\tau}\, x(\tau t + 1). \tag{5.19}$$

As a result, $\dot{\widetilde{x}}_1(t) = 1$ at $t \in (0,1)$, $\dot{\widetilde{x}}_2 = 1/\beta$ at $t \in (1, 1+\mu/\tau)$ and

$$\frac{\widetilde{x}_1\left(1+\dfrac{\mu}{\tau}\right)}{\widetilde{x}_1(0)} = \frac{-x_1(1+\mu+\tau)}{-x_1(1)} = \mu,$$

as was required. Thus, $\widetilde{x}(t)$ is also a **b.c.r.** solution of system (5.12) with the parameters $\widetilde{\beta} = 1/\beta$, $\widetilde{\tau} = \mu/\tau$, $\widetilde{\mu} = \mu$. Recall that $\nu = (\mu+\tau)/(1+\tau)$, hence

$$\widetilde{\nu} = \frac{\mu + \dfrac{\mu}{\tau}}{1 + \dfrac{\mu}{\tau}} = \frac{\mu}{\nu}.$$

Therefore, the coordinates of the vector $\widetilde{x}(0)$ can be obtained by the substitution $\widetilde{\mu} = \mu$, $\widetilde{\nu} = \mu/\nu$, $\widetilde{\beta} = 1/\beta$ into (5.17). On the other hand, it follows from (5.19) that $\widetilde{x}(0) = \mathfrak{g}_{1/\tau}\big(x(1)\big)/\beta$, or, equivalently, $x(1) = \mathfrak{g}_{\tau}\big(\beta\widetilde{x}(0)\big)$. This implies that the expression (5.18) for the coordinates of $x(1)$ can be obtained from (5.17) by means of the substitution $\mu' = \mu$, $\nu' = \mu/\nu$, $\beta' = 1/\beta$, $\tau' = \mu/\tau$ and consequent multiplying of the n–th coordinate by $\beta\tau^n$. Let

$$\phi_n(\mu,\nu) = \sum_{l=1}^{n}(-1)^l C_n^l \frac{1-\nu^l}{1-\mu^l};\qquad (5.20)$$

then (5.17) yields

$$\big[x(1)\big]_n = \beta\tau^n \frac{(-1)^{n+1}(1-1/\beta)}{n!(1-\mu/\nu)^n}\left(\frac{1/\beta}{1/\beta-1}+\phi_n(\mu,\mu/\nu)\right),$$

which coincides with (5.18). This is the relation we have been looking for.

5.5 Optimality of b.c.r. Solutions

It was proved in the previous section that the pair (μ,ν) determines a **b.c.r.**–solution of (5.12) by means of relations (5.17), (5.18). Let us clarify when the **b.c.r.** solution also meets the maximum condition (5.13). First of all we have the switching condition:

$$\big[x\big]_{2k} = \big[x(1)\big]_{2k} = 0;$$

hence,

$$\begin{cases} \phi_{2k}(\mu,\nu) = \dfrac{\beta}{1-\beta}, \\[2mm] -\phi_{2k}\big(\mu,\dfrac{\mu}{\nu}\big) - 1 = \dfrac{\beta}{1-\beta}. \end{cases} \qquad (5.21.a)$$

Besides that, we have

$$(-1)^k[x(0)]_{2k-1} < 0, \quad (-1)^k[x(1)]_{2k-1} > 0,$$

that implies

$$\begin{cases} (-1)^k\left(\dfrac{\beta}{\beta-1}+\phi_{2k-1}(\mu,\nu)\right) < 0, \\[2mm] (-1)^k\left(\dfrac{1}{\beta-1}-\phi_{2k-1}\big(\mu,\dfrac{\mu}{\nu}\big)\right) > 0. \end{cases} \qquad (5.21.b)$$

PROPOSITION 5.1. *The triple (β, μ, ν) determines some* **b.c.r.** *solution of system (5.12)–(5.13) iff the inequalities $\beta < 0$, $0 < \mu < 1$, $0 < \nu < 1$ and relations (5.21) hold.*

To prove this proposition we need some auxiliary construction. In this chapter we often deal with polynomials and their roots. Let us restate some results of polynomial algebra. Let $\mathcal{P}(t) = a_0 t^n + a_1 t^{n-1} + \ldots + a_n$, $a_0 \neq 0$, be an arbitrary polynomial of n–th power with real coefficients. We would like to determine how many roots of the equation

$$\mathcal{P}(t) = 0$$

there are on a semi–interval $(a, b]$. Denote this number by $N(a, b)$. The simplest way to the upper estimate $N(a, b)$ is given by the following

Budan–Fourier Theorem. *For any $s \in (a, b]$, consider the series of $(n + 1)$ numbers:*

$$\mathcal{P}(t), \mathcal{P}'(t), \ldots, \mathcal{P}^{(k)}(t), \ldots, \mathcal{P}^{(n)}(t) \equiv n!\, a_0. \qquad (5.22)$$

Let us replace the number $\mathcal{P}^{(k)}(s)$ in (5.22) by the sign "+" if $\mathcal{P}^{(k)}(s) > 0$ and by the sign "−" if $\mathcal{P}^{(k)}(s) < 0$. If $\mathcal{P}^{(k)}(s) = 0$ for some $0 \leqslant k \leqslant n$, then we omit this number in the sequence (5.22). Denote by $S(t)$ the number of the sign's changes for all pairs of neighboring members in sequence (5.22). Then

$$N(a, b) \leqslant S(a) - S(b).$$

For the proof, see [P.M. Cohn, 1974; Theorem 4, p. 149]. The Budan–Fourier criterion in the given form is too rough to give a precise estimation of the function $N(a, b)$. As a rule, we use it to prove that a certain polynomial has at most one root at some interval. Fortunately, the criterion is applicable to all the polynomials we treat below.

Besides the Budan–Fourier theorem, we need a technique for the determination of the common roots of two polynomials in one or two variables. Given two polynomials of powers n and k,

$$\mathcal{P}_1(t) = a_0 t^n + a_1 t^{n-1} + \ldots + a_n,$$
$$\mathcal{P}_2(t) = b_0 t^k + b_1 t^{k-1} + \ldots + b_k,$$

whose coefficients a_i, b_j are in their own turn polynomials of power m in the variable τ. We would like to determine the common roots of $\mathcal{P}_1(t)$ and $\mathcal{P}_2(t)$, i.e., to solve the system

$$\begin{cases} \mathcal{P}_1(t) = 0, \\ \mathcal{P}_2(t) = 0. \end{cases} \qquad (5.23)$$

Define the *resultant* R of the polynomials $\mathcal{P}_1$ and $\mathcal{P}_2$,

$$
R = \left.\left|
\begin{array}{cccccccc}
a_0 \, a_1 & \cdots & a_{n-1} \, a_n & & & & & \\
& a_0 \, a_1 & \cdots & a_{n-1} \, a_n & & & & \\
& & \multicolumn{4}{c}{\cdots\cdots\cdots\cdots\cdots} & \\
& & & & a_0 \, a_1 & \cdots & a_{n-1} \, a_n & \\
b_0 \, b_1 & \cdots & b_{k-1} \, b_k & & & & & \\
& b_0 \, b_1 & \cdots & b_{k-1} \, b_k & & & & \\
& & \multicolumn{4}{c}{\cdots\cdots\cdots\cdots\cdots} & \\
& & & b_0 \, b_1 & \cdots & b_{k-1} \, b_k & &
\end{array}
\right|\right.
\begin{array}{l}
\left.\vphantom{\begin{array}{c}1\\1\\1\\1\end{array}}\right\} k \text{ rows} \\[1.5em]
\left.\vphantom{\begin{array}{c}1\\1\\1\\1\end{array}}\right\} n \text{ rows}
\end{array}
$$

(with other terms equaling zero).

A Theorem on the Resultant of Polynomials

Let $a_0 \neq 0$, $b_0 \neq 0$. Then system (5.23) is compatible iff the resultant R of the polynomials $\mathcal{P}_1(t)$, $\mathcal{P}_2(t)$ equals zero.

For the proof, see any standard textbook on algebra, e.g., [P.M. Cohn, 1974].

In the particular case $\mathcal{P}_2(t) = \partial \mathcal{P}_1/\partial t$, the function $(-1)^{n(n-1)/2} a_0^{-1} R$ is called the *discriminant* of the polynomial $\mathcal{P}_1(t)$.

If the coefficients a_i, b_j of the polynomials $\mathcal{P}_1(t)$, $\mathcal{P}_2(t)$ depend on τ, then the resultant becomes a function of τ also. The roots of the resultant give all the values τ where system (5.23) is compatible (but now some roots of the resultant can be irrelevant). We see that the power of the resultant is not greater than m^{n+k}.

Now let us return to the sequence (5.22) and the function $S(t)$.

LEMMA 5.7. *The function $S(t)$ is decreasing.*

Proof. The function $S(t)$ has a finite number of values, so it can vary only at those very points at which the sequence (5.22) has some zero members. We shall say that a value t is a zero of (5.22) if for the given t there is any zero member in (5.22). We shall say that t is a multiple zero of sequence (5.22) if there are two or more zeros in succession at some place in (5.22).

Assume that $\mathcal{P}(t_0) \neq 0$. Let us consider the situation when $\mathcal{P}^{(k)}(t_0) = 0$, $\mathcal{P}^{(k+1)}(t_0) > 0$, $\mathcal{P}^{(k-1)}(t_0) < 0$, which gives the following subsequence of signs in (5.22): $\ldots - 0 + \ldots$. Then for $t > t_0$ in a sufficiently small neighborhood of t_0 we have the following distribution of signs in (5.22): $\ldots - + + \ldots$ and for $t < t_0$ the distribution of signs is the following: $\ldots - - + \ldots$. If $\mathcal{P}^{(k)}(t)$ is a single zero in (5.22) at $t = t_0$, that implies $S(t_0 - 0) = S(t_0) = S(t_0 + 0)$.

Assume that $\mathcal{P}^{(k+1)}(t)$ and $\mathcal{P}^{(k-1)}(t)$ have the same signs, e.g. $\ldots + 0 + \ldots$. Then in a sufficiently small neighborhood of t_0 for $t > t_0$

we have $\ldots + + + \ldots$, and for $t < t_0$ we have $\ldots + - + \ldots$. If $\mathcal{P}^{(k)}(t_0)$ is a single zero member in (5.22), it means that $S(t_0 - 0) - 2 = S(t_0) = S(t_0 + 0)$.

Assume that $\mathcal{P}(t_0) = 0$, $\mathcal{P}'(t_0) \neq 0$, e.g., $\mathcal{P}'(t_0) > 0$. Then for all neighboring $t < t_0$ we have $- + \ldots$, and for $t > t_0$ we have $+ + \ldots$. If $\mathcal{P}(t_0)$ is a single zero member, then $S(t_0 - 0) - 1 = S(t_0) = S(t_0 + 1)$. Thus, in a small neighborhood of a single nonmultiple zero of sequence (5.22) the function $S(t)$ is either constant or decreases by 1 or 2 (in accordance with the place of the zero member). This number (0, 1, or 2) we shall call the defect of the zero member. If for $t = t_0$ there are two or more zero members in (5.22), then the decreasing of function $S(t)$ equals the sum of the defects of the isolated zero members.

Let us consider the situation of multiple zeros. Assume that for $t = t_0$ there are two successive zeros and some positive member stands to their right in (5.22): $\ldots 0\, 0 + \ldots$. Then in (5.22) we have the sign distribution $\ldots + - + \ldots$ for all neighboring $t < t_0$ and the distribution $\ldots + + + \ldots$ for $t > t_0$. If there are no other zero members in (5.22), then $S(t_0 - 0) - 2 = S(t_0) = S(t_0 + 0)$. It is easy to see that for a zero of an arbitrary order p (i.e., if there exists a series of exactly p zeros) we have

$$S(t_0 - 0) - p = S(t_0) = S(t_0 + 0)$$

in the case that p is even $(p = 2l)$ and two different possibilities in the case that p is odd $(p = 2l + 1)$ and the zero series does not starts from $\mathcal{P}(t_0)$. The first possibility is

$$S(t_0 - 0) - p - 1 = S(t_0) = S(t_0 + 0).$$

if members of the same sign stand to the right and to the left of the zero–series; the second is

$$S(t_0 - 0) - p + 1 = S(t_0) = S(t_0 + 0),$$

if these signs are opposite. If the zero series starts from $\mathcal{P}(t_0)$, then we have

$$S(t_0 - 0) - p = S(t_0) = S(t_0 + 0).$$

This implies the statement of Lemma 5.7. **Q.E.D.**

COROLLARY 5.4. *If there are p zeros of the polynomial $\mathcal{P}(t)$ on $(a, b]$, then $S(a) \geqslant S(b) + p$.*

COROLLARY 5.5. *If there is at least one multiple zero of $\mathcal{P}(t)$ (i.e., $\mathcal{P}(t) = \mathcal{P}'(t) = 0$), then $S(a) \geqslant S(b) + 2$.*

Assume that $x(t)$ is a **b.c.r.** solution of system (5.12) with the parameters β, μ, τ at $x_0 \in \Pi \overset{\text{def}}{=} \{ x \in \mathbb{R}^d \mid x_{2k} = 0 \}$ such that $x(1) \in \Pi$. We do not assume that the maximum condition (5.13) is valid.

Lemma 5.8.

1. $x_{2k-1}(0) \neq 0$, $x_{2k-1}(1) \neq 0$.

2. *There are exactly three zeros of the function* $x_{2k}(t)$ *at* $t \in [0, 1 + \tau]$, *namely,* $t = 0$, $t = 1$, $t = 1 + \tau$.

Proof. For any $t \in (0, 1)$ and $t \in (1, 1 + \tau)$, the function $x_{2k}(t)$ is represented by two different polynomials $\mathcal{P}_1(t)$ and $\mathcal{P}_2(t)$ of order $2k$. Denote by $S_i(t)$ the number of sign changes in the sequences (5.22) for $\mathcal{P}_i(t)$, $i = 1, 2$. For both $\mathcal{P}_1(t)$ and $\mathcal{P}_2(t)$ the first $2k$ members in (5.22) are $2k$ current coordinates of $x(t)$ (i.e., $x_{2k}(t)$, $x_{2k-1}(t)$, $\ldots$, $x_1(t)$). The last $(2k+1)$–th member equals 1 for $\mathcal{P}_1(t)$ and β for $\mathcal{P}_2(t)$. Let us demonstrate that $S_2(1)$ is not greater than $S_1(0) + 1$. Let $l < 2k$ be the least power of the nonzero derivative $d^l x_{2k}(t)/dt^l = x_{2k-l}(t)$ at $t = 1$. If $x_{2k-l}(1) > 0$, then $S_1(1) + 1 = S_2(1)$ (the sequence of signs $\ldots + 00 \ldots 0+$ is converted at the instant $t = 1$ into the sequence $\ldots + 00 \ldots 0-$ because $t = 1$ is the switching moment). If $x_{2k-l} < 0$, we pass from the sequence $\ldots - 00 \ldots 0+$ to the sequence $\ldots - 00 \ldots 0-$. Hence, $S_1(1) - 1 = S_2(1)$, i.e., $S_2(1) < S_1(1)$. Let us show that the remainders $S_1(0) - S_1(1)$ and $S_2(1) - S_2(1 + \tau)$ are not greater than 1. Suppose the contrary, e.g., $S_1(0) - S_1(1) \geqslant 2$. In view of Lemma 5.7 and Corollary 5.4 we have $S_2(1 + \tau) \leqslant S_2(1) - 1 \leqslant S_1(1) \leqslant S_1(0) - 2$, i.e.,

$$S_2(1 + \tau) \leqslant S_1(0) - 2. \tag{5.24}$$

Recall that $S_2(1+\tau)$ is the number of sign changes in the sequence $x_{2k}(1+\tau)$, $x_{2k-1}(1 + \tau)$, $\ldots$, $x_1(1 + \tau)$, β. Since $x(1 + \tau) = \mathfrak{g}_\mu x(0)$, we have $\operatorname{sgn} x_l(0) = \operatorname{sgn} x_l(1+\tau)$ for all $l = 1, \ldots, 2k$. So $|S_2(1+\tau) - S_1(0)| \leqslant 1$, which contradicts (5.24). The inequality $S_2(1) - S_2(1 + \tau) \leqslant 1$ can be proved in the same manner.

Now suppose that $x_{2k-1}(0) = 0$ or $x_{2k-1}(1) = 0$. It follows from Corollary 5.5 that $S_1(0) - S_1(1) \geqslant 2$, which is impossible. If $x_{2k}(t) = 0$ at some internal point on $(0, 1)$ or $(1, 1+\tau)$ then on account of Corollary 5.4 we also have $S_1(0) - S_1(1) \geqslant 2$ or $S_2(1) - S_2(1+\tau) \geqslant 2$. **Q.E.D.**

Proof of Proposition 5.1. Lemma 5.8 provides the following sufficient condition for the existence of a **b.c.r.** trajectory: there are no zeros of odd order on $\left[x(t) \right]_{2k}$ at $t \in (0, 1)$ and $t \in (1, 1 + \tau)$. Thus, in the domain $\beta < 0$, $0 < \mu < 1$, $0 < \nu < 1$, the feasibility of equations (5.21) is a necessary and sufficient condition for the triple (β, μ, ν) to determine some **b.c.r.** solutions of system (5.12).

Q.E.D.

We would like to prove that the number of solutions of system (5.21) coincides for all $\beta < 0$. It is convenient to reformulate this problem as follows. Given the curve

$$\phi_{2k}(\mu, \nu) + \phi_{2k}\left(\mu, \frac{\mu}{\nu}\right) + 1 = 0, \tag{5.25}$$

defined in the domain $\mathcal{K} \overset{\text{def}}{=} \{\mu, \nu \mid 0 < \mu < \nu < 1\}$, we would like to understand how many points are there on the curve subject to the equation

$$\phi_{2k}(\mu, \nu) = p \tag{5.26}$$

for the given p.

Let us note that the inequality $\mu < \nu$ follows from the estimation

$$\nu - \mu = \frac{\mu + \tau}{1 + \tau} - \mu = \frac{\tau(1 - \mu)}{1 + \tau} > 0.$$

The set (5.25) is invariant under the mapping $i^* : (\mu, \nu) \to (\mu, \mu/\nu^2)$. The stationary points of the mapping i^* lie on the parabola $\mu = \nu^2$.

PROPOSITION 5.2. *The point* (μ_0, ν_0) *belongs to the intersection of the set (5.25) and the curve* $\mu = \nu^2$ *iff*

$$\phi_{2k}(\mu_0, \nu_0) = \phi_{2k}\left(\mu_0, \frac{\mu_0}{\nu_0}\right) = -\frac{1}{2}$$

(and hence $\beta = -1$*).*

Proof. The direct assertion is obvious, so we need to prove the inverse statement, namely, if $p = -1/2$ in (5.25)–(5.26), then $\mu = \nu^2$. Function $\phi_{2k}(\mu, \nu)$ can be represented in the form of an absolutely convergent series:

$$\phi_{2k}(\mu, \nu) = \sum_{\alpha=0}^{\infty} \left((1 - \mu^\alpha)^{2k} - (1 - \nu\mu^\alpha)^{2k}\right).$$

For all $\alpha \in \mathbb{N}$ and for all (μ, ν) in the domain $\mathcal{K}$ we have $(1 - \mu^\alpha)^{2k} < (1 - \nu\mu^\alpha)^{2k}$, hence $\phi_{2k} < 0$. Also,

$$\frac{\partial \phi_{2k}}{\partial \nu}(\mu, \nu) = 2k \sum_{\alpha=0}^{\infty} \mu^\alpha (1 - \nu\mu^\alpha)^{2k-1} > 0.$$

The restriction of ϕ_{2k} to the straight line $\mu = \text{Const}$ is an increasing function. If $p = -1/2$ and the pair (μ, ν) meets (5.25)–(5.26), then $\phi_{2k}(\mu, \nu) = \phi_{2k}(\mu, \mu/\nu) = -1/2$. The monotonicity of ϕ_{2k} implies $\nu = \mu/\nu$, i.e., $\mu = \nu^2$ as required. **Q.E.D.**

5.6 Numerical Verification of the Conjecture on the Number of Cycles in the Orbit Space

Proposition 5.2 shows that the number of points of (5.25) on the parabola $\mu = \nu^2$ equals the number of **b.c.r.** solutions in the symmetric case of Fuller's problem when $\beta = -1$. If we prove that for any $p \in (-1, 0)$ the curve (5.25) is transversal to (5.26) in $\mathcal{K}$, it would imply that the number of **b.c.r.** solutions to Fuller's problem does not depend on β and coincides with those in the symmetric Fuller's problem.

We have not obtained yet the full analytic solution of this problem. Nevertheless, there exists some direct procedure for the verification of the conjecture in any concrete dimension k. Set

$$\Delta_{2k}(\mu, \nu) \overset{\text{def}}{=} \phi_{2k}(\mu, \nu) + \phi_{2k}\left(\mu, \frac{\mu}{\nu}\right) + 1.$$

The function $\phi_{2k}(\mu, \nu)$ is a quotient of two polynomials depending on μ and ν. Therefore, within to nonzero multipliers, the equation $\Delta_{2k} = 0$ can be written as $P_{2k}(\mu, \nu) = 0$ where $P_{2k}(\mu, \nu)$ is a polynomial of power $k(2k+1)$ in μ and power $4k$ in ν. Furthermore, the equation

$$\begin{vmatrix} \dfrac{\partial \Delta_{2k}}{\partial \mu} & \dfrac{\partial \Delta_{2k}}{\partial \nu} \\[2ex] \dfrac{\partial \phi_{2k}}{\partial \mu} & \dfrac{\partial \phi_{2k}}{\partial \nu} \end{vmatrix} = 0$$

can be written as $Q_{2k}(\mu, \nu) = 0$ where Q_{2k} is a polynomial in μ and in ν. All that we need is to prove that the equations $P_{2k}(t) = 0$ and $Q_{2k}(t) = 0$ are not compatible in $\mathcal{K}$. To do that, we are to calculate the resultant of the polynomials P_{2k} and Q_{2k} (where P_{2k} and Q_{2k} are considered to be polynomials in ν). Therefore, the resultant is a polynomial in μ. The conjecture is that the resultant has no zeros at $0 < \mu < 1$. The appropriate numerical experiment has been performed for the first ten values of the dimension k: $k = 1, 2, \ldots, 10$. This allows us to state the following

CONJECTURE 5.6. *In the region* $\mathcal{K}$ *the set* $\Gamma_{2k} \overset{\text{def}}{=} \{\, \mu, \nu \mid \Delta_{2k} = 0 \}$ *consists of* k *analytic branches, call them* Γ^i_{2k}, $i = 1, 2, \ldots, k$. *The restriction of the function* ϕ_{2k} *to each of* Γ^i_{2k} *is a strictly monotonic function.*

For $k = 2$, the statement follows from Lemma 3.3. For $k = 3$, we give an analytic proof of the statement below. For $k = 4, 5, \ldots, 10$, the conjecture is confirmed by a numerical procedure. If the statement is true in its full volume, then it supplies the following algorithm of numerical determination of the number of **b.c.r.** solutions of systems (5.12)–(5.13).

We are to solve the equation $\phi_{2k}(\mu, \nu) = -1/2$ at points of the parabola $\mu = \nu^2$. In view of (5.20) this is equivalent to the equation

$$\sum_{l=1}^{2k} (-1)^l C_{2k}^l \frac{1}{1 + \nu^l} = -\frac{1}{2}.$$

There are not more than $k(2k + 1)$ solutions of the equation. We are to select those that lie in $(0, 1)$ and meet the maximum condition (5.13). In view of (5.21.b) this gives

$$(-1)^k \left[\phi_{2k-1}(\nu^2, \nu) + \frac{1}{2} \right] > 0,$$

or, equivalently,

$$(-1)^k \left[\sum_{l=1}^{2k-1} (-1)^l C_{2k-1}^l \frac{1}{1 + \nu^l} + \frac{1}{2} \right] > 0. \tag{5.27}$$

In view of Lemma 5.8, the $(2k - 1)$–th coordinate of the starting point on **b.c.r.** solution is not zero. Since function $\phi_{2k}(\mu, \nu)$ is continuous, the inequality (5.27) remains true for all other points of the branch.

The results of the numerical calculation of the number of **b.c.r.** solutions to the k–dimensional Fuller problem are presented in the following table:

k	2	3	4	5	6	7	8	9	10
$n(k)$	1	2	3	4	5	6	7	8	9
$m(k)$	1	1	2	2	3	3	4	4	5

Here k is the dimension of the Fuller problem, $n(k)$ is the number of all **b.c.r.** solutions of (5.12), and $m(k)$ is the number of **b.c.r.** solutions of (5.12) which meet the maximum condition (5.13). These results are in agreement with Conjecture 5.2 on the number of cycles in orbit space.

5.7 Three–Dimensional Fuller Problems

This section contains the main result of Chapter 5. Namely, we give a full description of optimal solutions with chattering arcs to some class of three–dimensional problems. Though plenty of two–dimensional problems have been successfully solved, there are only a few three–dimensional examples that have been analyzed completely.

Besides the usual difficulties connected with a study and representation of three–dimensional dynamic systems, a peculiar singularity is inherent in Hamiltonian systems at the Fuller point on a singular solution of third order. In the case of singular solutions of second order, there exists an analytic change of variables with a singularity (blowing–up procedure) that resolves the singularity of the Hamiltonian system at the switching curve of optimal chattering solutions. In the case of third order solutions, any blowing–up procedure remains at least a one–dimensional manifold on a two–dimensional optimal switching surface where the Hamiltonian system still has a singularity. Therefore, the standard technique of the invariant manifold theorem allows us to obtain only a part of the switching surface. To complete the study of the switching surface we need to apply some supplementary topological methods.

Among the k–dimensional analogs of Fuller's problem which were considered in Section 5.1 it is only the three-dimensional one in which the optimal synthesis was designed. Indeed, we solve a generalization of this problem, namely

PROBLEM 5.5. *The three–dimensional Fuller problem.*
Minimize

$$J\big(x(\cdot)\big) = \int_0^\infty \left(\frac{1}{2}x_1^2 + \frac{\epsilon}{3}\,|x_2|^3 \right) dt$$

subject to

$$\dot{x}_1 = x_2, \quad \dot{x}_2 = x_3, \quad \dot{x}_3 = u \in [\beta, 1] \quad (\beta < 0) \tag{5.28}$$

with initial conditions

$$x_1(0) = x_1^0, \quad x_2(0) = x_2^0, \quad x_3(0) = x_3^0.$$

Problem 5.5 is a particular case of Problem 5.4 with $n = 3$, $c_1 = 1/2$, $c_3 = 0$, $\sigma = 2$ and small $c_2 = \epsilon/3$. The symmetry group action is given by the mappings

$$\mathfrak{g}_\lambda : \mathbb{R}^3 \longrightarrow \mathbb{R}^3,$$
$$\mathfrak{g}_\lambda(x_1, x_2, x_3) = (\lambda^3 x_1, \lambda^2 x_2, \lambda x_3),$$

with $\lambda \in \mathbb{R}_+$ having been an arbitrary positive number.

It follows from Lemma 5.6 that the switching surface of optimal solutions to Problem 5.5 is a manifold $\Pi \cong \mathbb{R}^2$. All that we need is to establish the smoothness of the surface Π and to give a more complete characterization of the behavior of optimal solutions. It is convenient to imagine the optimal synthesis in Problem 5.5 by means of the following factorization procedure.

Let us consider the two–dimensional unit sphere S^2 in $\mathbb{R}^3$ as an orbit–space of $\mathbb{R}^3\backslash\{0\}$ under the action of the group $\mathfrak{g}_\lambda$. It is proven in Lemma 5.7 below that for any $x_0 \in \mathbb{R}^3\backslash\{0\}$ there exists a unique intersection of the orbit $\bigcup_{\lambda>0}\mathfrak{g}_\lambda(x_0)$ with S^2. Consider the mapping $\mathcal{P} : \mathbb{R}^3\backslash\{0\} \to S^2$ which transfers the point $x \in \mathbb{R}^3\backslash\{0\}$ to the point $\left\{\bigcup_{\lambda>0}\mathfrak{g}_\lambda(x)\right\}\bigcap S^2$. For any optimal trajectory in $\mathbb{R}^3$ we can consider its "trace" on the sphere S^2, i.e., the $\mathcal{P}$–image of the trajectory. This determines a quotient–system on S^2, call it $\mathfrak{A}$. It is easy to see that the system $\mathfrak{A}$ is specified up to the choice of the reference time (this will be described in detail later). Thus, the optimal synthesis in Problem 5.5 can be characterized in terms of $\mathfrak{A}$–trajectories on S^2. It appears that there exists a unique closed cycle on S^2. The cycle partitions S^2 into two two–dimensional disks with a common boundary. The behavior of $\mathfrak{A}$–trajectories inside each of these disks is represented in Fig. 16, on page 196. The final purpose of this section is to substantiate this picture.

Let us investigate the singular solutions to Problem 5.5. Pontryagin's maximum principle leads to the equations

$$
\begin{aligned}
\dot{\psi}_3 &= -\psi_2, & \dot{x}_1 &= x_2, \\
\dot{\psi}_2 &= -\psi_1 + \epsilon x_2^2 \,\mathrm{sgn}x_2, & \dot{x}_2 &= x_3, \\
\dot{\psi}_1 &= x_1, & \dot{x}_3 &= \widehat{u},
\end{aligned}
\qquad (5.29.\mathrm{a})
$$

where

$$
\widehat{u} = \begin{cases} 1, & \text{if } \psi_3 > 0, \\ \beta, & \text{if } \psi_3 < 0. \end{cases}
\qquad (5.29.\mathrm{b})
$$

Let us set $z = (x_3, x_2, x_1, \psi_1, -\psi_2, \psi_3)$, i.e., $z_1 = x_3$, $z_2 = x_2, \ldots, z_5 = -\psi_2$, $z_6 = \psi_3$. Let $Z(t, z^0)$ be a solution of system (5.29) with an initial point z^0. Denote by

$$
H_0 = \left\{ z \,\Big|\, \psi_1 x_2 + \psi_2 x_3 + \psi_3 \widehat{u} - \frac{1}{2}x_1^2 - \frac{\epsilon}{3}|x_2|^3 = 0 \right\}
\qquad (5.30)
$$

the zero–level surface of the Hamiltonian of system (5.29). Denote by

$$
S = \left\{ z \,\big|\, \psi_3 = 0 \right\}
$$

the switching surface of system (5.29). Let us ascertain that there are no singular solutions of system (5.29) on the surface H_0 except for the zero–solution.

LEMMA 5.9. *For any* $z^0 \in S \bigcap H_0 \backslash \{0\}$ *and for any sufficiently small* $\epsilon > 0$ *there exists* $t^0 = t^0(z^0, \epsilon) > 0$ *such that*

$$Z(t, z^0)\Big|_{(-t^0, t^0)} \bigcap S = z^0.$$

Proof. Let $z^0 \in S \bigcap H_0 \backslash \{0\}$. Differentiate a solution $Z(t, z^0)$ with respect to t at $t = 0$. We have:

$$\frac{d\psi_3}{dt} = -\psi_2, \qquad\qquad \frac{d^3\psi_3}{dt^3} = x_1 - 2\epsilon|x_2|x_3,$$

$$\frac{d^2\psi_3}{dt^2} = \psi_1 - \epsilon x_2^2 \operatorname{sgn} x_2, \qquad \frac{d^4\psi_3}{dt^4} = x_2 - 2\epsilon x_3^2 \operatorname{sgn} x_2 - 2\epsilon|x_2|\widehat{u}.$$

If any of these derivatives is not zero at z^0, then $Z(t, z^0)$ satisfies the statement of Lemma 5.9. Assume that all the derivatives are zero. If we take into account the condition that $z^0 \in S \bigcap H_0 \backslash \{0\}$, we obtain

$$\psi_3 = 0, \quad \psi_2 = 0, \quad \psi_1 = \epsilon x_2^2 \operatorname{sgn} x_2, \quad x_1 = 2\epsilon|x_2|x_3,$$

$$2\epsilon|x_2|\widehat{u} = x_2 - 2\epsilon x_3^2 \operatorname{sgn} x_2,$$

$$\psi_1 x_2 + \psi_2 x_3 + \psi_3\widehat{u} - \frac{1}{2}x_1^2 - \frac{\epsilon}{3}|x_2|^3 = 0.$$

If $x_2 = 0$, then $x_1 = \psi_1 = 0$ and hence $x_3 \neq 0$. Integration of (5.29) gives that at a sufficiently small interval $t \in (-t^0, t^0)$ the coordinates of the trajectory $Z(t, z^0)$ can be written as follows:

$$x_2(t) = x_3 t + o(t), \quad x_1(t) = \frac{x_3 t^2}{2} + o(t^2),$$

$$\psi_1(t) = \frac{x_3 t^3}{6} + o(t^3),$$

$$\psi_2(t) = -\frac{x_3 t^4}{24} + o(t^4) + \frac{\epsilon x_3^2 t^3}{3}\operatorname{sgn}(x_3 t) + \epsilon o(t^3),$$

$$\psi_3(t) = \frac{x_3 t^5}{120} + o(t^5) - \frac{\epsilon x_3^2 t^4}{12}\operatorname{sgn}(x_3 t) + \epsilon o(t^4).$$

It can readily be seen that if $x_3 \neq 0$, then nonzero roots of the equation $\psi_3(t) = 0$, or, in an equivalent form,

$$t - 10\epsilon x_3 \operatorname{sgn}(x_3 t) + o(t) + \epsilon o(1) = 0,$$

are separated from $t = 0$. That implies the statement of Lemma 5.9. **Q.E.D.**

System (5.29) is homogeneous under $\widetilde{\mathfrak{G}}$–action in $\mathbb{R}^6$, where

$$\widetilde{\mathfrak{G}} = \bigcup_{\lambda > 0} \widetilde{\mathfrak{g}}_\lambda, \quad \widetilde{\mathfrak{g}}_\lambda : \mathbb{R}^6 \to \mathbb{R}^6,$$

$$\widetilde{\mathfrak{g}}_\lambda(z) = (\lambda x_3, \lambda^2 x_2, \lambda^3 x_1, \lambda^4 \psi_1, -\lambda^5 \psi_2, \lambda^6 \psi_3).$$

As usual,

$$Z\big(t, \widetilde{\mathfrak{g}}_\lambda(z^0)\big) = \widetilde{\mathfrak{g}}_\lambda\big(Z(t/\lambda,\, z^0)\big).$$

Consider the Poincaré mapping $F^\epsilon : S \to S$, associated with system (5.29). Mapping F^ϵ transfers an arbitrary point $z \in S\backslash\{0\}$ to the point at which $Z(t, z)$ intersects the surface S for the first positive value of t. The homogeneity provides

$$F^\epsilon \cdot \widetilde{\mathfrak{g}}_\lambda = \widetilde{\mathfrak{g}}_\lambda \cdot F^\epsilon$$

for all $\lambda > 0$. Hence, one can define the *quotient–mapping* $F^\epsilon/\widetilde{\mathfrak{G}}$ that transfers the set of $\widetilde{\mathfrak{G}}$–orbits on the switching surface S to itself. Though the mapping F^ϵ is not continuous at any small neighborhood of the origin, the mapping $F^\epsilon/\widetilde{\mathfrak{G}}$ appears to be continuous and differentiable almost everywhere in some small neighborhood of the switching surface of the optimal synthesis in Problem 5.5. Hence, we can use the standard technique of the invariant manifold theorem.

In the previous section we defined the notion of **b.c.r.** solutions of system (5.12)–(5.13). Let us define its analog for system (5.29).

DEFINITION 5.3. *Let $Z(t, z^0)$ be a solution of system (5.29) and z^0, z^1, $z^2 \in S$ be three points of successive switches on it. Let $t = 0$, $t = t_1$, $t = t_2$ be the corresponding switching instants. We shall say that $Z(t, z^0)$ is a **b.c.r.** (bi–constant ratio) solution of system (5.29) if there exists a constant $\lambda > 0$ such that*

$$Z(t, z^0) = \widetilde{\mathfrak{g}}_\lambda\big(Z(t - t_2, z^0)\big)$$

for $t \geqslant t_2$.

The relation means that the "tail" of the trajectory $Z(t, z^0)$ at $t \geqslant t_2$ can be generated by contracting the original trajectory $Z(t, z^0)$ along $\widetilde{\mathfrak{G}}$–orbits with the coefficient of contraction λ. The immediate consequence of the definition is the following property of **b.c.r.** solutions. Assume that the point z^0 belongs to some **b.c.r.** solution of system (5.29) with the given coefficient λ^0. Then for any $z \in \bigcup_{\lambda > 0} \widetilde{\mathfrak{g}}_\lambda(z^0)$ the trajectory $Z(t, z)$ is a **b.c.r.** solution of system (5.29) with the same coefficient λ^0.

Let us set $\epsilon = 0$. Then system (5.29) is a particular case of systems (5.12)–(5.13) in Section 5.4. Hence, the results of Sections 5.4–5.5 on **b.c.r.**

solutions are applicable to system (5.29). In particular, for $\beta = -1$ there exists a single one–parameter family of **b.c.r.** solutions of system (5.29) and these solutions fill a two–dimensional surface in $\mathbb{R}^6$. (This fact was discovered by [A.T. Fuller, P.E. Grensted, 1965].) The switching curve of the family of **b.c.r.** solutions, call it ρ^0, can be regarded as a fixed point of the double–iterated quotient–mapping $F^0/\widetilde{\mathfrak{G}}$. We shall directly calculate this mapping for $\beta = -1$. It proves that $\left(F^0/\widetilde{\mathfrak{G}}\right)^2$ is a local diffeomorphism at ρ^0 and that there exists a one–dimensional contracting $\left(F^0/\widetilde{\mathfrak{G}}\right)^2$–invariant manifold. This manifold is a part of the optimal switching surface in Problem 5.5. The whole switching surface can be obtained by applying the inverse mapping $\left(F^0\right)^{-1}$ to the invariant manifold on a finite step. Then the result will be extended to all values of the parameters $\beta < 0$ and to sufficiently small $\epsilon > 0$.

The qualitative description of the optimal synthesis in Problem 5.5 is contained in the following Theorem 5.1. Denote as

$$\mathbf{Orb}\ (x) = \bigcup_{\lambda > 0} \mathfrak{g}_\lambda(x)$$

the $\mathfrak{G}$–orbit of a point x in $\mathbb{R}^3$. The same notation

$$\mathbf{Orb}\ (z) = \bigcup_{\lambda > 0} \widetilde{\mathfrak{g}}_\lambda(z)$$

is also used for $\widetilde{\mathfrak{G}}$–orbit in $\mathbb{R}^6$ (because it does not lead to misunderstandings).

THEOREM 5.1. **Synthesis in the three–dimensional Fuller problem.** *For all sufficiently small* $\epsilon \geqslant 0$ *the following statements hold for Problem 5.5.*

(1) *The optimal switching surface, call it* Π^ϵ, *is piecewise smooth and homeomorphic to* $\mathbb{R}^2$.

(2) *There is a one–dimensional curve* $\rho^* \in \mathbb{R}^3$ *consisting of two optimal solutions which come at the origin without switches under constant controls* $u \equiv 1$ *or* $u \equiv \beta$. *The curve* ρ^* *consists of two* $\mathfrak{G}$*–orbits,*

$$\rho^* = \mathbf{Orb}\ \left(-\frac{1}{6}, \frac{1}{2}, -1\right) \bigcup \mathbf{Orb}\ \left(\frac{1}{6\beta^2}, \frac{1}{2\beta}, 1\right).$$

For any $x_0 \in \mathbb{R}^3 \backslash \rho^*$ *the optimal solution* $x(t, x_0)$ *attains the origin in finite time with an infinite number of switches.*

(3) *There is exactly one one–parameter family of* **b.c.r.** *solutions and its trajectories fill some two–dimensional piecewise smooth surface, say* P^ϵ. *The switching set of* **b.c.r.** *solutions is a one–dimensional curve, call it* ρ^ϵ, *whose two smooth branches are represented by two* $\mathfrak{G}$*–orbits. Any optimal solution outside of the surface* P^ϵ *have a finite number of switches in the reverse time current.*

The rest of the section is devoted to proving Theorem 5.1.

PROPOSITION 5.3. *For all sufficiently small $\epsilon \geqslant 0$ there is a unique one–parameter family of optimal* **b.c.r.** *solutions to Problem 5.5.*

Proof. Let $z^0 = (z_1^0,\, z_2^0,\, \ldots,\, z_6^0 = 0)$ be the initial point of a **b.c.r.** solution $z^0(t) = Z(t, z^0)$ of system (5.29) with the parameters $t_1 = 1$, $t_2 = 1 + \tau$, $\lambda = \mu$. Assume that $z_5^0 > 0$ and the control $u(t)$ on the solution equals 1 at $t \in (0,1)$ and β at $t \in (1, 1+\tau)$. Integration of (5.29) gives

$$\mu z_1^0 = z_1^0 + 1 + \beta\tau, \tag{5.31}$$

$$\mu^2 z_2^0 = z_2^0 + z_1^0 + \frac{1}{2} + (z_1^0 + 1)\tau + \frac{\beta\tau^2}{2},$$

$$\mu^3 z_3^0 = z_3^0 + z_2^0 + \frac{z_1^0}{2} + \frac{1}{6} + \left(z_2^0 + z_1^0 + \frac{1}{2}\right)\tau + (z_1^0 + 1)\frac{\tau^2}{2} + \frac{\beta\tau^3}{6},$$

$$\mu^4 z_4^0 = z_4^0 + z_3^0 + \frac{z_2^0}{2} + \frac{z_1^0}{6} + \frac{1}{24} + \left(z_3^0 + z_2^0 + \frac{z_1^0}{2} + \frac{1}{6}\right)\tau$$
$$+ \left(z_2^0 + z_1^0 + \frac{1}{2}\right)\frac{\tau^2}{2} + (z_1^0 + 1)\frac{\tau^3}{6} + \frac{\beta\tau^4}{24},$$

$$\mu^5 z_5^0 = z_5^0 + z_4^0 + \frac{z_3^0}{2} + \frac{z_2^0}{6} + \frac{z_1^0}{24} + \frac{1}{120}$$
$$+ \left(z_4^0 + z_3^0 + \frac{z_2^0}{2} + \frac{z_1^0}{6} + \frac{1}{24}\right)\tau + \left(z_3^0 + z_2^0 + \frac{z_1^0}{2} + \frac{1}{6}\right)\frac{\tau^2}{2}$$
$$+ \left(z_2^0 + z_1^0 + \frac{1}{2}\right)\frac{\tau^3}{6} + (z_1^0 + 1)\frac{\tau^4}{24} + \frac{\beta\tau^5}{120} + \epsilon\Theta_1(\tau, z^0),$$

$$0 = \left(z_5^0 + z_4^0 + \frac{z_3^0}{2} + \frac{z_2^0}{6} + \frac{z_1^0}{24} + \frac{1}{120}\right)\tau$$
$$+ \left(z_4^0 + z_3^0 + \frac{z_2^0}{2} + \frac{z_1^0}{6} + \frac{1}{24}\right)\frac{\tau^2}{2} + \left(z_3^0 + z_2^0 + \frac{z_1^0}{2} + \frac{1}{6}\right)\frac{\tau^3}{6}$$
$$+ \left(z_2^0 + z_1^0 + \frac{1}{2}\right)\frac{\tau^4}{24} + (z_1^0 + 1)\frac{\tau^5}{120} + \frac{\beta\tau^6}{720} + \epsilon\Theta_2(\tau, z^0).$$

Here Θ_1, Θ_2 are some functions of (τ, z^0), specified by integration of the term $z_2^2(t)\,\mathrm{sgn}\,z_2(t)$. Smooth dependence in initial data of solutions of ordinary differential equations implies that if the solution $z^0 = z^0(\mu, \tau)$ of system (5.31) satisfies the inequalities $z_5^0 \neq 0$ and $z_5^0 + z_4^0 + \frac{1}{2}z_3^0 + \frac{1}{6}z_2^0 + \frac{1}{24}z_6^0 + \frac{1}{120} \neq 0$ (i.e. $z_5^0(t) \neq 0$ at $t = 0$ and $t = 1$), then functions Θ_1 and Θ_2 are C^1.

Let us set first $\epsilon = 0$. Then (5.31) coincides with (5.14) for $k = 3$ and we can apply the results of Sections 5.4–5.5 to estimate the number of solutions of the system

$$\begin{cases} \phi_6(\mu,\nu) + \phi_6\left(\mu, \dfrac{\mu}{\nu}\right) + 1 = 0, \\[2mm] \phi_6(\mu,\nu) = \dfrac{\beta}{\beta - 1}, \end{cases} \tag{5.32}$$

with $\phi_6(\mu,\nu) = \sum_{l=1}^{6} (-1)^l\, C_6^l\, (1-\nu^l)/(1-\mu^l)$ and $\nu = (\tau+\mu)/(\tau+1)$. For brevity, everywhere below we will omit the index 6 on the function ϕ_6. We are to confirm that

(1) in the domain $\mathcal{K} = \left\{\mu,\nu \mid 0 < \mu < \nu < 1\right\}$ the set

$$\phi(\mu,\nu) + \phi\left(\mu, \frac{\mu}{\nu}\right) + 1 = 0$$

    consists of two analytic nonintersecting smooth branches, call them Γ^1 and Γ^2;

(2) the restrictions $\phi\big|_{\Gamma^1}$ and $\phi\big|_{\Gamma^2}$ are both monotonic;

(3) each of the curves Γ^1 and Γ^2 intersects the parabola $\mu = \nu^2$ at a single point. The function

$$\sum_{l=1}^{5} (-1)^l\, C_5^l\, \frac{1-\nu^l}{1-\mu^l} + \frac{\beta}{\beta - 1}$$

    (see (5.21.b)) has opposite signs at those two points.

If all of these assertions are proved, then Proposition 5.1 implies that for $\epsilon = 0$ and any $\beta < 0$ there is a unique solution of (5.31) with $z_2^0 > 0$ and it determines some optimal **b.c.r.** solution of system (5.29). Moreover, if for any fixed $\beta < 0$ the point $(0,0)$ is a regular value of the mapping

$$(\mu,\nu) \rightarrow \left(\phi(\mu,\nu) + \phi\left(\mu, \frac{\mu}{\nu}\right) + 1,\ \phi(\mu,\nu) + \frac{\beta}{\beta - 1}\right),$$

then, in view of the implicit function theorem, the analogous assertions about **b.c.r.** solutions are valid for all sufficiently small $\epsilon > 0$.

Let us set

$$D(\mu,\nu) = \det \frac{D\big(\Delta(\mu,\nu), \phi(\mu,\nu)\big)}{D(\mu,\nu)}. \tag{5.33}$$

We wish to prove that

$$D(\mu,\nu)\big|_{\Delta=0} \neq 0. \tag{5.34}$$

In order to justify that the set $\Delta = 0$ is a smooth manifold, let us check that $\partial\Delta/\partial\nu \neq 0$ at $\Delta = 0$ except for the points of the intersection $\{\Delta = 0\} \cap \{\mu = \nu^2\}$ where $\partial\Delta/\partial\mu \neq 0$. By the definition, we have

$$\Delta = \sum_{l=1}^{6} (-1)^l C_6^l \frac{2 - \nu^l - (\mu/\nu)^l}{1 - \mu^l} + 1.$$

Since $2 - \nu^l - (\mu/\nu)^l = (1 - \nu^l) - \left((\mu/\nu)^l - \mu^l\right) - (\mu^l - 1)$ it follows that,

$$\sum_{l=1}^{6} (-1)^l C_6^l \frac{1 - \mu^l}{1 - \mu^l} = (1 - 1)^6 - 1 = -1.$$

Straightforward algebraic manipulations lead to the representation

$$\Delta = \sum_{l=1}^{6} (-1)^l C_6^l \frac{1 + \mu^l - (\nu^l + (\mu/\nu)^l)}{1 - \mu^l},$$

or, equivalently,

$$\Delta = \sum_{l=1}^{6} (-1)^l C_6^l \frac{(1/\sqrt{\mu})^l + (\sqrt{\mu})^l - \left((\nu/\sqrt{\mu})^l + (\sqrt{\mu}/\nu)^l\right)}{(1/\sqrt{\mu})^l - (\sqrt{\mu})^l}.$$

If we multiply the equation $\Delta = 0$ by $1/\sqrt{\mu} - \sqrt{\mu}$ and take $\sigma = \nu/\sqrt{\mu} + \sqrt{\mu}/\nu$, $\theta = \sqrt{\mu} + 1/\sqrt{\mu}$ as the coordinates in $\mathcal{K}\backslash\{\mu = \nu^2\}$, then after the reduction to the common denominator within to some nonzero multipliers we obtain

$$(\sigma - \theta)^2 R_1(\sigma, \theta) = 0,$$

where

$$R_1(\sigma, \theta) \overset{\text{def}}{=} (\theta^6 - 5\theta^4 + 7\theta^2 - 2)\sigma^4$$
$$+ (-4\theta^7 + 26\theta^5 - 52\theta^3 + 32\theta)\sigma^3$$
$$+ (6\theta^8 - 54\theta^6 + 159\theta^4 - 171\theta^2 + 57)\sigma^2$$
$$+ (-4\theta^9 + 56\theta^7 - 250\theta^5 + 416\theta^3 - 186\theta)\sigma$$
$$+ \theta^{10} - 29\theta^8 + 220\theta^6 - 677\theta^4 + 849\theta^2 - 288.$$

It is convenient to let $q = \sigma/\theta$, $p = 1/\theta^2$. If we cancel out the term $(\sigma - \theta)^2$ and divide the equation by θ^{10}, we obtain

$$R_2(q, p) = 0, \tag{5.35}$$

where

$$R_2(q,p) \overset{\text{def}}{=} q^4(-2p^3 + 7p^2 - 5p + 1)$$
$$+ q^3(32p^3 - 52p^2 + 26p - 4)$$
$$+ q^2(57p^4 - 171p^3 + 159p^2 - 54p + 6)$$
$$+ q(-186p^4 + 416p^3 - 250p^2 + 56p - 4)$$
$$- 288p^5 + 849p^4 - 677p^3 + 220p^2 - 29p + 1.$$

Let us check that the equation (5.35) has no multiple roots in the domain

$$\widetilde{\mathcal{K}} = \left\{ q,p \mid 2\sqrt{p} < q < 1, \quad 0 < p < 1/4 \right\}.$$

The domain $\widetilde{\mathcal{K}}$ is the image of the region $\mathcal{K}$ under the mapping

$$(\mu, \nu) \rightarrow (q, p),$$
$$q = \frac{\nu^2 + \mu}{\nu(1 + \mu)}, \quad p = \frac{\mu}{(1 + \mu)^2}.$$

To solve the system

$$\begin{cases} R_2(q,p) = 0, \\ \dfrac{\partial R_2}{\partial q}(q,p) = 0, \end{cases}$$

one must calculate the roots of the discriminant of the polynomial R_2,

$$\Xi_1(p)$$
$$= \quad 585252.875\,p^{28} - \quad 25817242\,p^{27} + \quad 367595552\,p^{26}$$
$$- \quad 2839178752\,p^{25} + \quad 14265834496\,p^{24} - \quad 51084296192\,p^{23}$$
$$+ \quad 137628368898\,p^{22} - \quad 289074806784\,p^{21} + 485286215680\,p^{20}$$
$$- \quad 663052877824\,p^{19} + \quad 7474074768\,p^{18} - \quad 702254022656\,p^{17}$$
$$+ \quad 554286907392\,p^{16} - \quad 368632706560\,p^{15} + 209091657728\,p^{14}$$
$$- \quad 100573519872\,p^{13} + \quad 41172488192\,p^{12} - \quad 14336442368\,p^{11}$$
$$+ \quad 4236572160\,p^{10} - \quad 1058176064\,p^{9} + \quad 222000272\,p^{8}$$
$$- \quad 38769452\,p^{7} + \quad 5565034\,p^{6} - \quad 644983.3125\,p^{5}$$
$$+ \quad 58830.4023\,p^{4} - \quad 4063.5457\,p^{3} + \quad 199.68615\,p^{2}$$
$$- \quad 6.21752\,p + \quad 0.00921 \quad = \quad 0.$$

Similarly, the system

$$\begin{cases} R_2(q,p) = 0, \\ \dfrac{\partial R_2}{\partial p}(q,p) = 0 \end{cases}$$

leads to the equation

$$
\begin{aligned}
\Xi_2(q) \\
= {}&\ 2.5502\,q^{64} - 57.7723\,q^{63} + 563.227\,q^{62} - 2822.02\,q^{61} \\
+{}&\ 6914.33\,q^{60} - 2074.11\,q^{59} - 34564.0\,q^{58} + 84552.6\,q^{57} \\
-{}&\ 40467.5\,q^{56} - 124290.0\,q^{55} + 162374.9\,q^{54} + 116813.5\,q^{53} \\
-{}&\ 299539.3\,q^{52} - 169928.1\,q^{51} + 658276.5\,q^{50} - 275090.0\,q^{49} \\
-{}&\ 566450.8\,q^{48} - 117482.1\,q^{47} + 1279775.9\,q^{46} + 806615.9\,q^{45} \\
+{}&\ 489826\,q^{44} + 19470557.9\,q^{43} + 5651940.1\,q^{42} + 62597224.5\,q^{41} \\
+{}&\ 3923407.2\,q^{40} + 60178714.9\,q^{39} + 10000000\,q^{38} + 8878708.5\,q^{37} \\
+{}&\ 5083934.1\,q^{36} + 5135686.4\,q^{35} + 5199169.5\,q^{34} + 3058953.3\,q^{33} \\
+{}&\ 1393238.6\,q^{32} + 293128.8\,q^{31} - 886950.1\,q^{30} - 1263169.3\,q^{29} \\
-{}&\ 785533.3\,q^{28} - 799230.5\,q^{27} - 773541.8\,q^{26} - 483290.5\,q^{25} \\
-{}&\ 161643.4\,q^{24} + 70692.0\,q^{23} + 123761.1\,q^{22} + 97286.43\,q^{21} \\
+{}&\ 58980.8\,q^{20} + 26831.56\,q^{19} - 21872.85\,q^{18} - 9992.982\,q^{17} \\
+{}&\ 1433.623\,q^{16} - 6762.175\,q^{15} + 255.350\,q^{14} + 2446.003\,q^{13} \\
-{}&\ 223.359\,q^{12} - 597.640\,q^{11} + 722.2505\,q^{10} - 239.6378\,q^{9} \\
-{}&\ 194.6962\,q^{8} + 219.5920\,q^{7} - 105.2866\,q^{6} + 17.3650\,q^{5} \\
+{}&\ 15.6924\,q^{4} - 13.3771\,q^{3} + 5.5670\,q^{2} - 1.5233\,q \\
+{}&\ 0.201693 = 0.
\end{aligned}
$$

Any standard criterion (e.g., the Budan–Fourier theorem) gives that there are no roots of $\Xi_1(p)$ at $p \in (0, 1/4)$ and of $\Xi_2(q)$ at $q \in (0, 1)$.

It follows that

$$
\left. \frac{\partial \Delta}{\partial \nu} \right|_{\Delta=0} = m(\sigma - \theta)^2 \frac{\partial R_2}{\partial \nu},
$$
$$
\left. \frac{\partial \Delta}{\partial \mu} \right|_{\Delta=0} = m(\sigma - \theta)^2 \frac{\partial R_2}{\partial \mu},
$$

(5.36.a)

where $m = m(\sigma, \theta)$ is some positive multiplier, smooth in σ, θ.

Furthermore,

$$\frac{\partial R_2}{\partial \nu} = \frac{\partial p}{\partial \nu}\frac{\partial R_2}{\partial p} + \frac{\partial q}{\partial \nu}\frac{\partial R_2}{\partial q} = \frac{\nu^2 - \mu}{\nu^2(1+\mu)}\frac{\partial R_2}{\partial q} \qquad (5.36.\text{b})$$

$$\frac{\partial R_2}{\partial \mu} = \frac{\partial p}{\partial \mu}\frac{\partial R_2}{\partial p} + \frac{\partial q}{\partial \mu}\frac{\partial R_2}{\partial q} = \frac{1-\mu}{(1+\mu)^3}\frac{\partial R_2}{\partial p} + \frac{1-\nu^2}{\nu(1+\mu)^2}\frac{\partial R_2}{\partial q}.$$

Since $\dfrac{\partial R_2}{\partial q} \neq 0$, it follows that $\left.\dfrac{\partial \Delta}{\partial \nu}\right|_{\Delta=0} = 0$ iff $\mu = \nu^2$, so the set $\Delta = 0$ is a smooth manifold in $\mathcal{K}\backslash\{\mu = \nu^2\}$. It is left to check the points where the curve $\Delta = 0$ intersects the parabola $\mu = \nu^2$. In this case we have $\sigma = 2$ and θ is the root of the polynomial

$$(\theta - 2)^2(\theta + 1)^2(\theta^2 + \theta - 1)(\theta^4 - 7\theta^3 - 6\theta^2 + 29\theta + 23) = 0.$$

We search for the roots lying in the interval $2 < \theta < \infty$, that is, either $\theta_1 = 2,31126428\ldots$ or $\theta_2 = 7,2131538\ldots$. This gives $\mu_1 = 0,33147229\ldots$ and $\mu_2 = 0,019996165\ldots$. A straightforward calculation gives

$$\frac{\partial \Delta}{\partial \mu}\left(\mu_1, \sqrt{\mu_1}\right) \neq 0 \quad \text{and} \quad \frac{\partial \Delta}{\partial \mu}\left(\mu_2, \sqrt{\mu_2}\right) \neq 0.$$

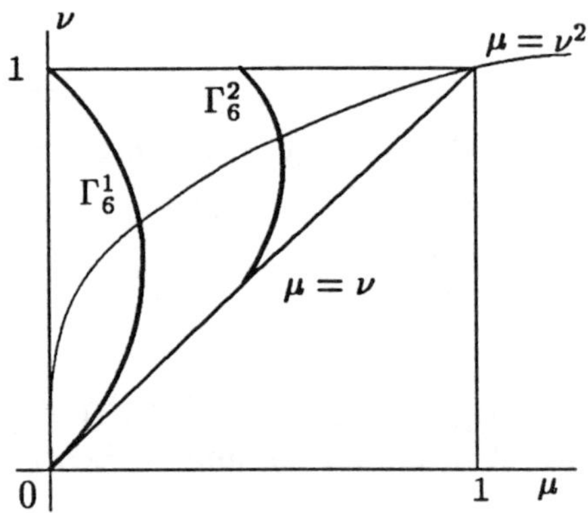

FIG. 12: NUMBER OF FAMILIES OF SELF–SIMILAR SOLUTIONS

Let us show that the set $\Delta = 0$ consists of exactly two smooth branches in the region $\mathcal{K}$ as it is shown in Fig. 12. To do this let us examine the behavior of the set $\Delta = 0$ at the boundary of $\mathcal{K}$. If $\mu = \nu$ or $\nu = 1$, then we have $\sigma = 0$ and $q = 1$. It follows from (5.36) that all points of the straight lines $\mu = \nu$ and $\nu = 1$ are the double roots of the equation $\Delta = 0$. The substitution $q = 1$ into (5.35) gives

$$-6p(4p - 1)^2(3p^2 - 6p + 1) = 0.$$

Thus, $p = 1/4$ is the double root of the equation $\Delta = 0$. This corresponds to the corner point $\mu = 1$, $\nu = 1$ of the triangle $\mathcal{K}$. There is also the unique simple root $p_0 = 0,18350342\ldots$ in the region $0 < p < 1/4$ that gives $\mu_0 = 0,31948873\ldots$. It follows from the implicit function theorem that at some neighborhood of the points $\mu = \nu = \mu_0$ and $\mu = \mu_0$, $\nu = 1$, the set $\Delta = 0$ is a smooth one–dimensional manifold, transversal to the boundaries $\mu = \nu$ and $\nu = 1$.

If $\mu = 0$, then a straightforward calculation gives that $\nu = 1$ or $\nu = 0$. Hence, we are left with examining the corner points of the triangle $\mathcal{K}$ where the set $\Delta = 0$ has a singularity.

Let us begin with the points $\mu = 0$, $\nu = 0$ and $\mu = 0$, $\nu = 1$. If $\mu \to 0$, then $p \to 0$. We have

$$R_2(q,p) = (q - 1)^4 - 6p + (q - 1)pR_3(q,p) + p^2R_4(q,p),$$

R_3, and R_4 being some polynomials in q, p. Using the standard technique of resolution of singularities of analytic sets we obtain that for all small p in $\mathcal{K}$ the equation $R_2(q,p) = 0$ gives the curve

$$q = 1 - \sqrt[4]{6p} + o(\sqrt[4]{p}).$$

Returning to the variables μ, ν, we obtain the following representation of the set $\Delta = 0$:

$$\nu = \mu + \sqrt[4]{6}\mu^{5/4} + o(\mu^{5/4})$$

near the point $\mu = 0, \nu = 0$ and the representation

$$\nu = 1 - \sqrt[4]{6}\mu^{1/4} + o(\mu^{1/4})$$

near the point $\mu = 0, \nu = 1$.

Now let us deal with the point $\mu = 1$, $\nu = 1$. For the corresponding values of variables (q,p) we have $p_0 = 1/4, q_0 = 1$. The polynomial $R_2(q,p)$ can be written as follows:

$$R_2(q,p) = -6p(3p^2 - 6p + 1)\Big((4p - 1)^2 + (4p - 1)(q - 1)\Big)$$
$$+ (q - 1)^2p^2(57p^2 - 87p + 45) + (q - 1)^3R_5(q,p),$$

where R_5 is a polynomial in q, p. In the vicinity of the point $p = 1/4$, $q = 1$, Taylor's formula gives

$$R_2(q,p) = \frac{15}{2}\left(p - \frac{1}{4}\right)^2 + \frac{15}{8}\left(p - \frac{1}{4}\right)(q-1) + \frac{429}{64}(q-1)^2 + \dots .$$

Since the discriminant of the quadratic polynomial is negative, the point $p = 1/4$, $q = 1$ is an isolated zero of R_2.

To complete the study of the structure of the set $\Delta = 0$ in $\mathcal{K}$, let us show that this set does not intersect the straight lines $\mu = 0.3$ and $\mu = 0.4$. For $\mu = 0.3$ a straightforward calculation gives $p = p_1 = 0.17751479\dots$ and

$$R_2(q,p_1) = 0.321819q^4 - 0.8442133q^3$$
$$+ 0.52457q^2 + 0.2052628q - 0.2101022.$$

For $\mu = 0.4$ we have $p = p_2 = 0.20408163\dots$ and

$$R_2(q,p_2) = 0.2541373q^4 - 0.5876464q^3$$
$$+ 0.2472324q^2 + 0.2295374q - 0.1391488.$$

Using the Budan–Fourier theorem one can check that in both cases there are no roots of the polynomials $R_2(q,p_i)$ $(i = 1, 2)$ at $q \in (0.3, 1)$.

It follows that the first branch of the set $\Delta = 0$, call it Γ^1, is disposed on the left–hand side of the line $\mu = 0.3$. The second one, call it Γ^2, is disposed in the region $0.3 < \mu < 0.4$. Let us show that only the second branch Γ^2 satisfies the maximum principle. In view of Section 5.6 it is sufficient to calculate the value of the function z_2 at the points $\Gamma^1 \bigcap \{\mu = \nu^2\}$ and $\Gamma^2 \bigcap \{\mu = \nu^2\}$, i.e., for $\mu = \mu_1 = 0.019996165\dots$ and $\mu = \mu_2 = 0.33147229\dots$ respectively. One can check that $z_2 < 0$ in the first case and $z_2 > 0$ in the second. Hence, we need to confirm the inequality (5.34) only for the curve Γ^2.

Let us show first that $\left.\frac{\partial \Delta}{\partial \mu}\right|_{\Gamma^2} > 0$. Because of (5.36), it is enough to demonstrate that $\partial R_2/\partial p > 0$ and $\partial R_2/\partial q > 0$ at Γ^2. Since $\partial R_2/\partial p$ and $\partial R_2/\partial q$ does not equal zero anywhere on Γ^2, let us calculate their values at the point $\Gamma^2 \bigcap \{\mu = \nu^2\}$, i.e., for $p = 0.18697474\dots$, $q = 0.86481146\dots$. We have

$$\frac{\partial R_2}{\partial p} = 0.5542024\dots , \qquad \frac{\partial R_2}{\partial q} = 0.02926808\dots ,$$

which implies $\partial \Delta/\partial \mu > 0$.

Let us prove that $\partial\phi/\partial\mu < 0$ in the region $0.3 < \mu < 0.4$, $\mu < \nu < 1$. Recall that the function ϕ can be expanded in the following absolutely convergent series

$$\phi = \sum_{\alpha=0}^{\infty} \left((1 - \mu^{\alpha})^6 - (1 - \nu\mu^{\alpha})^6 \right),$$

hence

$$\frac{1}{6} \frac{\partial\phi}{\partial\mu} = \sum_{\alpha=0}^{\infty} \alpha\mu^{\alpha-1} \left(\nu(1 - \nu\mu^{\alpha})^5 - (1 - \mu^{\alpha})^5 \right).$$

We see that $\dfrac{\partial\phi}{\partial\mu}(\mu, 1) \equiv 0$, so it is enough to prove that $\dfrac{\partial}{\partial\nu} \dfrac{\partial\phi}{\partial\mu} > 0$ at $0.3 < \mu < 0.4$. We have

$$\frac{1}{6} \frac{\partial^2\phi}{\partial\mu\partial\nu} = \sum_{\alpha=1}^{\infty} \alpha\mu^{\alpha-1}(1 - \nu\mu^{\alpha})^4(1 - 6\nu\mu^{\alpha}).$$

If $\alpha \geqslant 2$, then $1 - 6\nu\mu^{\alpha} \geqslant 1 - 6\mu^2 \geqslant 1 - 6 \cdot 0.4^2 > 0$. Therefore, only the first term in the series can be negative. It is easy to see that the minimum of the function $(1 - t)^4(1 - 6t)$ for $t \in (0, 1)$ is attained at $t = 1/3$. A straightforward estimation of the members in the series gives

$$\frac{1}{6} \frac{\partial^2\phi}{\partial\mu\partial\nu} > -\left(\frac{2}{3}\right)^4 + 2 \cdot 0.3\,(1 - 0.4^2)^4(1 - 6 \cdot 0.4^2)$$
$$+ 3 \cdot 0.3^2(1 - 0.4^3)^4(1 - 6 \cdot 0.4^3) + \ldots .$$

The sum of the first four positive terms is greater than the absolute value of the negative one, so $\partial^2\phi/\partial\mu\partial\nu > 0$, which implies that $\partial\phi/\partial\mu < 0$.

Let us note that the derivative $\partial\phi/\partial\nu = 6 \sum_{\alpha=0}^{\infty} \mu^{\alpha}(1 - \nu\mu^{\alpha})^5$ is positive.

Now everything has been prepared for proving (5.34) at points of Γ^2. Since $\Delta = \phi + \tilde{\phi} + 1$, it follows from (5.33) that

$$D(\mu, \nu) = \det \frac{D(\phi, \tilde{\phi})}{D(\mu, \nu)}$$

where

$$\tilde{\phi}(\mu, \nu) = \phi\left(\mu, \frac{\mu}{\nu}\right).$$

We have

$$\frac{\partial\tilde{\phi}}{\partial\mu} = \frac{\partial\hat{\phi}}{\partial\mu} + \frac{1}{\nu} \frac{\partial\hat{\phi}}{\partial\nu}, \quad \frac{\partial\tilde{\phi}}{\partial\nu} = -\frac{\mu}{\nu^2} \frac{\partial\hat{\phi}}{\partial\nu}.$$

The hat over the letter means that the function has been calculated at the point $(\mu, \mu/\nu)$. Thus,

$$
D\left(\mu, \frac{\mu}{\nu}\right) = \det
\begin{pmatrix}
\dfrac{\partial \widehat{\phi}}{\partial \mu} & \dfrac{\partial \widehat{\phi}}{\partial \nu} \\[2ex]
\dfrac{\partial \phi}{\partial \mu} + \dfrac{\nu}{\mu}\dfrac{\partial \phi}{\partial \mu} & -\dfrac{\nu^2}{\mu}\dfrac{\partial \phi}{\partial \mu}
\end{pmatrix}
$$

$$
= -\frac{\nu^2}{\mu}\det
\begin{pmatrix}
\dfrac{\partial \widehat{\phi}}{\partial \mu} + \dfrac{1}{\nu}\dfrac{\partial \widehat{\phi}}{\partial \nu} & -\dfrac{\mu}{\nu^2}\dfrac{\partial \widehat{\phi}}{\partial \nu} \\[2ex]
\dfrac{\partial \phi}{\partial \mu} & \dfrac{\partial \phi}{\partial \nu}
\end{pmatrix}
= \frac{\nu^2}{\mu} D(\mu, \nu).
$$

It follows that if $D(\mu_0, \nu_0) = 0$ at some point $(\mu_0, \nu_0) \in \mathcal{K}$, then $D(\mu_0, \mu_0/\nu_0) = 0$ also. According to what has been said, three terms of the matrix $D(\mu, \nu)$ (namely, $\partial\Delta/\partial\mu$, $\partial\phi/\partial\mu$, $\partial\phi/\partial\nu$) have the same signs for both (μ, ν) and $(\mu, \mu/\nu)$ at all points of the curve Γ^2. The fourth term, $\partial\Delta/\partial\nu$, has opposite signs at points (μ, ν) and $(\mu, \mu/\nu)$. This implies $D(\mu, \nu)\big|_{\Gamma^2} \neq 0$.

The consideration concerning the branch Γ^1 is completely analogous, but it is not used in the following, so it is omitted here.

Q.E.D.

Let $z^0 \in S$. Define the function $t^\epsilon(z^0)$ as the minimal positive root of the equation $z_6(t, z^0) = 0$. By the definition of the Poincaré mapping we have $F^\epsilon(z^0) = Z(t^\epsilon(z^0), z^0)$. What we wish to do now is to give an explicit form of the quotient–mapping $(F^\epsilon)^2/\widetilde{\mathfrak{G}}$.

The homogeneity of system (5.29) under $\widetilde{\mathfrak{G}}$–action yields

$$
t^\epsilon\left(\widetilde{\mathfrak{g}}_\lambda(z^0)\right) = \lambda\, t^\epsilon(z^0), \tag{5.37}
$$

hence, it is natural to consider the surface

$$
L^\epsilon = \left\{ z \in S \mid t^\epsilon(z) = 1, \quad z_5 > 0 \right\}
$$

as an orbit–space of $\mathbb{R}^6\backslash\{0\}$ under $\widetilde{\mathfrak{G}}$–action (at least, as its coordinate map in a sufficiently large region of space). Let $M^\epsilon = L^\epsilon \bigcap \rho^\epsilon$, where ρ^ϵ is the switching curve of **b.c.r.** solutions to Problem 5.5. Now the quotient–mapping $\widetilde{F}^\epsilon \overset{\text{def}}{=} (F^\epsilon)^2/\widetilde{\mathfrak{G}}$ at some neighborhood of the point M^ϵ can be defined as follows:

$$
\widetilde{F}^\epsilon : L^\epsilon \longrightarrow L^\epsilon,
$$

$$
\widetilde{F}^\epsilon(z) = \widetilde{\mathfrak{g}}_{1/\mu}\left(Z\big(\tau, F^\epsilon(z)\big)\right),
$$

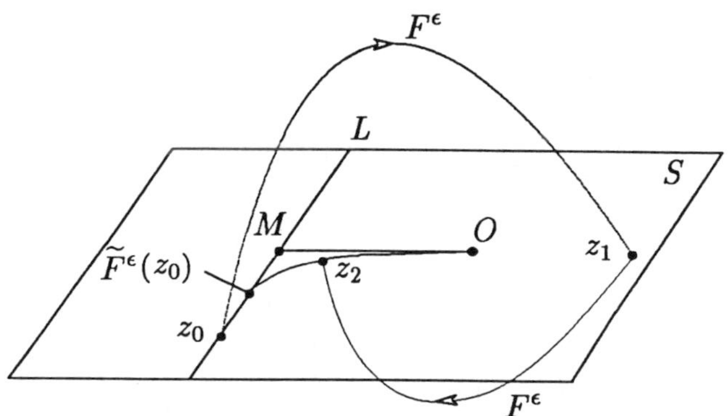

FIG. 13: POINCARÉ MAPPING IN THE PHASE SPACE

where $\tau = \tau(z)$ equals $t^\epsilon\big(F^\epsilon(z)\big)$ and $\mu = \mu(z)$ equals $t^\epsilon\big(Z(\tau, F^\epsilon(z))\big)$ (see Fig. 13).

Some comments are necessary here. It follows from the definition of $t^\epsilon(\cdot)$ that the trajectory with the initial point F^ϵ intersects the switching surface at the instant $\tau = t^\epsilon\big(F^\epsilon(z)\big)$. In view of (5.37) we have

$$t^\epsilon\left(\widetilde{\mathfrak{g}}_{1/\mu}\big(Z(\tau, F^\epsilon(z))\big)\right) = \frac{1}{\mu}\, t^\epsilon\left(Z\big(\tau, F^\epsilon(z)\big)\right) = 1,$$

so $\widetilde{F}^\epsilon(z) \in L^\epsilon$. Furthermore, $\widetilde{F}^\epsilon(M^\epsilon) = M^\epsilon$ and z_5–coordinates of points M^ϵ and $F(M^\epsilon)$ do not equal zero (Lemma 5.8)), so solutions of system (5.29) transversally intersect S at some neighborhood of M^ϵ and $F^\epsilon(M^\epsilon)$. It follows that mapping $\widetilde{F}^\epsilon$ is well–defined and a C^1–diffeomorphism at some neighborhood of M^ϵ on L^ϵ.

Let us show that mapping $\widetilde{F}^\epsilon$ has a saddle structure at its fixed point M^ϵ.

LEMMA 5.10. *Eigenvalues of the matrix $D\widetilde{F}^\epsilon\Big|_{M^\epsilon}$ are not equal to 1 or -1.*

Proof. Assume first that $\epsilon = 0$. Then in the vicinity of the point M^0 the surface L^0 is given by the equations

$$\psi_3 = 0, \quad -\psi_2 + \frac{\psi_1}{2} + \frac{x_1}{6} + \frac{x_2}{24} + \frac{x_3}{120} + \frac{1}{720} = 0 \qquad (5.38)$$

(see (5.30)). It follows that the values ψ_1, x_1, x_2, x_3 can be taken as coordinates on the surface L^0.

Let us demonstrate that $\widetilde{F^0} - E$ is a diffeomorphism at M^0 (here E is the identity mapping on L^0, so $(\widetilde{F^0} - E)(z)$ means $\widetilde{F^0}(z) - z$).

By the definition, the mapping $\widetilde{F}^0 - E$ can be written at the point $z \in L^0$ as follows:

$$(\widetilde{F}^0 - E)(z) = \mathcal{K}_{1/\mu(z)}\left(e^{(\tau(z)+1)M}z + p(\tau(z))\right) - z,$$

where

$$M = \begin{pmatrix} 0 & 0 & 0 & 0 & 0 & 0 \\ 1 & 0 & 0 & 0 & 0 & 0 \\ 0 & 1 & 0 & 0 & 0 & 0 \\ 0 & 0 & 1 & 0 & 0 & 0 \\ 0 & 0 & 0 & 1 & 0 & 0 \\ 0 & 0 & 0 & 0 & 1 & 0 \end{pmatrix},$$

$$p(\tau) = \begin{pmatrix} 1 + \beta\tau \\ \dfrac{1}{2} + \tau + \dfrac{\beta\tau^2}{2} \\ \dfrac{1}{6} + \dfrac{\tau}{2} + \dfrac{\tau^2}{2} + \dfrac{\beta\tau^3}{6} \\ \dfrac{1}{24} + \dfrac{\tau}{6} + \dfrac{\tau^2}{4} + \dfrac{\tau^3}{6} + \dfrac{\beta\tau^4}{24} \\ \dfrac{1}{120} + \dfrac{\tau}{24} + \dfrac{\tau^2}{12} + \dfrac{\tau^3}{12} + \dfrac{\tau^4}{24} + \dfrac{\beta\tau^5}{120} \\ \dfrac{1}{720} + \dfrac{\tau}{120} + \dfrac{\tau^2}{48} + \dfrac{\tau^3}{36} + \dfrac{\tau^4}{48} + \dfrac{\tau^5}{120} + \dfrac{\beta\tau^6}{720} \end{pmatrix},$$

$$\mathcal{K}_{1/\mu} = \begin{pmatrix} 1/\mu & 0 & 0 & 0 & 0 & 0 \\ 0 & 1/\mu^2 & 0 & 0 & 0 & 0 \\ 0 & 0 & 1/\mu^3 & 0 & 0 & 0 \\ 0 & 0 & 0 & 1/\mu^4 & 0 & 0 \\ 0 & 0 & 0 & 0 & 1/\mu^5 & 0 \\ 0 & 0 & 0 & 0 & 0 & 1/\mu^6 \end{pmatrix}.$$

The function $\tau = \tau(z)$ denotes the interval between the first and second switches on the trajectory, emanating from the point z; the function $1/\mu(z)$ equals the coefficient of expansion along the $\mathfrak{G}$–orbit at which the point of the second switch is carried into the surface L^0.

We need to prove that for any y in an open neighborhood of zero on the plane (5.38) there exists a unique solution $z = z(y)$ of the equation

$$(\widetilde{F}^0 - E)(z) = y$$

or, in coordinate notation,

$$\mathcal{K}_{1/\mu(z)}\left(e^{(\tau(z)+1)M}z + p(\tau(z))\right) - z = y, \tag{5.39}$$

and $y \to z(y)$ is a C^∞–mapping. Let us note that the plane (5.38) contains the full image of the surface L^0 under the $(\widetilde{F}^0 - E)$–mapping.

Let us fix for an instance the values of μ and τ and resolve (5.39) with respect to z as a function of μ, ν, y. Since the operator $\mathcal{K}_{1/\mu} e^{(\tau+1)M} - E$ is invertible, $z = z(\mu, \tau, y)$ is C^1. It is easy to see that the parameters μ and τ are solutions of the equations

$$
\begin{aligned}
\left[z\right]_6 &= 0, \\
\left[e^M z + w(1)\right]_6 &= 0,
\end{aligned}
\tag{5.40}
$$

(recall that the notation $\left[\,\cdot\,\right]_m$ means the m–th coordinate of the vector). If $y = 0$, then until redenotation we obtain that the equations (5.32) specify the **b.c.r.** solution of system (5.31). It follows from (5.34) that for any $\beta < 0$ the functions on the left–hand sides of equations (5.40) are functionally independent at the point (μ_0, ν_0) corresponding to the stationary point M^0. Hence, at some neighborhood of (μ_0, ν_0), the Jacobian of the functions on the left–hand sides of (5.40) does not equal zero. If y is sufficiently close to zero on the surface (5.38), then the Jacobian is still nonzero and there exists a unique C^∞–solution $z = z(y) \in L^0$, $\mu = \mu(y)$, $\tau = \tau(y)$ of (5.39)–(5.40) that has been required. It has thereby been proved that the number 1 is not an eigenvalue of the matrix $D\widetilde{F}^0(M^0)$.

To complete the case $\epsilon = 0$ and show that -1 is not an eigenvalue of the matrix $D\widetilde{F}^0(M^0)$ also, we can use a similar method. It is sufficient to replace the mapping $\widetilde{F}^0 - E$ in (5.39) by $\left(\widetilde{F}^0\right)^2 - E$ and literally repeat the performance above.

For any sufficiently small ϵ the statement being proved follows from the implicit function theorem. **Q.E.D.**

LEMMA 5.11. *For any fixed $\beta < 0$ and for all sufficiently small $\epsilon = \epsilon(\beta) > 0$ there exists a contracting $\widetilde{F}^\epsilon$–invariant manifold in some neighborhood of M^ϵ on the zero–level surface of the Hamiltonian H_0.*

Proof. The surface $L^\epsilon \cap H_0$ is a three–dimensional $\widetilde{F}^\epsilon$-invariant manifold. Let us prove that the matrix

$$
Q \stackrel{\text{def}}{=} D\left(\widetilde{F}^\epsilon\Big|_{L^\epsilon \cap H_0}\right)(M^\epsilon)
$$

has exactly one eigenvalue in the unit ball $|z| < 1$ in $\mathbb{C}^1$. This eigenvalue is real and belongs to the interval $(0, 1)$. Consider the characteristic equation

$$
f(\lambda) \stackrel{\text{def}}{=} \det\left(Q - \lambda E\right) = 0,
\tag{5.41}
$$

where E is the unit (3×3)–matrix. Let us verify the statement first for $\epsilon = 0$ and $\beta = -1$. Most of the remaining proof is devoted to the explicit calculation of the matrix Q in this case. On account of (5.38) the values of $\psi_1,\, x_1,\, x_2,\, x_3$ can be taken to be coordinates on L^0. As follows from (5.31), the mapping $\widetilde{F}^0$ can be written in the form

$$
\psi_1 \mapsto \mu^{-4}\Big(\psi_1 + x_1 + \frac{x_2}{2} + \frac{x_3}{6} + \frac{1}{24} + \big(x_1 + x_2 + \frac{x_3}{2} + \frac{1}{6}\big)\tau
$$
$$
+ \big(x_2 + x_3 + \frac{1}{2}\big)\frac{\tau^2}{2} + \big(x_3 + 1\big)\frac{\tau^3}{6} - \frac{\tau^4}{24}\Big),
$$
$$
x_1 \mapsto \mu^{-3}\Big(x_1 + x_2 + \frac{x_3}{2} + \frac{1}{6} + \big(x_2 + x_3 + \frac{1}{2}\big)\tau \qquad\qquad (5.42.a)
$$
$$
+ \big(x_3 + 1\big)\frac{\tau^2}{2} - \frac{\tau^3}{6}\Big),
$$
$$
x_2 \mapsto \mu^{-2}\Big(x_2 + x_3 + \frac{1}{2} + \big(x_3 + 1\big)\tau - \frac{\tau^2}{2}\Big),
$$
$$
x_3 \mapsto \mu^{-1}\big(x_3 + 1 - \tau\big).
$$

Here $\mu > 0$ and $\tau > 0$ are functions of $\psi_1,\, x_1,\, x_2,\, x_3$ defined as follows:

$$
-\Big(\psi_2 - \psi_1 - \frac{x_1}{2} - \frac{x_2}{6} - \frac{x_3}{24} - \frac{1}{120}\Big)\tau
$$
$$
+ \Big(\psi_1 + x_1 + \frac{x_2}{2} + \frac{x_3}{6} + \frac{1}{24}\Big)\frac{\tau^2}{2}
$$
$$
+ \Big(x_1 + x_2 + \frac{x_3}{2} + \frac{1}{6}\Big)\frac{\tau^3}{6} + \Big(x_2 + x_3 + \frac{1}{2}\Big)\frac{\tau^4}{24} \qquad (5.42.b)
$$
$$
+ \big(x_3 + 1\big)\frac{\tau^5}{120} - \frac{\tau^6}{720} = 0
$$

and

$$
-\Big(\psi_2 - \psi_1 - \frac{x_1}{2} - \frac{x_2}{6} - \frac{x_3}{24} - \frac{1}{120} - \big(\psi_1 + x_1 + \frac{x_2}{2} + \frac{x_3}{6}\big)\tau
$$
$$
- \big(x_1 + x_2 + \frac{x_3}{2} + \frac{1}{6}\big)\frac{\tau^2}{2} - \big(x_2 + x_3 + \frac{1}{2}\big)\frac{\tau^3}{6} \qquad (5.42.c)
$$
$$
- \big(x_3 + 1\big)\frac{\tau^4}{24} + \frac{\tau^5}{120}\Big)\mu + \Big(\psi_1 + x_1 + \frac{x_2}{2} + \frac{x_3}{6} + \frac{1}{24}
$$
$$
+ \big(x_1 + x_2 + \frac{x_3}{2} + \frac{1}{6}\big)\tau + \big(x_2 + x_3 + \frac{1}{2}\big)\frac{\tau^2}{2} + \big(x_3 + 1\big)\frac{\tau^3}{6} - \frac{\tau^4}{24}\Big)\frac{\mu^2}{2}
$$
$$
+ \Big(x_1 + x_2 + \frac{1}{2}x_3 + \frac{1}{6} + \big(x_2 + x_3 + \frac{1}{2}\big)\tau + \big(x_3 + 1\big)\frac{\tau^2}{2} - \frac{\tau^3}{6}\Big)\frac{\mu^3}{6}
$$
$$
+ \Big(x_2 + x_3 + \frac{1}{2} + \big(x_3 + 1\big)\tau - \frac{\tau^2}{2}\Big)\frac{\mu^4}{24}
$$

$$+ \left(x_3 + 1 - \tau\right)\frac{\mu^5}{120} + \frac{\mu^6}{720} = 0.$$

The value ψ_2 in (5.42.b) and (5.42.c) is to be replaced by the expression

$$\psi_2 = \frac{1}{2}\psi_1 + \frac{1}{6}x_1 + \frac{1}{24}x_2 + \frac{1}{120}x_3 + \frac{1}{720}$$

(see (5.38)). Relations (5.42) look too cumbersome, but make sense: we have simply integrated the linear system (5.29.a) first with $u = 1$ on the interval $t \in (0, 1)$, secondly with $u = -1$ on the interval $t \in (1, 1 + \tau)$, and finally with $u = 1$ on the interval $t \in (1 + \tau, 1 + \tau + \mu)$. Equations (5.42.b) and (5.42.c) correspond to the conditions $\psi_3(1 + \tau) = 0$ and $\psi_3(1 + \tau + \mu) = 0$ respectively. Using the symmetry properties of Fuller's problem with $\beta = -1$, we could slightly simplify equations (5.42), but this is not essential. Be it as it may, we have a concrete mapping at a concrete fixed point and so we can numerically estimate (with some degree of accuracy) the spectrum of its Jacobian.

Let us agree to use the subscript 0 for any function which has been calculated at the point M^0. It follows from (5.42) that

$$D_0 = \frac{D\widetilde{F}^0}{D(\psi_1, x_1, x_2, x_3)}$$

$$= \begin{pmatrix} \mu_0^{-4} & \mu_0^{-4}(1 + \tau_0) & \mu_0^{-4}\left(\frac{1}{2} + \tau_0 + \frac{\tau_0^2}{2}\right) & \mu_0^{-4}\left(\frac{1}{6} + \frac{\tau_0}{2} + \frac{\tau_0^2}{2} + \frac{\tau_0^3}{6}\right) \\ 0 & \mu_0^{-3} & \mu_0^{-3}(1 + \tau_0) & \mu_0^{-3}\left(\frac{1}{2} + \tau_0 + \frac{\tau_0^2}{2}\right) \\ 0 & 0 & \mu_0^{-2} & \mu_0^{-2}(1 + \tau_0) \\ 0 & 0 & 0 & \mu_0^{-1} \end{pmatrix}$$

$$+ \begin{pmatrix} \mu_0^{-4}\,(\mu_0^3 x_{10})\,\mathbf{grad}\,\tau_0 \\ \mu_0^{-3}\,(\mu_0^2 x_{20})\,\mathbf{grad}\,\tau_0 \\ \mu_0^{-2}\,(\mu_0 x_{30})\,\mathbf{grad}\,\tau_0 \\ -\mu_0^{-1}\,\mathbf{grad}\,\tau_0) \end{pmatrix} + \begin{pmatrix} (\mu_0^4 \psi_{10})\,(-4\mu_0^{-5})\,\mathbf{grad}\,\mu_0 \\ (\mu_0^3 x_{10})\,(-3\mu_0^{-4})\,\mathbf{grad}\,\mu_0 \\ (\mu_0^2 x_{20})\,(-2\mu_0^{-3})\,\mathbf{grad}\,\mu_0 \\ (\mu_0 x_{30})\,(-\mu_0^{-2})\,\mathbf{grad}\,\mu_0 \end{pmatrix},$$

where the rank of the last two matrices equals 1 and the rows are proportional to the vectors

$$\mathbf{grad}\,\tau_0 = \frac{-1}{-\mu_0^5 \psi_{20}} \left(\frac{\tau_0}{2} + \frac{\tau_0^2}{2},\ \frac{\tau_0}{3} + \frac{\tau_0^2}{2} + \frac{\tau_0^3}{6},\ \frac{\tau_0}{8} + \frac{\tau_0^2}{4} + \frac{\tau_0^3}{6} + \frac{\tau_0^4}{24},\right.$$
$$\left.\frac{\tau_0}{30} + \frac{\tau_0^2}{12} + \frac{\tau_0^3}{12} + \frac{\tau_0^4}{24} + \frac{\tau_0^5}{120}\right)$$

and

$$\mathbf{grad}\ \mu_0 = \frac{-1}{(\mu_0\tau_0)^5\psi_{20}} \left[\left(\left(\frac{1}{2} - \tau_0\right)\mu_0 + \frac{\mu_0^2}{2},\right.\right.$$

$$\left(\frac{1}{3} - \tau_0 - \frac{\tau_0^2}{2}\right)\mu_0 + (1 + \tau_0)\frac{\mu_0^2}{2} + \frac{\mu_0^3}{6},$$

$$\left(\frac{1}{8} - \frac{\tau_0}{2} - \frac{\tau_0^2}{2} - \frac{\tau_0^3}{6}\right)\mu_0 + \left(\frac{1}{2} + \tau_0 + \frac{\tau_0^2}{2}\right)\frac{\mu_0^2}{2} + (1 + \tau_0)\frac{\mu_0^3}{6} + \frac{\mu_0^4}{24},$$

$$\left(\frac{1}{30} - \frac{\tau_0}{6} - \frac{\tau_0^2}{4} - \frac{\tau_0^3}{6} - \frac{\tau_0^4}{24}\right)\mu_0 + \left(\frac{1}{6} + \frac{\tau_0}{2} + \frac{\tau_0^2}{2} + \frac{\tau_0^3}{6}\right)\frac{\mu_0^2}{2}$$

$$+ \left(\frac{1}{2} + \tau_0 + \frac{\tau_0^2}{2}\right)\frac{\mu_0^3}{6} + (1 + \tau_0)\frac{\mu_0^4}{24} + \frac{\mu_0^5}{120}\right)$$

$$\left. + \mu_0^5\left(\psi_{10} + \frac{x_{10}}{2} + \frac{x_{20}}{6} + \frac{x_{30}}{24} - \frac{1}{120}\right)\mathbf{grad}\ \tau_0\right]$$

We have used that in the symmetrical case

$$F^0(M^0) = -\mathfrak{g}_{\tau_0}(M^0),$$
$$(F^0)^2(M^0) = \mathfrak{g}_{\mu_0}(M^0).$$

If we substitute the values

$$\tau_0 = 0.5757363, \qquad\qquad x_{10} = 0.04164337,$$
$$\mu_0 = (\tau_0)^2 = 0.33147229, \quad \psi_{20} = -0.00538198,$$
$$x_{30} = -0.63462392, \qquad\quad \psi_{10} = -0.02527156,$$
$$x_{20} = 0.10110916,$$

we obtain

$$D_0 = \frac{1}{k_0} \times \begin{pmatrix} -11.016174 & -9.920351 & -5.161861 & -1.961579 \\ 13.806120 & 12.421936 & 6.456702 & 2.450631 \\ 21.279190 & 19.191702 & 10.003331 & 3.808584 \\ -70.874651 & -63.744496 & -33.120007 & -12.56568 \end{pmatrix}$$

where $k_0 = -0.00002157\ldots$.

The matrix D_0 has four real eigenvalues $\lambda_0 \approx 0.0495$, $\lambda_2 \approx 753.9$, $\lambda_3 \approx 2274.4$, $\lambda_4 \approx 50599.1$. We are to select those of the eigenvalues whose eigenvectors belong to the tangent space of the surface H_0. One can check that λ_3 does not meet the last condition.

Thus, $\widetilde{F}^0$ is a C^1–hyperbolic diffeomorphism at some neighborhood of M^0. The function $x_2^2 \, \text{sgn} \, x_2$ on the right–hand side of the equation (5.29) is C^1 everywhere and C^∞ at points of the region $x_2 \neq 0$. However, if for some $z^0 \in S \backslash \{0\}$ the trajectory $Z(t, z^0)$ intersects the surface $z_2 = 0$ under a nonzero angle and with nonzero phase velocity, then F^0 is C^∞–mapping at z_0. Solutions of system (5.29), tangential to the surface $x_2 = 0$, have to pass through the plane $x_3 = 0$. If we resolve the system

$$x_{30} + t = 0, \quad x_{20} + x_{30}t + \frac{1}{2}t^2 = 0$$

or, respectively, the system

$$x_{30} - t = 0, \quad x_{20} + x_{30}t - \frac{1}{2}t^2 = 0,$$

with respect to t, then we obtain $x_{20} = \frac{1}{2}x_{30}^2$ or $x_{20} = -\frac{1}{2}x_{20}^2$. A straightforward calculation yields that at $\epsilon = 0$ **b.c.r.** solutions of system (5.29) do not meet these relations. (The same can be proved for any $\beta < 0$.) It follows that , F^ϵ is a C^∞–mapping at some neighborhood of M^ϵ for all sufficiently small $\epsilon > 0$.

Now the existence of $\widetilde{F}^0$–contracting manifold follows from the invariant manifold theorem. Namely, the single eigenvalue of the derivative $D\widetilde{F}(M^0)$ at the interval $(0, 1)$ determines this invariant contracting curve. Let us denote the eigenvalue by $\lambda(0)$ and the curve by γ^0. Hence, the case $\epsilon = 0$, $\beta = -1$ is completed. Let us prove the same for an arbitrary value $\beta < 0$.

The matrix Q is continuous in β. It follows from Lemma 5.10 that the polynomial (5.41) has the same signs at $\lambda = 0$, $\lambda = 1$, and $\lambda = -1$. Therefore, there are at least two real eigenvalues of $f(\lambda)$, one on the interval $(0, 1)$ and one on the interval $(1, \infty)$. Suppose that for some $\beta_0 < 0$ there are two or more eigenvalues of matrix Q in the unit circle $|z| < 1$ in $\mathbb{C}$. It is easy to see that then there is another value $\beta_1 < 0$ such that $Q = Q(\beta_1)$ has two complex–conjugated eigenvalues in $|z| < 1$ and hence its spectrum is simple (there are no double roots of (5.41)). In this situation the statement of the invariant manifold theorem holds and for $\beta = \beta_1$ there exists a contracting three–dimensional $\widetilde{F}^0$–invariant manifold, call it $\mathfrak{R}^*$.

It can readily be proved (see Lemma 5.12 below) that solutions of system (5.29) with initial conditions in **Orb** $\mathfrak{R}^*$ hit the origin in finite time. It follows from the sufficiency theorem that they are optimal solutions to Problem 5.5 with $\beta = \beta_1$. But these trajectories fill some five–dimensional manifold in $\mathbb{R}^6$ that contradicts Lemma 5.4.

Now for all sufficiently small $\epsilon > 0$ the statement of Lemma 5.11 holds in view of the implicit function theorem. **Q.E.D.**

Let us set $\Gamma_0^\epsilon = \textbf{Orb} \, \gamma^\epsilon; \quad \Gamma_n^\epsilon = F^\epsilon(\Gamma_{n-1}^\epsilon), \quad n = 1, 2 \ldots .$

LEMMA 5.12.

(1) $\Gamma^\epsilon_{n+2} \subset \Gamma^\epsilon_n$ *for any* $n \in \mathbb{N}$.

(2) *For any* $z^0 \in \Gamma^\epsilon_0$, *the solution* $Z(t, z^0)$ *attains the origin in finite time.*

Proof. Consider a neighborhood U of M^ϵ on γ^ϵ and an arbitrary $z^0 \in U$. By the definition of mapping $\widetilde{F}^\epsilon$ we have

$$\widetilde{F}^\epsilon(z^0) = \widetilde{\mathfrak{g}}_{1/\mu(z^0)}\Big(F^\epsilon\big(F^\epsilon(z^0)\big)\Big),$$

where $\mu(z^0)$ is specified by (5.39)–(5.40). The tangent plane to γ^ϵ at M^ϵ is associated with the eigenvalue $\lambda(\epsilon) \in (0,1)$ of the differential $D\widetilde{F}^\epsilon(M^\epsilon)$. Hence, if the neighborhood U is small enough, then $\widetilde{F}^\epsilon(z^0) \in U$, hence

$$F^\epsilon(F^\epsilon(\mathbf{Orb}\,)) \subset U. \tag{5.43}$$

Denote by $\tau(z^0)$ the second switching instant on $Z(t, z^0)$. The homogeneity of (5.29) under $\widetilde{\mathfrak{G}}$–action provides that $\tau\big(\widetilde{\mathfrak{g}}_\lambda(z^0)\big) = \lambda\tau(z^0)$. Since the function $\mu(z^0)$ is continuous and $\mu(M^\epsilon) \in (0,1)$, there exists

$$\mu_0 = \sup_{z^0 \in U} \mu(z^0) \in (0,1).$$

It follows from (5.43) that intervals between successive switches with even numbers on $Z(t, z^0)$, $z^0 \in \Gamma^\epsilon_0$, can be upper estimated by members of a decreasing geometric progression whose denominator is less than μ_0. It follows from Lemma 5.4 and the sufficiency theorem that $Z(t, z^0)$ is an optimal solution to Problem 5.5.   **Q.E.D.**

Denote by $v(x)$ the section of the space $\Gamma(T\mathbb{R}^3)$ of all limited vector–fields corresponding to the optimal feedback control $\widehat{u}(x)$ of Problem 5.5,

$$v(x) = (x_2,\, x_3,\, \widehat{u}(x)).$$

In view of Lemma 5.3, for any $\lambda > 0$ the transformation

$$x(t) \to \mathfrak{g}_\lambda x(t/\lambda)$$

unites any optimal solution to Problem 5.5 with an optimal one under appropriate boundary conditions. Hence, the system

$$\dot{x} = v(x)$$

with a discontinuous right–hand side can be associated with a dynamic system on $\mathbb{R}^3/\mathfrak{G}$ ($\mathbb{R}^3/\mathfrak{G}$ is the orbit–space of $\mathbb{R}^3\backslash\{0\}$ with respect to $\mathfrak{G}$–action). As $\mathbb{R}^3/\mathfrak{G}$ we can take the manifold S^2 defined as follows:

$$S^2 = \big\{x \in \mathbb{R}^3 \mid x_1^4 + x_2^6 + x_3^{12} = 1\big\}.$$

It is easy to see that for any $x^0 \in \mathbb{R}^3\backslash\{0\}$ there exists a unique intersection of $\mathbf{Orb}\,x^0 = \bigcup_{\lambda > 0}\mathfrak{g}_\lambda(x^0)$ with the manifold S^2. The quotient–system on S^2 can be specified up to orbital equivalence. This means that it is uniquely defined only in the direction of the velocity of the quotient–system, but not in its magnitude. Let us consider the mapping

$$\mathbf{g} : \mathbb{R}^3 \backslash \{0\} \to S^2$$

where

$$\mathbf{g}(x) = \mathfrak{g}_{\Lambda(x)}(x) = \left(\Lambda^3(x)x_1, \, \Lambda^2(x)x_2, \, \Lambda(x)x_3 \right),$$

$$\Lambda(x) \stackrel{\text{def}}{=} \left(x_1^4 + x_2^6 + x_3^{12} \right)^{-1/12}.$$

For any $s \in S^2$ let us define the vector–field v^* on S^2 as follows:

$$v^*(s) = \frac{1}{\Lambda(x)} D\mathbf{g}\big|_x v(x),$$

x being any point in **Orb** s. The matrix $D\mathbf{g}(x)\big|_x$ being the differential of the mapping $\mathbf{g}$ at x,

$$D\mathbf{g}\big|_x = \begin{pmatrix} \Lambda^3(x) & 0 & 0 \\ 0 & \Lambda^2(x) & 0 \\ 0 & 0 & \Lambda(x) \end{pmatrix} -$$

$$- \frac{\Lambda^{13}}{12} \begin{pmatrix} 12\Lambda^2 x_1^4 & 18\Lambda^2 x_1 x_2^5 & 36 x_1 x_3^{11} \\ 8\Lambda x_1^3 x_2 & 12\Lambda x_2^6 & 24\Lambda x_2 x_3^{11} \\ 4 x_1^3 x_2 & 6 x_2 x_3 & 12 x_3^{12} \end{pmatrix}.$$

Let us prove that the definition of vector v^* does not depend on the choice of the point $x \in \mathbf{Orb}\ (s)$. Since for any $\lambda_0 > 0$ we have

$$\Lambda\big(\mathfrak{g}_{\lambda_0}(x)\big) = \frac{1}{\lambda_0}\Lambda(x),$$

it follows that

$$D\mathbf{g}\Big|_{\mathfrak{g}_{\lambda_0}(x)} \cdot v\big(\mathfrak{g}_{\lambda_0}(x)\big) = \left[\begin{pmatrix} \Lambda^3(x)/\lambda_0^3 & 0 & 0 \\ 0 & \Lambda^2(x)/\lambda_0^2 & 0 \\ 0 & 0 & \Lambda(x)/\lambda_0 \end{pmatrix} \right.$$

$$\left. - \frac{\Lambda^{13}}{12} \begin{pmatrix} 12\Lambda^2(x)\dfrac{x_1^4}{\lambda_0^3} & 18\Lambda^2(x)\dfrac{x_1 x_2^5}{\lambda_0^2} & 36\Lambda^2(x)\dfrac{x_1 x_3^{11}}{\lambda_0} \\ 8\Lambda(x)\dfrac{x_1^3 x_2}{\lambda_0^3} & 12\Lambda(x)\dfrac{x_2^6}{\lambda_0^2} & 24\Lambda(x)\dfrac{x_2 x_3^{11}}{\lambda_0} \\ \dfrac{4 x_1^3 x_3}{\lambda_0^3} & \dfrac{6 x_2^5 x_3}{\lambda_0^2} & \dfrac{12 x_3^{12}}{\lambda_0} \end{pmatrix} \right] \cdot \begin{pmatrix} \lambda_0^2 x_2 \\ \lambda_0 x_3 \\ \widehat{u} \end{pmatrix}$$

$$= \frac{1}{\lambda_0} D\mathbf{g}\big|_x \cdot v(x).$$

The vector–field $v^*(x)$ is thereby well–defined. We take v^* as the right–hand side of the quotient–system associated with the vector–field v. What we wish to do is to characterize the singularities (fixed points and closed cycles) of v^* on S^2.

Set $T \overset{\text{def}}{=} \Pi^\epsilon \cap S^2$, i.e., we denote as T the "trace" of the optimal switching surface. It follows from Lemma 5.6 that T is homeomorphic to S^1 (the one–dimensional nonself–intersecting circle). The field v^* is analytic at points of $S^2 \backslash T$ and has a discontinuity at T.

Recall that in view of Proposition 5.3 there exists a single family of **b.c.r.** solutions to Problem 5.5; P^ϵ denotes the two–dimensional surface in $\mathbb{R}^3$, constituted by the solutions. Consider the curve $\mathcal{E} \overset{\text{def}}{=} S^2 \cap P^\epsilon$, the "trace" of **b.c.r.** solutions on the orbit–space S^2. Let $S^+ \subset S^2$ be any two half–spheres separated by the curve $\mathcal{E}$, and $\rho \overset{\text{def}}{=} \left\{ x \in \mathbb{R}^3 \middle| x_1 = x_3^3/6, \ x_2 = -x_3^2/2 \ \text{sgn} \ x_3 \right\}$.

It is easy to see that there is a unique solution of system (5.29), invariant relative to $\widetilde{\mathfrak{G}}$–action, namely, the trajectory

$$\psi_3 = -\frac{1}{720} x_3^6 \, \text{sgn} \, x_3, \quad \psi_2 = -\frac{1}{120} x_3^5, \quad \psi_1 = -\frac{1}{24} x_3^4 \, \text{sgn} \, x_3,$$

$$x_1 = \frac{1}{6} x_3^3, \quad x_2 = -\frac{1}{2} x_3^2 \text{sgn} \, x_3.$$

Its projection on x–space is the curve ρ. Let the point Q be the intersection of S^+ and this curve. It follows from the uniqueness of an invariant solution that Q is the unique fixed point of the vector–field v^* on S^+.

PROPOSITION 5.4. *For all sufficiently small $\epsilon \geqslant 0$ there is a unique closed cycle of the field v^* on S^2, namely, the curve $\mathcal{E}$.*

Proof. Let us prove first that there are no closed v^*–cycles with constant control $\widehat{u}(t) \equiv 1$ or $\widehat{u}(t) \equiv \beta$. Indeed, any such cycle Z^* would determine an infinite number of trajectories of system (5.29) with constant control on the surface **Orb** Z^*. This contradicts the fact that the origin is not a fixed point of system (5.29) with constant control $u \neq 0$.

Assume that there is some closed trajectory Z^* of the field v^* on S^2 that is different from $\mathcal{E}$. Let us agree to imagine S^+ as the two–dimensional disk with the boundary $\mathcal{E}$. The curve T (the "trace" of the switching surface) partitions S^+ into two half–disks, call them S_1^*, S_2^*, where $\widehat{u} = 1$ and $\widehat{u} = \beta$, respectively. It follows from Lemma 5.8 that any v^*–trajectory intersects T at a discrete set of points. Hence, in view of the uniqueness theorem for ordinary differential equations there is a unique trajectory of v^* through any $x \in S^2$ for both $t \geqslant 0$ and $t \leqslant 0$.

Let us demonstrate that the closed cycle Z^* is to be "two–linked", i.e., to consist of two arcs where the control $\widehat{u}$ equals 1 and β respectively. Assume that Y_1, Y_2, Y_3 are three points of successive switches on Z^*. We assert that $Y_1 = Y_3$. Assume the contrary: $Y_1 \neq Y_3$. Then either Y_3 lies between Y_2 and Y_1 on the segment T, or it lies on a supplement of T to the subset $[Y_1, Y_2]$ (see Figs. 14, 15).

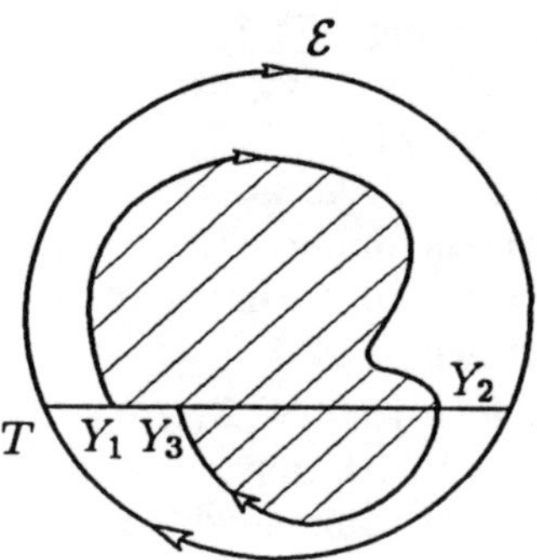

FIG. 14: POINCARÉ MAPPING IN THE ORBIT SPACE, A

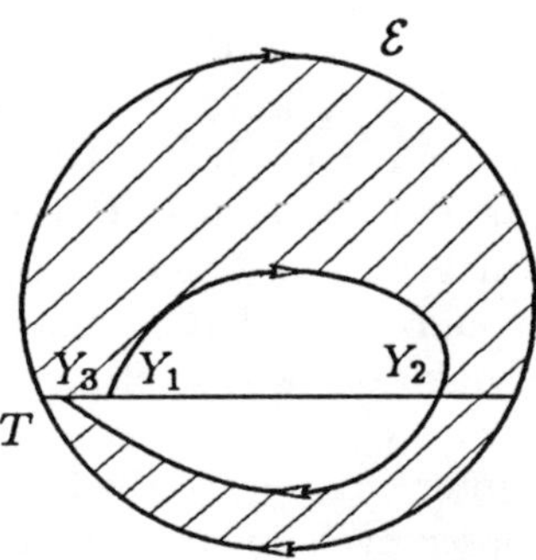

FIG. 15: POINCARÉ MAPPING IN THE ORBIT SPACE, B

In both cases, passing through Y_3, the trajectory Z^* is to belong to the domain on S^+ not containing the point Y_3 (this domain is hatched in Figs. 14, 15). Hence, Z^* is not a cycle in spite of the presumption.

It follows that there are two $\widetilde{\mathfrak{G}}$–orbits on the switching surface $\psi_3 = 0$, mutually permuted by the Poincaré mapping $F^{\mathfrak{e}}$. In other words, the cycle Z^* determines a new family of **b.c.r.** solutions of system (5.29). This contradicts Proposition 5.3. **Q.E.D.**

Everything has now been prepared for the final part of the proof of Theorem 5.1.

Recall that F^ϵ is the Poincaré mapping of the surface $\psi_3 = 0$ along the trajectories of system (5.29). Consider the restriction of F^ϵ on the switching surface of optimal solutions to Problem 5.5. It defines the mapping of Π^ϵ into itself. Recall that Π^ϵ is the projection of the switching surface of the extended state space into x–space. That, in turn, induces the mapping $F^* : T \to T$, transferring an orbit on the switching surface Π^ϵ in $\mathbb{R}^3$ to the corresponding orbit. Let us set $\pi(\psi, x) = x$, $\tau_0 = \pi\Gamma_0 \bigcap T_0$, $\tau_{-n-1} = (F^*)^{-1}\tau_{-n}$, $\Gamma_{-n-1} = (F^\epsilon)^{-1}\Gamma_{-n}$, $n = 0, 1, 2, \ldots$. We assert that the sequence of the preimages of Γ_0 with respect to the mapping F^ϵ is finite, i.e., there is some $n_0 < \infty$ such that $\bigcup_{n=1}^{n_0} \tau_{-n} = T$. (In turn, the union $\bigcup_{n=1}^{n_0} \Gamma_{-n}$ coincides with the whole switching surface in $\mathbb{R}^6$.)

It is clear that if the trajectory $Z(t, z^0)$ of system (5.29) intersects S under a nonzero angle and with nonzero phase velocity at $z^0 \in S$, then the mapping F^ϵ is C^1 at some neighborhood of z^0. Since $\dot{z}_1 = z_2$ in (5.29), it follows that if the set $\Gamma^*_{-n} \overset{\text{def}}{=} \Gamma_{-n} \bigcap \{z \mid z_2 = 0\}$ is empty, then the restriction $F^\epsilon\big|_{\Gamma_{-n-1}}$ is a diffeomorphism. Assume that $\Gamma^*_{-n} = \varnothing$ for any natural n. Then we assert that the ω–limit set of the trajectories of the vector–field v^* is represented by a closed cycle that is different from $\mathcal{E}$. Indeed, let us consider an arbitrary point $\mathcal{A}_0 \in T_0$ and the corresponding sequence of preimages of $\mathcal{A}_0$ under the iterations of the mapping $(F^*)^{-1}$, i.e., the sequence $\mathcal{A}_n = (F^*)^{-1}\mathcal{A}_{n-1}$, $n = 1, 2 \ldots$. It has been assumed that for any n the restriction of $(F^*)^{-1}$ to the segment $[\mathcal{A}_n, \mathcal{A}_{n+2}] \subset T$ is a continuous one–to–one correspondence. Hence, the sequences $\{\mathcal{A}_{2k}\}_{k=1}^\infty$ and $\{\mathcal{A}_{2k+1}\}_{k=1}^\infty$ are naturally ordered on T, and there are two limit points, call them $\mathcal{A}^+$ and $\mathcal{A}^-$, of the sequences. If $\mathcal{A}^+ = \mathcal{A}^- = \mathcal{A}$, then $\mathcal{A} \in T$ is a fixed point of the trajectories on S^+. This is impossible, because $\Psi_3(Q) \neq 0$, so $\mathcal{A}^+ \neq \mathcal{A}^-$. In this case the trajectory of v_S emanating from $\mathcal{A}^+$ intersects T again in $\mathcal{A}^-$, and vice versa. This gives the appropriate closed cycle in spite of Proposition 5.4.

Thus, there is some n_0 such that $\Gamma^*_{-n_0} \neq \varnothing$. In view of Lemma 5.5 we have $\operatorname{grad} \omega(x^0) = (-\Psi_1(x^0), 0, 0)$ at some point $x^0 \in \pi\Gamma^*_{-n}$. Hence, the surface $x_1 = x_1^0$ is tangent to the level–surface of the Bellman function $\omega(x) = \omega(x_0)$. But ω is strictly convex, so there are not more than two points with this property. It follows that $\bigcup_{n=0}^\infty \Gamma^*_{-n_0}$ is to contain exactly two orbits of the symmetry group $\widetilde{\mathfrak{G}}$. Hence, there is a single point, call it $\mathcal{B}_0$, on $T_0 \overset{\text{def}}{=} S^+ \bigcap T$, where $\psi_2(\mathcal{B}_0) = 0$. Denote also by $\mathcal{B}_1$ a point on the opposite semisphere S^-, where $\psi_2(\mathcal{B}_1) = 0$. Since vector $\Psi(x) = -\operatorname{grad} \omega(x)$ is zero only at the origin, we have $\Psi_1(\mathcal{B}_0) \neq 0$. It follows from Corollary 5.2 that optimal trajectories $Z(t, z^0)$ are continuous in z^0, so the mapping $(F^*)^{-1}$ can be continuously extended to the closure of Γ_{-n}. It follows that $(F^*)^{-1}\Gamma_{-n_0} = T\backslash\Gamma_{-n_0}$. It was noted above that there is a unique v_S–trajectory through any point of S. Hence, v_S–trajectories through the segment $[\mathcal{B}_0, F^*(\mathcal{B}_0)]$ of the curve T cannot have any switch at the backward time current. All of them are thus to approach Q, which is the unique fixed point of v_S. The phase portrait of v_S–trajectories is depicted in Fig. 16.

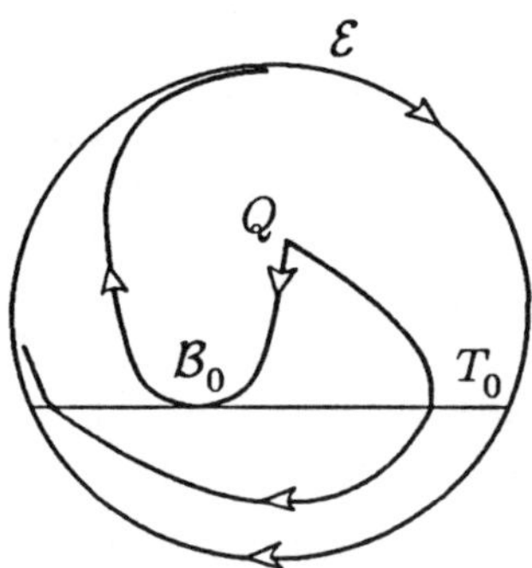

FIG. 16: SYNTHESIS OF FULLER'S PROBLEM IN THREE DIMENSIONS

To complete the proof of Theorem 5.1, it remains to show that Π^ϵ is a smooth surface. By the construction, the lifting of Π^ϵ to the extended state space, i.e., $\widetilde{\Pi}^\epsilon \stackrel{\text{def}}{=} \pi^{-1}(\Pi^\epsilon)$, is a C^1–surface everywhere outside the points of the curve

$$\xi \stackrel{\text{def}}{=} \left\{ (\psi, x) \,\middle|\, x \in \mathbf{Orb}\,(\mathcal{B}_0 \textstyle\bigcup \mathcal{B}_1),\, \psi = \Psi(x) \right\}.$$

LEMMA 5.13. *The restriction of π to $\widetilde{\Pi}^\epsilon\backslash\xi$ is a C^1–diffeomorphism.*

Proof. Denote by $\mathcal{D}_+$ the region in $\mathbb{R}^3$ where the optimal control $\widehat{u}$ equals 1 and by $\mathcal{D}_-$ the region where $\widehat{u}$ equals β. Denote by S^* the subset of $\mathcal{D}_+\bigcup\mathcal{D}_-$ such that π is not regular on $\pi^{-1}(S^*)$. It follows from the Sard theorem [S. Sternberg, 1964; Theorem 3.1, p. 45] that the Lebesgue measure of S^* equals 0. Hence, the sets $\mathcal{D}_+\backslash S^*$ and $\mathcal{D}_-\backslash S^*$ are open and everywhere dense in $\mathcal{D}_+$ and $\mathcal{D}_-$. It can readily be seen that the restriction of π to the integral variety of system (5.29), consisting of optimal trajectories of Problem 5.5, has an inverse continuous mapping, namely

$$x \mapsto \big(x, \Psi(x)\big).$$

In view of Lemma 5.5, function $\Psi(x) = -\mathrm{grad}\,\omega(x)$ is uniquely defined and continuous. So if $x^0 \in S^*$, then

$$\lim_{\substack{x \to x_0 \\ x \in \mathcal{D}_\pm\backslash S^*}} \|\Omega(x)\| = \infty,$$

where $\Omega(x) = D\Psi(x)/Dx$, and $\|\Omega\|$ denotes the operator norm

$$\|\Omega\| = \sup_{\|y\|=1} \|\Omega \cdot y\|.$$

It remains to demonstrate that the matrix $\Omega = -D^2\omega/Dx^2$ has a restricted norm. One can check that the eigenvector of the matrix Ω, tangential to the invariant curve γ^ϵ of the mapping $\widetilde{F}^\epsilon$, has a nonzero projection into x–space. Hence, for some small neighborhood of the switching curve ρ^ϵ, the projections of trajectories of system (5.29) are transversal to Π^ϵ. Without loss of generality we can suppose that this neighborhood coincides with Γ_0^ϵ. It is clear that the function $\Psi(x)$ is smooth at the subsets of $\mathcal{D}_+$ and $\mathcal{D}_-$ which are filled by solutions of system (5.29) emanating from Γ_0^ϵ with $u = \widehat{u}(x)$.

Let us consider the part of the region $\mathcal{D}_+$ consisting of the arcs of optimal solutions that are escaping from Γ_0^ϵ at the backward time current with $u = 1$ and ending at Γ_{-1}^ϵ. If $\Gamma_{-1}^\epsilon \bigcap \xi = \varnothing$, then the restriction $\pi\big|_{\Gamma_{-1}^\epsilon}$ is regular and the surface $\pi\Gamma_{-1}^\epsilon$ is smooth. We are to show that function $\Psi(x)$ can be smoothly prolonged from the region $\mathcal{D}_-$ to Γ_{-1}^ϵ. Assume that there is $x^0 \in \pi\Gamma_{-1}^\epsilon \bigcap \mathrm{cl}\, S^*$, where cl means the closure of the set. Since $\|\Omega(x^0)\| = \infty$, there is some $x^1 \in \Pi^\epsilon \bigcap \mathrm{cl}\,(\mathcal{D}_-\backslash S^*)$, such that $\|\Omega^-\| > \|\Omega^+\|$, where

$$\Omega^- = \lim_{\substack{x \to x_1 \\ x \in \mathcal{D}_-\backslash S^*}} \Omega(x),$$

$$\Omega^+ = \lim_{\substack{x \to x_1 \\ x \in \mathcal{D}_+\backslash S^*}} \Omega(x).$$

Let us denote by $\widetilde{\omega}$ the smooth extension of the Bellman function ω from $\mathcal{D}_+$ into the region $\mathcal{D}_-$ at some neighborhood of x^1 along the trajectories of system (5.28) with $u \equiv 1$. Then $\widetilde{\omega} \geqslant \omega$. Function ω is strictly convex, so the matrices Ω^- and Ω^+ are positive definite. Consider the function $\Delta = \widetilde{\omega} - \omega \geqslant 0$. We have $\Delta(x^1) = 0$, $\quad$ grad $\Delta(x^1) = 0$ but the matrix

$$\frac{D^2\Delta}{Dx^2}(x_1) = \Omega^+ - \Omega^-$$

is not positive definite. That contradicts the presumption that $\Delta \geqslant 0$ at some neighborhood of x^1.

Thus, the restriction $\pi\big|_{\mathcal{D}_-}$ is also regular at Γ_{-1}^ϵ. Repeating the preceding consideration we arrive to the conclusion that the restriction $\pi\big|_{\widetilde{\Pi}^\epsilon\backslash\xi}$ is regular at points of Γ_{-2}^ϵ, Γ_{-3}^ϵ, and so on. This implies the statement of Lemma 5.14 and Theorem 5.1.

Q.E.D.

Chapter 6

APPLICATIONS

6.1 Fibrations in Three–Dimensional Space

The simplest nontrivial case admitting visualization is a three-dimensional space with a curve as a singular manifold. We consider two such examples. The first one (Problem 6.1) is a modification of Fuller's problem. This is required to minimize the weighted mean–square deviation of a massive object from the origin. The role of the control belongs to a bounded force and the weight is the discounted factor $e^{-\alpha t}$. The second example (Problem 6.2) is a time–optimal problem for some control system of third order.

PROBLEM 6.1. *Minimize*

$$\int_0^\infty e^{-\alpha t} x^2 \, dt \quad (\alpha \geqslant 0)$$

subject to

$$\dot{x} = y, \quad \dot{y} = u, \quad u \in [-1, 1], \tag{6.1}$$

$$x(0) = x_0, \quad y(0) = y_0.$$

The proof of the existence and uniqueness of solutions to Problem 6.1 is completely analogous to that in Lemmas 5.1 and 5.2.

PROPOSITION 6.1. *Let* $\big(\widehat{x}(t), \widehat{y}(t), \widehat{u}(t)\big)$ *be a solution to Problem 6.1 with an initial point* (x_0, y_0). *Then for any* $\tau > 0$ *the triple* $\big(\widehat{x}(t+\tau), \widehat{y}(t+\tau), \widehat{u}(t+\tau)\big)$ *is a solution to Problem 6.1 with the initial point* $\big(\widehat{x}(\tau), \widehat{y}(\tau)\big)$.

Proof. Assume that the trajectory $\widehat{x}(t+\tau)$, $\widehat{y}(t+\tau)$ is not optimal. Let $(x^*(t), y^*(t), u^*(t))$ be a solution to Problem 6.1 with the initial point $(\widehat{x}(\tau), \widehat{y}(\tau))$. Consider the trajectory $(x(t), y(t))$ of system (6.1) of the form

$$(x(t), y(t)) = \begin{cases} (\widehat{x}(t), \widehat{y}(t)) & \text{if } t \in [0, \tau], \\ (x^*(t-\tau), y^*(t-\tau)) & \text{if } t \in [\tau, \infty). \end{cases}$$

We have

$$\int_0^\infty e^{-\alpha t} x^2 \, dt = \int_0^\tau e^{-\alpha t} x^2 \, dt + \int_\tau^\infty e^{-\alpha t} x^2 \, dt,$$

where

$$\int_\tau^\infty e^{-\alpha t} x^2 \, dt = \int_\tau^\infty e^{-\alpha t} \big(x^*(t-\tau)\big)^2 \, dt$$

$$= e^{-\alpha \tau} \int_0^\infty e^{-\alpha t} \big(x^*(t)\big)^2 \, dt$$

$$< e^{-\alpha \tau} \int_0^\infty e^{-\alpha t} \big(\widehat{x}(t+\tau)\big)^2 \, dt = \int_\tau^\infty e^{-\alpha t} \widehat{x}^2(t) \, dt.$$

It follows that

$$\int_0^\infty e^{-\alpha t} x^2 \, dt < \int_0^\tau e^{-\alpha t} \widehat{x}^2 \, dt + \int_\tau^\infty e^{-\alpha t} \widehat{x}^2 \, dt = \int_0^\infty e^{-\alpha t} \widehat{x}^2 \, dt,$$

which implies that $\widehat{x}(t)$ is not optimal.   **Q.E.D.**

Usually, we seek the optimal control as a function of the reference position of the trajectory in the phase space, i.e., as a feedback control. In nonautonomous problems, the feedback control depends both on the state variable and on the time.

COROLLARY 6.1. *An optimal feedback control in Problem 6.1, call it* $u(t, x, y)$, *is independent of* t, *i.e.,*

$$u(t, x, y) = \widehat{u}(x, y).$$

Let us demonstrate now that in the vicinity of the origin the optimal feedback control to Problem 6.1 is given as follows:

$$\widehat{u}(x, y) = -\text{sgn}\,(x + \lambda_\alpha(y)\, y^2 \,\text{sgn}\, y),$$

where $\lambda_\alpha(\cdot) \in C^1$ and $0 < \lambda_\alpha(0) < 0.5$.

It can be derived from the last inequality (for details, see Lemma 3.4), that intervals between switches on the optimal trajectories in Problem 6.1 are upper estimated by members of a convergent geometric progression.

Hence, the trajectories hit the origin at finite time with an infinite number of control switches.

Let us define a new state coordinate ϕ related to the time by the equations $\dot\phi = 1$, $\phi(0) = 0$. The Pontryagin's function for Problem 6.1 is written as $H = \psi_\phi + \psi_x y + \psi_y u - \psi_0 e^{-\alpha\phi} x^2$. It can readily be proved that $\psi_0 \neq 0$, so we can take $\psi_0 = 1/2$. Pontryagin's maximum principle leads to the equations

$$
\begin{aligned}
\dot\psi_\phi &= -\alpha e^{-\alpha\phi} x^2/2, & \dot x &= y, \\
\dot\psi_x &= x e^{-\alpha\phi}, & \dot y &= u, \\
\dot\psi_y &= -\psi_x, & \dot\phi &= 1,
\end{aligned}
\tag{6.2}
$$

where $u = \operatorname{sgn}\psi_y$ for $\psi_y \neq 0$. For singular solutions, we have $\psi_y = 0$ at some time–interval. If $\psi_y(t) \equiv 0$, then $\psi_x(t) \equiv 0$, hence $x(t) \equiv y(t) \equiv 0$. It follows that all singular solutions of system (6.2) fill a two–dimensional surface in the extended space.

Let us put

$$
\begin{aligned}
z_1 &= \psi_y, & w_1 &= \psi_\phi, \\
z_2 &= -\psi_x, & w_2 &= \phi. \\
z_3 &= -x e^{-\alpha\phi}, & & \\
z_4 &= -y e^{-\alpha\phi}, & &
\end{aligned}
$$

In these coordinates the system (6.2) is written as follows:

$$
\begin{aligned}
\dot z_1 &= z_2, & \dot z_4 &= -e^{-\alpha w_2} u - \alpha z_4, \\
\dot z_2 &= z_3, & \dot w_1 &= -\alpha e^{\alpha w_2} z_3^2/2, \\
\dot z_3 &= z_4 - \alpha z_3, & \dot w_2 &= 1.
\end{aligned}
\tag{6.3}
$$

Since the right–hand side of (6.3) does not depend on w_1, the fifth equation of (6.3) can be canceled out. This follows from Theorem 3.1, on bundles, that in the five–dimensional space $(z_1, z_2, z_3, z_4, w_2)$ there exists a fibration with the base $S_0 = \{(z, w) \mid z_1 = z_2 = z_3 = z_4 = 0\}$ and with two–dimensional piecewise smooth fibres, filled by the chattering trajectories of system (6.3). It follows from Theorem 3.3 that the projections of the chattering arcs into (x, y, ϕ)–space are optimal for the following problem:

$$
\int_0^\infty e^{-\alpha\phi} x^2\, dt \to \min
$$

subject to

$$\dot{x} = y, \quad \dot{y} = u, \quad \dot{\phi} = 1,$$

$$u \in [-1, 1],$$

$$x(0) = x_0, \quad y(0) = y_0, \quad \phi(0) = \phi_0.$$

According to (3.36), the switching curves inside the two–dimensional fibres have the form

$$x = \lambda_i y^2, \quad \phi = \phi_0 + \nu_i y, \quad i = 0, 1,$$

(up to higher order terms with respect to the y–variable.)

Consider an arbitrary fibre in (x, y, ϕ)–space and the trajectories whose initial points belong to the intersection of the fibre with the plane $\phi = 0$. It is clear that the trajectories are optimal for Problem 6.1. In view of Proposition 6.1 the switching surface in the space (x, y, ϕ) is projected on the optimal switching curve in (x, y)–space (see Fig. 17). The inequality $0 < \lambda_i < 0.5$ follows from (3.32).

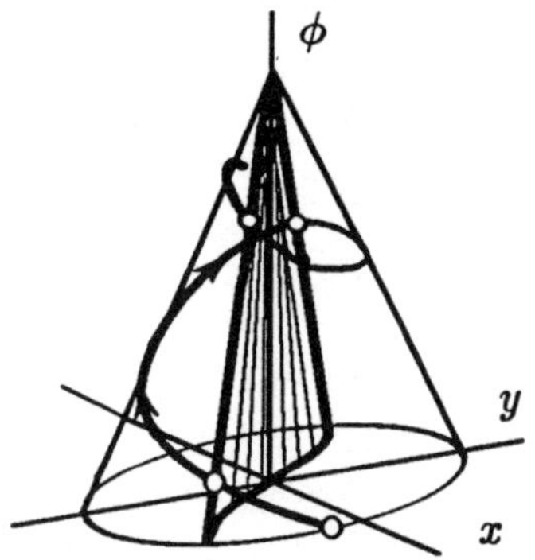

FIG. 17: CHATTERING SYNTHESIS IN $\mathbb{R}^3$

The second example of chattering synthesis is provided by the time–optimal problem for the following third order control system.

PROBLEM 6.2. $T \to \min$

subject to

$$\dot{x} = y, \quad \dot{y} = u, \quad \dot{z} = e^{-\alpha x^2},$$

$$u \in [-1, 1], \quad \alpha > 0,$$

with boundary conditions

$$x(0) = x_0, \quad y(0) = y_0, \quad z(0) = z_0 < 1,$$
$$x(T) = y(T) = 0, \quad z(T) = 1.$$

Pontryagin's maximum principle gives

$$\dot{\psi}_1 = 2\alpha x \psi_3 e^{-\alpha x^2}, \qquad \dot{x} = y,$$
$$\dot{\psi}_2 = \psi_1, \qquad\qquad \dot{y} = u,$$
$$\dot{\psi}_3 = 0, \qquad\qquad \dot{z} = e^{-\alpha x^2},$$

$$u = \operatorname{sgn} \psi_2.$$

This can readily be proved that $\psi_3 \neq 0$ on solutions to Problem 6.2, so we put $\psi_3 = 1/(2\alpha)$. The manifold of singular solutions is given by the equations $\psi_2 = \psi_1 = x = y = 0$. By means of change of variables

$$z_1 = \psi_2, \quad z_2 = -\psi_1, \quad z_3 = -x, \quad z_4 = -y, \quad w = z,$$

the system of the maximum principle can be derived to the form

$$\dot{z}_1 = z_2, \qquad\qquad \dot{z}_4 = -u,$$
$$\dot{z}_2 = z_3 + z_3(e^{-\alpha z_3^2} - 1), \quad \dot{w} = e^{-\alpha z_3^2},$$
$$\dot{z}_3 = z_4, \qquad\qquad u = \operatorname{sgn} z_1.$$

Since the projection of the singular manifold into (x, y, z)–space is one–dimensional, all the conditions of Theorems 3.1 and 3.3 are satisfied, and the optimal control for Problem 6.2 is chattering. The one–parameter family of chattering arcs hitting a point of the straight line $x = y = 0$ constitutes a two–dimensional piecewise smooth surface. It follows from Lemma 3.5 that the switching surface of all the chattering trajectories is a two–dimensional C^1–surface.

6.2 Stabilization of a Rigid Body

Consider the control problem of rotation of a rigid body around its gravity center (Eulerian case)

$$J_{11}\dot{x} = (J_{22} - J_{33})yz + u_1 a_1,$$
$$J_{22}\dot{y} = (J_{33} - J_{11})xz + u_2 a_2, \qquad (6.4)$$
$$J_{33}\dot{z} = (J_{11} - J_{22})xy + u_3 a_3.$$

Here $(x, y, z) \in \mathbb{R}^3$ are three components of the angle velocity of the body, $(J_{11}, J_{22}, J_{33}) \in \mathbb{R}^3$ are three principal moments of inertia, $(u_1, u_2, u_3) \in \mathbb{R}^3$ are independent components of control, a_1, a_2, a_3 are some positive constants. The controls are bounded:

$$|u_1| \leqslant 1, \quad |u_2| \leqslant 1, \quad |u_3| \leqslant 1.$$

The Case of an Axially Symmetric Body $(J_{22} = J_{33})$.

The scale transformation $x = \tilde{x}J_{22}/(J_{22} - J_{11})$ leads to the system

$$\begin{aligned}
\dot{x} &= u_1 b_1, \\
\dot{y} &= xz + u_2 b_2, \\
\dot{z} &= -xy + u_3 b_3.
\end{aligned} \tag{6.5}$$

For brevity, we assume that $b_1 = b_2 = b_3 = 1$. Consider the following

    PROBLEM 6.3. **Stabilization of an axially symmetric rigid body.**

$$T \to \min$$

subject to (6.5) and boundary conditions

$$\begin{aligned}
x(0) &= x_0, \quad y(0) = y_0, \quad z(0) = z_0, \\
x(T) &= y(T) = z(T) = 0.
\end{aligned}$$

The existence of a chattering mode for the symmetric problem was proved in [M.Z. Borschevskii, I.V. Ioslovich, 1985]. Our technique allows us to prove that the optimal chattering synthesis is piecewise smooth.

    Let us consider the system of Pontryagin's maximum principle with $\psi_0 = 1$ and check the order of singular solutions of the system. We have

$$H = \psi_1 u_1 + \psi_2(xz + u_2) + \psi_3(-xy + u_3) \equiv 1,$$

$$\begin{aligned}
\dot{\psi}_1 &= -\psi_2 z + \psi_3 y, \\
\dot{\psi}_2 &= \psi_3 x, \\
\dot{\psi}_3 &= -\psi_2 x.
\end{aligned} \tag{6.6}$$

    The maximum condition gives

$$u_i = \operatorname{sgn} \psi_i \quad \text{if } \psi_i \neq 0 \quad (i = 1, 2, 3).$$

Consider u_1–singular solutions such that $\psi_1(t) \equiv 0$ and $\psi_2(t) \neq 0$, $\psi_3(t) \neq 0$ at $t \in (t_0, t_1)$. It follows that the components u_2, u_3 of the control are constant. In this case we have

$$\frac{d\psi_1}{dt} = -\psi_2 z + \psi_3 y = 0,$$

$$\frac{d^2\psi_1}{dt^2} = -\psi_2(-xy + \mathrm{sgn}\,\psi_3) + \psi_3(xz + \mathrm{sgn}\,\psi_2) - \psi_3 xz + (-\psi_2 x)y$$

$$= -\psi_2\,\mathrm{sgn}\,\psi_3 + \psi_3\,\mathrm{sgn}\,\psi_2 \equiv 0,$$

$$\frac{d^3\psi_1}{dt^3} = -\psi_3 x\,\mathrm{sgn}\,\psi_3 - \psi_2 x\,\mathrm{sgn}\,\psi_2 = -\left(|\psi_3| + |\psi_2|\right) x \equiv 0,$$

$$\frac{d^4\psi_1}{dt^4} = -\left(|\psi_3| + |\psi_2|\right) u_1 - \left(-\psi_2\,\mathrm{sgn}\,\psi_3 + \psi_3\,\mathrm{sgn}\,\psi_2\right) x^2.$$

It follows that the singular solution meets the equation

$$\psi_1 = 0, \qquad |\psi_2| = |\psi_3|, \tag{6.7}$$
$$\psi_2 z = \psi_3 y, \qquad x = 0.$$

The component u_1 of the control is zero, $u_1 \equiv 0$. We see that u_1–singular solutions have second intrinsic order and the conditions of Theorem 3.1, on bundles, are satisfied. This yields the existence of the chattering structure in the six–dimensional extended space in some neighborhood of the two–dimensional surface (6.7). The projection of (6.7) into (x, y, z)–space is the curve

$$\left|\frac{z}{y}\right| = 1, \quad x = 0.$$

Consider the branch $z = y$ of the singular curve. To generate a Lagrangian manifold with chattering arcs (of dimension 3), let us define the field of ψ–variables on the curve

$$x = 0, \quad z = y$$

as follows:

$$\psi_1 = 0, \quad |\psi_2| = |\psi_3|, \quad |\psi_2| + |\psi_3| = 1$$

(the latter condition implies that $H \equiv 1$). Hence,

$$\psi_2 = \pm\frac{1}{2}, \quad \psi_3 = \pm\frac{1}{2}.$$

If $y_0 > 0$, then to hit the endpoint along the singular solution one needs $u_2 = -1$, $u_1 = -1$, and hence

$$\psi_2 = -\frac{1}{2}, \quad \psi_3 = -\frac{1}{2}.$$

Now conditions of the chattering optimality theorem (Theorem 3.3) are trivially fulfilled. Indeed, we put

$$
\begin{aligned}
Z_1 &= \psi_1, & W_1 &= -(\psi_2 u_2 + \psi_3 u_3), \\
Z_2 &= \psi_3 y - \psi_2 z, & W_2 &= \psi_3 y + \psi_2 z. \\
Z_3 &= \psi_3 u_2 - \psi_2 u_3, \\
Z_4 &= -x(\psi_2 u_2 + \psi_3 u_3),
\end{aligned}
$$

System (6.5)–(6.6) yields

$$
\begin{aligned}
\dot{Z}_1 &= Z_2, & \dot{Z}_4 &= -Z_3 \frac{Z_4^2}{W_1^2} + u_1 W_1, \\
\dot{Z}_2 &= Z_3, & \dot{W}_1 &= -\frac{Z_3 Z_4}{W_1}, \\
\dot{Z}_3 &= Z_4, & \dot{W}_2 &= \frac{2Z_4(u_2 u_3(Z_3^2 + W_1^2) + 2Z_3 W_1 W_2)}{W_1(Z_3^2 - W_1^2)} - u_2 u_3 W_1.
\end{aligned}
$$

Let us check the regularity of the projection of the chattering bundle. Denote by Π the projection of the six–dimensional vectors to (x, y, z)–space. Since the singular manifold in (x, y, z)–space is a curve, in view of Corollary 3.1, it is enough to verify that vectors

$$
D\pi \cdot \left(\frac{\partial}{\partial Z_3}\right), \ D\pi \cdot \left(\frac{\partial}{\partial Z_4}\right), \ D\pi \cdot \left(\frac{\partial}{\partial W_1}\right) = \lambda D\pi \cdot \left(\frac{\partial}{\partial W_2}\right)
$$

are linearly independent at points of the singular manifold.

One can check that

$$
\begin{aligned}
x &= \frac{Z_4}{W_1}, & \psi_1 &= Z_1, \\
y &= \frac{Z_2 + W_2}{u_2 Z_3 - u_3 W_1}, & \psi_2 &= -\frac{u_3 Z_3 + u_2 W_1}{2}, \\
z &= \frac{Z_2 - W_2}{u_3 Z_3 + u_2 W_1}, & \psi_3 &= \frac{u_2 Z_3 - u_3 W_1}{2}.
\end{aligned}
$$

Hence,

$$
\begin{aligned}
D\pi \cdot \left(\frac{\partial}{\partial Z_3}\right) &= \frac{\partial x}{\partial Z_3}\frac{\partial}{\partial x} + \frac{\partial y}{\partial Z_3}\frac{\partial}{\partial y} + \frac{\partial z}{\partial Z_3}\frac{\partial}{\partial z} \\
&= \frac{W_2}{W_1^2}\left(-u_2\frac{\partial}{\partial y} + u_3\frac{\partial}{\partial z}\right).
\end{aligned}
$$

Similarly,

$$D\pi \cdot \left(\frac{\partial}{\partial Z_4}\right) = \frac{1}{W_1}\frac{\partial}{\partial x},$$
$$D\pi \cdot \left(\frac{\partial}{\partial W_1}\right) = \frac{W_2}{W_1^2}\left(u_3\frac{\partial}{\partial y} + u_2\frac{\partial}{\partial z}\right).$$

It is readily seen that these vectors constitute a basis of $\mathbb{R}^3$, hence the conditions of Theorem 3.4, on projection, are satisfied.

It must be proved separately that any optimal solution to the problem of interest is to enter into the curve $x = 0$, $z = y$ at a small time τ if the initial points are sufficiently close to this curve. The time τ is continuous in initial data. For details, see [M.Z. Borschevckii, I.V. Ioslovich, 1985] or the robot control problem below. It follows that chattering fibrations are locally optimal to the problem of stabilization of a rigid body. The behavior of trajectories in the vicinity of the branch $x = 0$, $z = -y$ is the same.

The General (Asymmetric) Case

Now let us consider the general case of the system (6.4)

$$\begin{aligned}
\dot{x} &= A_1 yz + u_1 a_1, \\
\dot{y} &= A_2 xz + u_2 a_2, \\
\dot{z} &= A_3 xy + u_3 a_3,
\end{aligned} \qquad (6.8)$$

$$A_1 = \frac{J_{22} - J_{33}}{J_{11}}, \; A_2 = \frac{J_{33} - J_{11}}{J_{22}}, \; A_3 = \frac{J_{11} - J_{22}}{J_{33}},$$

$$|u_i| \leqslant 1, \quad a_i > 0 \quad (i = 1,2,3).$$

Without loss of generality we can assume that

$$J_{22} > J_{33} > J_{11}, \quad \text{i.e.,} \quad A_1 > 0, A_2 > 0, A_3 < 0.$$

LEMMA 6.1. *System (6.8) can be reduced to the system*

$$\begin{aligned}
\dot{x} &= \beta^2 yz + u_1 b_1, \\
\dot{y} &= \quad xz + u_2 b_2, \quad (b_i > 0). \\
\dot{z} &= -xy + u_3 b_3,
\end{aligned} \qquad (6.9)$$

Proof. Let us introduce the scale parameters λ, μ, ν for the angular velocities

$$x = \lambda\widetilde{x}, \quad y = \mu\widetilde{y}, \quad z = \nu\widetilde{z}. \tag{6.10}$$

To obtain (6.9) we put

$$\frac{A_1\mu\nu}{\lambda} = \beta^2, \quad A_2\frac{\nu\lambda}{\mu} = 1, \quad A_3\frac{\lambda\mu}{\nu} = -1. \tag{6.11}$$

Multiplication of two first equations of (6.11) gives $\nu^2 = \beta^2/(A_1 A_2)$. Similarly, $\lambda^2 = -1/(A_2 A_3)$, $\mu^2 = -\beta^2/(A_1 A_3)$. So the change of variables (6.10) with the values

$$\lambda = \frac{1}{\sqrt{-A_2 A_3}}, \quad \mu = \frac{\beta}{\sqrt{-A_1 A_3}}, \quad \nu = \frac{\beta}{\sqrt{A_1 A_2}}$$

gives (6.9).   **Q.E.D.**

Hence, the general case is reduced to the following

PROBLEM 6.4.   *Minimize the time T of hitting the origin*

$$x(T) = y(T) = z(T) = 0$$

starting from the point

$$x(0) = x_0, \quad y(0) = y_0, \quad z(0) = z_0$$

subject to (6.9) under a measurable bounded control (u_1, u_2, u_3), $|u_i| \leqslant 1$, $i = 1, 2, 3$.

Pontryagin's maximum principle gives

$$H = \psi_1\beta^2 yz + \psi_2 xz - \psi_3 xy + \psi_1 u_1 b_1 + \psi_2 u_2 b_2 + \psi_3 u_3 b_3,$$

$$u_i = \operatorname{sgn} \psi_i, \quad \text{if} \quad \psi_i \neq 0 \quad (i = 1, 2, 3.)$$

$$\begin{aligned}
\dot{\psi}_1 &= \psi_3 y - \psi_2 z, \\
\dot{\psi}_2 &= \psi_3 x - \psi_1\beta^2 z, \\
\dot{\psi}_3 &= -\psi_2 x - \psi_1\beta^2 y.
\end{aligned}$$

Let us find u_1–singular solutions. We have

$$\psi_1 = 0. \tag{6.12}$$

Assume that $\psi_2 \neq 0$ and $\psi_3 \neq 0$ at $t \in (t_0, t_1)$ (it follows that the components u_2 and u_3 of the control are constant). While differentiating the identity (6.12) we successively introduce the new variables

$$Z_1 = \psi_1,$$

$$\dot{Z}_1 = \psi_3 y - \psi_2 z,$$

$$Z_2 = \psi_3 y - \psi_2 z,$$

$$\dot{Z}_2 = u_2 b_2 \psi_3 - u_3 b_3 \psi_2 + Z_1 \beta^2 (z^2 - y^2),$$

$$Z_3 = u_2 b_2 \psi_3 - u_3 b_3 \psi_2,$$

$$\dot{Z}_3 = -u_2 b_2 \psi_2 x - u_3 b_3 \psi_3 x + Z_1 \beta^2 (u_3 b_3 z - u_2 b_2 y),$$

$$Z_4 = -u_2 b_2 \psi_2 x - u_3 b_3 \psi_3 x$$

$$= -x(b_2|\psi_2| + b_3|\psi_3|),$$

$$\dot{Z}_4 = -(b_1 u_1 + \beta^2 yz)(b_2|\psi_2| + b_3|\psi_3|)$$
$$+ Z_1 \beta^2 x(u_3 b_3 y + u_2 b_2 z) - x^2 Z_3.$$

The singular trajectory has the (intrinsic) order 2 and satisfies the equations

$$S = \left\{ \begin{array}{ll} \psi_1 = 0, & \psi_3 y - \psi_2 z = 0, \\ b_2|\psi_3| - b_3|\psi_2| = 0, & x = 0 \end{array} \right\}.$$

The singular component of the control is

$$u_1 = -\frac{\beta^2 yz}{b_1}.$$

To meet the constraint $|u_1| \leqslant 1$ we need $|yz| \leqslant b_1/\beta^2$.

To complete the system of variables Z_i, $i = 1, 2, 3, 4$, we can take $W_1 = -b_2 u_2 \psi_2 - b_3 u_3 \psi_3$, $W_2 = \psi_3 y + \psi_2 z$. One can check that

$$x = \frac{Z_4}{W_1}, \qquad\qquad\qquad \psi_1 = Z_1,$$

$$y = \frac{Z_2 + W_2}{2} \cdot \frac{b_3^2 + b_2^2}{b_2 u_2 Z_3 - b_3 u_3 W_1}, \qquad \psi_2 = -\frac{b_3 u_3 Z_3 + b_2 u_2 W_1}{b_3^2 + b_2^2},$$

$$z = \frac{Z_2 - W_2}{2} \cdot \frac{b_3^2 + b_2^2}{b_3 u_3 Z_3 + b_2 u_2 W_1}, \qquad \psi_3 = \frac{b_2 u_2 Z_3 - b_3 u_3 W_1}{b_3^2 + b_2^2}.$$

Let us verify the conditions of Theorem 3.4, on the regular projection. The singular manifold in (x, y, z)–space is given by the equations

$$x = 0, \qquad y = -\frac{b_3^2 + b_2^2}{2 b_3 u_3} \cdot \frac{W_2}{W_1}, \qquad z = -\frac{b_3^2 + b_2^2}{2 b_2 u_2} \cdot \frac{W_2}{W_1}.$$

Eliminating the (W_1, W_2)–terms between the last two equations, we obtain

$$b_3 u_3 y = b_2 u_2 z.$$

It is readily seen that to approach the origin along the singular arc one must set $u_2 = -\mathrm{sgn}\, y, \quad u_3 = -\mathrm{sgn}\, z,$ so we have

$$\left| \frac{z}{y} \right| = \frac{b_3}{b_2}.$$

Since the singular manifold is one–dimensional, we are left with verifying whether the vectors

$$D\pi \cdot \left(\frac{\partial}{\partial Z_3} \right),\; D\pi \cdot \left(\frac{\partial}{\partial Z_4} \right),\; D\pi \cdot \left(\frac{\partial}{\partial W_1} \right) = \lambda D\pi \cdot \left(\frac{\partial}{\partial W_2} \right)$$

are independent (π being the projection of the six–dimensional vectors to (x, y, z)–space at S). We have

$$D\pi \cdot \left(\frac{\partial}{\partial Z_3} \right) = \frac{b_3^2 + b_2^2}{2} \frac{W_2}{W_1} \left(-\frac{b_2 u_2}{b_3^2} \right) \frac{\partial}{\partial z},$$

$$D\pi \cdot \left(\frac{\partial}{\partial Z_4} \right) = \frac{1}{W_1} \frac{\partial}{\partial x},$$

$$D\pi \cdot \left(\frac{\partial}{\partial W_1} \right) = \frac{b_3^2 + b_2^2}{2} \frac{W_2}{W_1^2} \left(\frac{1}{b_3 u_3} \frac{\partial}{\partial y} + \frac{1}{b_2 u_2} \frac{\partial}{\partial z} \right).$$

We see that $D\pi \cdot (\partial/\partial Z_3)$ is not proportional to $D\pi \cdot (\partial/\partial W_1)$. It follows from Corollary 3.1 that Theorems 3.1, 3.3 and 3.4 are applicable.

6.3 The Resource Allocation Problem

Let us consider the problem of the optimal resource allocation in an economical model of the controllable scientific and technological progress. For simplicity, assume that there are exactly two factors that influence the development of the production. The first one can be measured by the quantity x_1, and the second one by the quantity x_2. The flow of the product (the velocity of the production) depends on x_1, x_2. The flow is denoted by the function $Q(x_1, x_2)$. This function is called a production function of the model. Assume that a fixed part $sQ(x_1, x_2)$ of the flow of the product is invested in expansion of the production. A portion $usQ(x_1, x_2)$ of this part is invested in the production of factor x_1, and the remaining portion $(1 - u)sQ(x_1, x_2)$ is invested in the production of factor x_2. Here the control u is subjected to the restriction $u \in [0, 1]$. As usual in the continuous–time models of this sort, we presuppose the instantaneous turnover of funds.

There exist models simulating the situations in which the investments are assigned to an extensive growth of the production (see, e.g., [L.Ph. Zelikina, 1985]. In these models, control influences the first derivative of the factors x_1, x_2. We aim to consider a situation in which the investments are assigned to improve the production in an intensive way by changing technologies (the model of the scientific and technological progress). In our model we use the following suggestion. If there are no investments, then the growth rate of the corresponding factor x_i remains constant, i.e., the second derivative of x_i equals zero. Otherwise, $\ddot{x}_i$ is proportional to the value of the investment.

Thus, we consider the system

$$\ddot{x}_1 = usQ(x_1, x_2),$$
$$\ddot{x}_2 = (1 - u)sQ(x_1, x_2), \qquad u \in [0, 1].$$

Since the coefficient s is a fixed number, it will be included below in the function Q itself. As a result, we obtain

$$\ddot{x}_1 = uQ(x_1, x_2),$$
$$\ddot{x}_2 = (1 - u)Q(x_1, x_2), \qquad u \in [0, 1], \tag{6.13}$$

$$x_1 \geqslant 0, \quad x_2 \geqslant 0, \quad \dot{x}_1 \geqslant 0, \quad \dot{x}_2 \geqslant 0.$$

The production function Q is subjected to the following natural constraints:

$$Q(x_1, x_2) > 0, \quad \frac{\partial Q}{\partial x_1}(x_1, x_2) > 0, \quad \frac{\partial Q}{\partial x_2}(x_1, x_2) > 0.$$

The inequalities guarantee that the bounds $x_1 > 0$, $x_2 > 0$, $\dot{x}_1 > 0$, $\dot{x}_2 > 0$ hold at any positive instant in the case that they hold at the initial instant. So, in fact, our problem is not with the phase restrictions.

Let us consider the time optimal problem of hitting the target manifold $\mathcal{M} \subset \mathbb{R}_+^4$ from the initial point $x_i(0) = x_{i0}$, $\dot{x}_i(0) = \dot{x}_{i0}$ $(i = 1, 2)$ along a trajectory of system (6.13).

PROBLEM 6.5. $T \to \min$ *subject to*

$$\dot{x}_1 = y_1, \qquad\qquad \dot{x}_2 = y_2,$$
$$\dot{y}_1 = u_1 Q(x_1, x_2), \qquad \dot{y}_2 = u_2 Q(x_1, x_2),$$

$$u_1 \geqslant 0, \quad u_2 \geqslant 0, \quad u_1 + u_2 = 1,$$

$$x_i(0) = x_{i0}, \quad y_i(0) = y_{i0}, \quad i = 1, 2$$

with boundary conditions

$$\big(x_1(T), y_1(T), x_2(T), y_2(T)\big) \in \mathcal{M}.$$

The particular case of Problem 6.5 where $Q = x_1 x_2$ was analyzed in Chapter 1. Now we shall prove that singular solutions to Problem 6.5 have second intrinsic order and hence the conditions of Theorems 3.1–3.3 are satisfied.

Let us set

$$x = (x_1 - x_2)/2, \qquad p = (x_1 + x_2)/2,$$
$$y = (y_1 - y_2)/2, \qquad q = (y_1 + y_2)/2,$$
$$u = u_1 - u_2.$$

In the new coordinates we have

$$
\begin{aligned}
\dot{x} &= y, & \dot{p} &= q, \\
\dot{y} &= \frac{u}{2} Q(p + x, p - x), & \dot{q} &= \frac{1}{2} Q(p + x, p - x), \\
& & u &\in [-1, 1].
\end{aligned}
\tag{6.14}
$$

Set $F(x, p) = \frac{1}{2} Q(p + x, p - x)$. Pontryagin's maximum principle gives

$$H = \psi_1 y + u\psi_2 F(x, p) + \psi_3 q + \psi_4 F(x, p) - 1 \equiv 0,$$
$$\dot{\psi}_1 = -(\psi_4 + u\psi_2) F'_x, \qquad \dot{\psi}_3 = -(\psi_4 + u\psi_2) F'_p,$$
$$\dot{\psi}_2 = -\psi_1, \qquad\qquad\quad \dot{\psi}_4 = -\psi_3,$$

where

$$u = \begin{cases} 1, & \text{if } \psi_2 > 0, \\ -1, & \text{if } \psi_2 < 0, \end{cases}$$

(recall that $F > 0$).

If $\psi_2(t) \equiv 0$ on an interval (t_0, t_1), then it follows that $\psi_1(t) \equiv 0$ and $\psi_4 F'_x \equiv 0$. In the case $\psi_4(t) \equiv 0$, we have $\psi_3(t) \equiv 0$, which contradicts Pontryagin's maximum principle. Hence, $F'_x \equiv 0$. The equation does not contain the control variable u in an explicit form, so the order of the singular extremal is greater than 1. If we differentiate this identity with respect to t twice, we successively obtain

$$F''_{xx} y(t) + F''_{xp} q(t) \equiv 0$$

and

$$F'''_{xxx}\, y^2(t) + 2F'''_{xxp}\, y(t)q(t) + F'''_{xpp}\, q^2(t) + F''_{xx}\, uF + F''_{xp}\, F \equiv 0.$$

Assume that $F''_{xx} \neq 0$. If we let

$$\Delta = F'''_{xxx}y^2 + 2F'''_{xxp}yq + F'''_{xpp}q^2,$$

then

$$\widehat{u} = -\frac{\Delta + F''_{xp}F}{F''_{xx}F}.$$

Hence,

$$\frac{\partial}{\partial u}\frac{d^k}{dt^k}\Big(\psi_2 F(x,p)\Big) \equiv 0, \quad k = 0,1,2,3,$$

$$\frac{\partial}{\partial u}\frac{d^4}{dt^4}\Big(\psi_2 F(x,p)\Big) = \psi_4 F''_{xx}F^2 \quad (\mathrm{mod}\ \psi_1 = \psi_2 = x = y = 0),$$

and the singular extremal has second (local) order. To fulfill Kelley's condition, we need $\psi_4 F''_{xx} < 0$. To fulfill the restriction $u \in [-1,1]$, we need

$$\big|\Delta + F''_{xp}F\big| \leqslant \big|F''_{xx}F\big|.$$

A wide class of functions meets these conditions, e.g., the function $F_0 = p^2 - x^2$ and any function in some open neighborhood of F_0 in C^3-topology.

The singular manifold S_0 in Problem 6.5 is given by the equations

$$\psi_2 = 0, \quad \psi_1 = 0, \quad F'_x = 0, \quad F''_{xx}y + F''_{xp}q = 0.$$

To ensure that S_0 is a smooth submanifold in $\mathbb{R}^4$, we need

$$\mathrm{rk}\ \frac{D(F'_x, F''_{xx}y + F''_{xp}q)}{D(x,y,p,q)} = 2,$$

i.e.,

$$\mathrm{rk}\ \begin{pmatrix} F''_{xx} & 0 & F''_{xp} & 0 \\ F'''_{xxx}y + F'''_{xxp}q & F''_{xx} & F'''_{xxp}y + F'''_{xpp}q & F''_{xp} \end{pmatrix} = 2.$$

By the assumption, we have $F''_{xx} \neq 0$ and the first two columns of the matrix are independent. In particular, the values ψ_3, ψ_4, p, q can be taken as parameters on S_0. Let us reduce the Hamiltonian system for Problem 6.5 to the semi–canonical form. We have

$$z_1 = \psi_2 F,$$

$$\dot{z}_1 = -\psi_1 F + \psi_2(F'_x y + F'_p q),$$

$$z_2 = -\psi_1 F,$$

$$\dot{z}_2 = (\psi_4 + u\psi_2)F'_x F - \psi_1(F'_x y + F'_p q),$$

$$z_3 = \psi_4 F'_x F,$$

$$\dot{z}_3 = -\psi_3 F'_x F + \psi_4(F''_{xx} y + F''_{xp} q)F + \psi_4 F'_x (F'_x y + F'_p q),$$

$$z_4 = \psi_4(F''_{xx} y + F''_{xp} q)F,$$

$$\dot{z}_4 = -\psi_3(F''_{xx} y + F''_{xp} q)F + \psi_4(F'''_{xxx} y^2 + 2F'''_{xxp} yq + F'''_{xpp} q^2)F$$

$$+ \psi_4(F''_{xx} y + F''_{xp} q)(F'_x y + F'_p q) + \psi_4(F''_{xx} uF + F''_{xp} F)F.$$

Let us complement the functions z_1, z_2, z_3, z_4 with $w_1 = \psi_3$, $w_2 = \psi_4$, $w_3 = p$, $w_4 = q$. At points of the singular manifold S_0 the Jacobian of the change equals

$$\mathcal{J} = -F^4 \psi_4^2 \left(F''_{xx}\right)^2.$$

Now the conditions of Theorem 3.1 are satisfied and yield the existence of the chattering extremals for Problem 6.5. To prove optimality we are to design a Lagrangian manifold with the chattering trajectories.

Assume that the target manifold $\mathcal{M}$ belongs to the projection of S_0 into (x, y, p, q)–space, i.e., that $F'_x \equiv 0$, $F''_{xx} y + F''_{xp} q \equiv 0$ on $\mathcal{M}$. Let us take as $\mathcal{M}$ a one–dimensional manifold,

$$q = f(p),$$

with $f(\,\cdot\,)$ having been some C^3–function which will be chosen below. Let us take the manifold $\mathcal{M}$ to be transversal to the projection of the singular trajectories. The tangent vectors to $\mathcal{M}$ are directed along the vector

$$v_1 = \left(\frac{\partial x}{\partial q}\frac{df}{dp} + \frac{\partial x}{\partial p}, \ \frac{\partial y}{\partial q}\frac{df}{dp} + \frac{\partial y}{\partial p}, \ 1, \ \frac{df}{dp}\right),$$

where

$$\frac{\partial x}{\partial p} = -\frac{F''_{xp}}{F''_{xx}} = \frac{y}{q}; \quad \frac{\partial x}{\partial q} = 0;$$

$$\frac{\partial y}{\partial p} = -\frac{1}{F''_{xx}}\left(\left(F'''_{xxx}\frac{\partial x}{\partial p} + F'''_{xxp}\right)y + \left(F'''_{xxp}\frac{\partial x}{\partial p} + F'''_{xpp}\right)q\right)$$

$$= -\frac{1}{F''_{xx}}\left(F'''_{xxx}\frac{y^2}{q} + 2F'''_{xxp}y + F'''_{xpp}q\right) = -\frac{\Delta}{qF''_{xx}};$$

$$\frac{\partial y}{\partial q} = -\frac{F''_{xp}}{F''_{xx}} = \frac{y}{q}.$$

Thus,

$$v_1 = \left(\frac{y}{q}, \frac{y}{q}f' - \frac{\Delta}{qF''_{xx}}, 1, f'\right).$$

The (x, y, p, q)–component of the velocity on the singular trajectories equals

$$v_2 = \left(y, \widehat{u}F, q, F\right)$$

where

$$\widehat{u} = -\frac{1}{F''_{xx}F}\left(\Delta + F''_{xp}F\right) = -\frac{\Delta}{F''_{xx}F} + \frac{y}{q}.$$

To provide transversality, the vectors v_1 and v_2 are required not to be proportional. It is enough to choose the function f such that

$$\frac{1}{q} \neq \frac{f'}{F}, \quad \text{i.e.,} \quad qf' - F \neq 0.$$

Now the manifold $\mathcal{M}$ is to be lifted into the extended $\mathbb{R}^8$–space in such a way that the image, call it $\mathcal{N}$, would belong to both the zero–level surface of the Hamiltonian and the singular manifold S_0. Besides that, the differential form $\psi_1\,dx + \psi_2\,dy + \psi_3\,dp + \psi_4\,dq$ needs to be identically zero on every tangent vector to $\mathcal{N}$. This gives $\psi_3 + \psi_4 f' = 0$. We have

$$F'_x = 0, \quad F''_{xx}p + F''_{xp}q = 0, \quad q = f(p),$$

$$\psi_1 = \psi_2 = 0, \quad \psi_3 q + \psi_4 F = 1, \quad \psi_3 + \psi_4 f' = 0.$$

If one resolves the system with respect to ψ_3, ψ_4, he obtains

$$\psi_3 = \frac{-f'}{F - qf'}, \quad \psi_4 = \frac{1}{F - qf'}.$$

Kelley's condition $\psi_4 F''_{xx} < 0$, gives the inequality $F''_{xx}/(F - qf') < 0$. To apply the chattering optimality theorem 3.3 we are to verify the regular projecting of the chattering bundles to (x, y, p, q)–space. Let $D\pi$ be the projection of an eight–dimensional vector to (x, y, p, q)–space at points of S_0. We have

$$D\pi \cdot \left(\frac{\partial}{\partial z_3}\right) = \frac{\partial x}{\partial z_3}\frac{\partial}{\partial x} + \frac{\partial y}{\partial z_3}\frac{\partial}{\partial y} + \frac{\partial p}{\partial z_3}\frac{\partial}{\partial p} + \frac{\partial q}{\partial z_3}\frac{\partial}{\partial q}.$$

It follows from the relations $w_3 = p$, $w_4 = q$ that

$$\frac{\partial p}{\partial z_3} = \frac{\partial q}{\partial z_3} = 0.$$

Differentiating the identity $z_3 = w_2 F_x' F$ and $z_4 = w_2(F_{xx}'' y + F_{xp}'' q)$ with respect to z_3 at $z = 0$, we obtain

$$\begin{cases} 1 = w_2 F_{xx}'' \dfrac{\partial x}{\partial z_3} F, \\[2ex] 0 = w_2 \left(F_{xxx}''' \dfrac{\partial x}{\partial z_3} y + F_{xx}'' \dfrac{\partial y}{\partial z_3} + F_{xxp}''' \dfrac{\partial x}{\partial z_3} q \right) F. \end{cases}$$

Hence

$$\frac{\partial x}{\partial z_3} = \frac{1}{w_2 F_{xx}'' F},$$

$$\frac{\partial y}{\partial z_3} = -\frac{F_{xxx}''' y + F_{xxp}''' q}{w_2 F (F_{xx}'')^2}.$$

Thus

$$D\pi \cdot \left(\frac{\partial}{\partial z_3} \right) = \frac{1}{w_2 F_{xx}'' F} \left\{ \frac{\partial}{\partial x} - \frac{F_{xxx}''' y + F_{xxp}''' q}{F_{xx}''} \frac{\partial}{\partial y} \right\}.$$

Similarly,

$$D\pi \cdot \left(\frac{\partial}{\partial z_4} \right) = \frac{1}{w_2 F F_{xx}''} \frac{\partial}{\partial y}, \quad D\pi \cdot \left(\frac{\partial}{\partial w_1} \right) = D\pi \cdot \left(\frac{\partial}{\partial w_2} \right) = 0,$$

$$D\pi \cdot \left(\frac{\partial}{\partial w_3} \right) = \frac{\partial}{\partial p}, \qquad D\pi \cdot \left(\frac{\partial}{\partial w_4} \right) = \frac{\partial}{\partial q}.$$

In view of Corollary 3.1, the manifold $\mathcal{N}$ satisfies all the conditions of Theorem 3.3 and therefore generates a locally optimal synthesis for the resource allocation problem.

The simplest analysis of the solution indicates that analogous results can be obtained for the following $(n + 2)$–dimensional generalization of system (6.14).

PROBLEM 6.6.

$$T \rightarrow \min$$

subject to

$$\begin{aligned} \dot{x} &= y, \\ \dot{y} &= \alpha(x, y, r) + u\beta(x, y, r), \\ \dot{r} &= \gamma(x, r), \end{aligned} \tag{6.15}$$

with boundary conditions

$$x(0) = x_0, \quad y(0) = y_0, \quad r(0) = r_0,$$
$$\big(x(T), y(T), r(T)\big) \in \mathcal{M},$$

where $x, y \in \mathbb{R}$, $r \in \mathbb{R}^n$, $u \in [-1, 1]$, $\alpha, \beta : \mathbb{R}^{n+2} \to \mathbb{R}^1$, $\gamma : \mathbb{R}^{n+1} \to \mathbb{R}^n$.

Functions α, β, γ are smooth enough. We use the notation

$$r = \begin{pmatrix} r_1 \\ \vdots \\ r_n \end{pmatrix}, \quad \gamma = \begin{pmatrix} \gamma_1 \\ \vdots \\ \gamma_n \end{pmatrix}$$

for the column vectors r, γ.

Pontryagin's maximum principle gives

$$\dot{\psi}_1 = -\psi_2(\alpha_x' + u\beta_x') - \psi\gamma_x',$$
$$\dot{\psi}_2 = -\psi_2(\alpha_y' + u\beta_y') - \psi_1, \qquad (6.16)$$
$$\dot{\psi} = -\psi_2(\alpha_r' + u\beta_r') - \psi\gamma_r',$$

$$u = \operatorname{sgn}(\beta\psi_2).$$

Here $\psi_1, \psi_2 \in \mathbb{R}^1$, $\psi = (\psi^1, \ldots, \psi^n) \in (\mathbb{R}^n)^*$ (ψ is a row-vector),
$\alpha_r' = \left(\dfrac{\partial\alpha}{\partial r_1}, \ldots, \dfrac{\partial\alpha}{\partial r_n} \right)$, $\beta_r' = \left(\dfrac{\partial\beta}{\partial r_1}, \ldots, \dfrac{\partial\beta}{\partial r_n} \right)$,

$$\gamma_r' = \begin{pmatrix} \dfrac{\partial\gamma_1}{\partial r_1} & \cdots & \dfrac{\partial\gamma_1}{\partial r_n} \\ \vdots & \ddots & \vdots \\ \dfrac{\partial\gamma_n}{\partial r_1} & \cdots & \dfrac{\partial\gamma_n}{\partial r_n} \end{pmatrix} \qquad (\gamma_r' \text{ is } (n \times n))\text{--matrix}.$$

Let us consider the region where $\beta \neq 0$ and look for singular solutions of system (6.15)–(6.16). If we put $z_1 = \beta\psi_2$ and differentiate z_1 along a solution of (6.15)–(6.16), we obtain

$$\dot{z}_1 = -\beta\psi_1 + (\beta_x' y + \beta_y'\alpha + \beta_r'\gamma - \alpha_y'\beta)\psi_2.$$

Set $z_2 = -\beta\psi_1$. Then

$$\dot{z}_2 = \beta\psi\gamma_x' - (\beta_x' y + \beta_y'(\alpha + u\beta) + \beta_r'\gamma)\psi_1 + \beta(\alpha_x' + u\beta_x')\psi_2.$$

Set $z_3 = \beta\psi\gamma'_x$. Then

$$\dot{z}_3 = \left(\beta'_x y + \beta'_y(\alpha + u\beta) + \beta'_r\gamma\right)\psi\gamma'_x + \beta\left(-\psi_2(\alpha'_r + \beta'_r) - \psi\gamma'_r\right)\gamma'_x$$
$$+ \beta\psi\left(\gamma''_{xx}y + \gamma''_{xr}\gamma\right),$$

where

$$\gamma''_{xr} = \begin{pmatrix} \dfrac{\partial^2\gamma_1}{\partial x\partial r_1} & \cdots & \dfrac{\partial^2\gamma_1}{\partial x\partial r_n} \\ \vdots & \ddots & \vdots \\ \dfrac{\partial^2\gamma_n}{\partial x\partial r_1} & \cdots & \dfrac{\partial^2\gamma_n}{\partial x\partial r_n} \end{pmatrix}.$$

Set $z_4 = \beta\psi\left(-\gamma'_r\gamma'_x + \gamma''_{xr}\gamma + \gamma''_{xx}y\right)$. If we differentiate z_4 along a solution of (6.15)–(6.16), we obtain

$$\frac{\partial}{\partial u}\frac{dz_4}{dt} = \beta^2\psi\gamma''_{xx} \quad (\mathrm{mod} \quad z_1 = z_2 = z_3 = z_4 = 0).$$

Assume that $\psi\gamma''_{xx}\big|_{S_0} < 0$, where S_0 denotes the surface $S_0 = \{z_1 = z_2 = z_3 = z_4 = 0\} \subset \mathbb{R}^{2(n+2)}$. Then

$$\mathrm{rk}\,\frac{D(z_1, z_2, z_3, z_4)}{D(x, y, r, \psi_1, \psi_2, \psi)}\bigg|_{S_0}$$

$$= \mathrm{rk}\begin{pmatrix} 0 & 0 & 0 & 0 & \beta & 0 \\ 0 & 0 & 0 & -\beta & 0 & 0 \\ \beta\psi\gamma''_{xx} & 0 & \beta\psi\gamma''_{xr} & 0 & 0 & \beta(\gamma'_x)^T \\ * & \beta\psi\gamma''_{xx} & * & 0 & 0 & * \end{pmatrix} = 4$$

(the asterisks denote the quantities of no significance). Hence, S_0 is a smooth submanifold of dimension $2n$ in $\mathbb{R}^{2(n+2)}$. If we complement the functions $z_1, \ldots, z_4$ with r and ψ up to a new basis in $\mathbb{R}^{2(n+2)}$, system (6.15)–(6.16) can be written as follows:

$$\begin{aligned} \dot{z}_1 &= z_2 + \Theta_4, & \dot{\psi} &= \Psi(z, r, \psi), \\ \dot{z}_2 &= z_3 + \Theta_3, & \dot{r} &= R(z, r, \psi) \\ \dot{z}_3 &= z_4 + \Theta_2, & u &= \mathrm{sgn}\, z_1. \\ \dot{z}_4 &= A(z, r, \psi) + u\beta^2\psi\gamma''_{xx}, \end{aligned}$$

As usual, Θ_i denote functions meeting the conditions

$$\varlimsup_{\kappa \to +0} \frac{\Theta_i\big(\mathfrak{g}_\kappa(z), r, \psi\big)}{\kappa^i} < C,$$

$$\mathfrak{g}_\kappa(z) = (\kappa^4 z_1, \kappa^3 z_2, \kappa^2 z_3, \kappa z_4).$$

In the domain $\psi\gamma''_{xx} < 0$, $\quad |A| < -\beta^2\psi\gamma''_{xx}$, all conditions of Theorem 3.1 are satisfied. Therefore, for any point $\sigma \in S_0$ there exists a one–parameter family of chattering trajectories through σ. The construction of a locally optimal synthesis for Problem 6.6 is completely analogous (with the correction in dimension) to that for Problem 6.5.

6.4 Control of Two Interdependent Oscillators

Let us consider two interdependent oscillators which define the bilinear system

$$\begin{aligned}
\dot{x}_1 &= x_2, & \dot{x}_3 &= x_4, \\
\dot{x}_2 &= -x_1 + ux_4, & \dot{x}_4 &= -x_3 - ux_2,
\end{aligned} \tag{6.17}$$

with initial conditions

$$x(0) = x_0.$$

Here u is a scalar control which transfers energy from one oscillator to another. The sign of u corresponds to the direction of the transference. Suppose that $|u| \leqslant 1$. The criterion is

$$\frac{1}{2}\int_0^\infty x_1^2(t)\,dt \to \min. \tag{6.18}$$

It means that we wish to damp the first oscillator by minimizing the mean square value of its amplitude.

Pontryagin's maximum principle (see Remark 2.1) yields

$$H = \psi_1 x_2 - \psi_2 x_1 + \psi_3 x_4 - \psi_4 x_3 + u\,(\psi_2 x_4 - \psi_4 x_2) - \frac{1}{2}x_1^2 \equiv 0;$$

$$u = \mathrm{sgn}\,(\psi_2 x_4 - \psi_4 x_2), \quad \text{for } \psi_2 x_4 - \psi_4 x_2 \neq 0;$$

$$\begin{aligned}
\dot{\psi}_1 &= \psi_2 + x_1, & \dot{\psi}_3 &= \psi_4, \\
\dot{\psi}_2 &= -\psi_1 + u\psi_4, & \dot{\psi}_4 &= -\psi_3 - u\psi_2.
\end{aligned} \tag{6.19}$$

Let us define the new variables in system (6.17),(6.19) as follows:

$$z_1 = \psi_2 x_4 - \psi_4 x_2,$$

$$\frac{dz_1}{dt} = \psi_4 x_1 - \psi_1 x_4 + \psi_3 x_2 - \psi_2 x_3,$$

$$z_2 = \psi_4 x_1 - \psi_1 x_4 + \psi_3 x_2 - \psi_2 x_3,$$

$$\frac{dz_2}{dt} = -x_1 x_4 + 2(\psi_1 x_3 - \psi_3 x_1) + 2(\psi_4 x_2 - \psi_2 x_4)$$

$$+ u(\psi_1 x_2 - \psi_2 x_1 + \psi_3 x_4 - \psi_4 x_3).$$

Later on it will be shown that, on the zero–level surface of the Hamiltonian, the term $\psi_1 x_2 - \psi_2 x_1 + \psi_3 x_4 - \psi_4 x_3$, which is equal to $\frac{1}{2} x_1^2 + |z_1|$, and the term $\psi_1 x_3 - \psi_3 x_1$ have a higher order in the sense of generalized homogeneity than the term $-x_1 x_4$. Thus we set

$$z_3 = -x_1 x_4,$$

and continue differentiating.

$$\frac{dz_3}{dt} = -x_2 x_4 + x_1 x_3 + u x_1 x_2.$$

We show below that the terms $x_1 x_3$ and $x_1 x_2$ have higher orders than the term $x_2 x_4$, and we set

$$z_4 = -x_2 x_4.$$

Hence,

$$\frac{dz_4}{dt} = x_1 x_4 - u x_4^2 + x_2 x_3 + u x_2^2$$

$$= u(x_2^2 - x_4^2) + x_1 x_4 + x_2 x_3.$$

Finally,

$$\begin{aligned}
\dot{z}_1 &= z_2, \\
\dot{z}_2 &= z_3 + 2(\psi_1 x_3 - \psi_3 x_1) - 2z_1 \\
&\qquad + u(\psi_1 x_2 - \psi_2 x_1 + \psi_3 x_4 - \psi_4 x_3), \\
\dot{z}_3 &= z_4 + x_1 x_3 + u x_1 x_2, \\
\dot{z}_4 &= -z_3 + x_2 x_3 + u(x_2^2 - x_4^2).
\end{aligned} \qquad (6.20)$$

Let us take

$$w_1 = x_3, \quad w_2 = x_4, \quad w_3 = \psi_3, \quad w_4 = \psi_4,$$

then

$$\det \frac{D(z,w)}{D(x,\psi)} = \frac{D(z)}{D(x_1,x_2,\psi_1,\psi_2)} = \begin{vmatrix} 0 & -\psi_4 & 0 & x_4 \\ \psi_4 & \psi_3 & -x_4 & -x_3 \\ -x_4 & 0 & 0 & 0 \\ 0 & -x_4 & 0 & 0 \end{vmatrix}$$

$$= x_4^4.$$

Thus, the mapping $(x,\psi) \to (z,w)$ is a diffeomorphism in the domain $x_4 \neq 0$. The manifold of singular trajectories is given by the equations

$$z_1 = z_2 = z_3 = z_4 = 0,$$

or, equivalently,

$$x_1 = x_2 = \psi_1 = \psi_2 = 0.$$

One might expect that these singular arcs are optimal because it is natural to arrest the first oscillator (i.e., to keep $x_1 = x_2 = 0$) for minimizing the functional (6.18). To apply Theorem 3.1 we are to verify that the additional terms of the right–hand sides of (6.20) have a higher order than z_i (in the sense of the generalized homogeneity $z_i \mapsto \lambda^{5-i} z_i$).

Denote by $\mathcal{H}(\psi, x) = \max_{u \in [-1,1]} H$ the Hamiltonian of Problem (6.17)–(6.19). The optimal trajectories of system (6.20) lie on the zero–level surface of the Hamiltonian,

$$\mathcal{H}(\psi, x) = 0. \tag{6.21}$$

It is readily seen that the expression $\psi_1 x_2 - \psi_2 x_1 + \psi_3 x_4 - \psi_4 x_3 - \frac{1}{2}x_1^2$ equals $|z_1|$ at points that satisfy the equation (6.21). Now replace system (6.20) by another system, coinciding with (6.20) at points of the surface (6.21). Namely, let us

(i) substitute $|z_1| + \frac{1}{2}x_1^2$ for $\psi_1 x_2 - \psi_2 x_1 + \psi_3 x_4 - \psi_4 x_3$;

(ii) replace the term $\psi_1 x_3 - \psi_3 x_1$ by the following expression (6.22).

$$\psi_1 x_3 - \psi_3 x_1 = \frac{|z_1|(x_2 x_3 - x_1 x_4) - z_2(x_3 x_4 + x_1 x_2)}{x_2^2 + x_4^2}$$

$$- \frac{z_1(x_1^2 + x_3^2) + \frac{1}{2}x_1^2(x_2 x_3 - x_1 x_4)}{x_2^2 + x_4^2}. \tag{6.22}$$

LEMMA 6.2. *The trajectories of system (6.20) lying on the zero–level surface of the Hamiltonian (6.21) coincide with trajectories of the system, obtained from (6.20) by means of the replacements (i)–(ii).*

Proof. At points of (6.21) we have

$$\begin{cases} \psi_1 x_2 - \psi_2 x_1 + \psi_3 x_4 - \psi_4 x_3 - \frac{1}{2} x_1^2 = |z_1|, \\[2mm] \psi_4 x_1 - \psi_1 x_4 + \psi_3 x_2 - \psi_2 x_3 = z_2. \end{cases} \qquad (6.23)$$

Let us resolve (6.23) with respect to ψ_1, ψ_3. Straightforward manipulations lead to the result

$$\psi_1 = \frac{|z_1| x_2 - z_2 x_4 + x_1(\psi_2 x_2 + \psi_4 x_4) + \frac{1}{2} x_1^2 x_2}{x_2^2 + x_4^2},$$

$$\psi_3 = \frac{|z_1| x_4 + z_2 x_2 + x_3(\psi_2 x_2 + \psi_4 x_4) + \frac{1}{2} x_1^2 x_4}{x_2^2 + x_4^2}.$$

To derive (6.22), it remains to combine the equations with coefficients x_3 and $-x_1$.   **Q.E.D.**

The homogeneity group action furnishes the variable z_1 with weight 4, z_2 with weight 3, z_3 with weight 2, and z_4 with weight 1. We consider the region $w_2 \neq 0$, so the weight of the function $x_2 = -z_4/w_2$ equals 1 and the weight of the function $x_1 = -z_3/w_2$ equals 2. It follows from (6.22) that the function $\psi_3 x_1 - \psi_1 x_3$ has not less than third order. Now it is readily seen that the right–hand side of the modified system (3.18) has the proper form.

If $x_4 \neq 0$, then the singular manifold is given by the equations

$$x_1 = x_2 = \psi_1 = \psi_2 = 0.$$

The singular control u is zero. For the rest of the variables we have

$$\begin{aligned} \dot{\psi}_3 &= \psi_4, & \dot{x}_3 &= x_4, \\ \dot{\psi}_4 &= -\psi_3, & \dot{x}_4 &= -x_3. \end{aligned}$$

Integrals of the system correspond to the circles $x_3^2 + x_4^2 = R_1^2$ and $\psi_3^2 + \psi_4^2 = R_2^2$, R_i being some nonnegative constants. Theorem 3.1, on bundles, provides the existence of chattering bundles. To obtain a Lagrangian submanifold, the values ψ_3, ψ_4 on the singular manifold can be specified as follows: $\psi_3 = \psi_4 = 0$.

Let us verify the projectibility of the subbundle to x–space. In view of Corollary 3.1 it is enough to check that the projections of the vectors $\partial/\partial z_3$ and $\partial/\partial z_4$ are not tangent to the surface $x_1 = x_2 = 0$. We have

$$D\pi \cdot \left(\frac{\partial}{\partial z_3} \right) = \frac{\partial x_1}{\partial z_3} \frac{\partial}{\partial x_1} + \frac{\partial x_2}{\partial z_3} \frac{\partial}{\partial x_2} + \frac{\partial x_3}{\partial z_3} \frac{\partial}{\partial x_3} + \frac{\partial x_4}{\partial z_3} \frac{\partial}{\partial x_4},$$

$$D\pi \cdot \left(\frac{\partial}{\partial z_4} \right) = \frac{\partial x_1}{\partial z_4} \frac{\partial}{\partial x_1} + \frac{\partial x_2}{\partial z_4} \frac{\partial}{\partial x_2} + \frac{\partial x_3}{\partial z_4} \frac{\partial}{\partial x_3} + \frac{\partial x_4}{\partial z_4} \frac{\partial}{\partial x_4}.$$

Since $x_1 = -z_3/w_2$, $x_2 = -z_4/w_2$, $x_3 = w_1$, $x_4 = w_2$, differentiation gives

$$\frac{\partial x_1}{\partial z_3} = -\frac{1}{w_2}, \quad \frac{\partial x_1}{\partial z_4} = 0, \quad \frac{\partial x_2}{\partial z_3} = 0, \quad \frac{\partial x_2}{\partial z_4} = -\frac{1}{w_2},$$

$$\frac{\partial x_3}{\partial z_3} = \frac{\partial x_3}{\partial z_4} = \frac{\partial x_4}{\partial z_3} = \frac{\partial x_4}{\partial z_4} = 0.$$

Thus

$$D\pi \cdot \frac{\partial}{\partial z_3} = -\frac{1}{x_4}\frac{\partial}{\partial x_1},$$

$$D\pi \cdot \frac{\partial}{\partial z_4} = -\frac{1}{x_4}\frac{\partial}{\partial x_2}.$$

The tangent plane to the singular manifold is the span of $\partial/\partial x_3$ and $\partial/\partial x_4$. These four vectors are independent.

Remark 6.1. *System (6.17) has an integral $\sum_{i=1}^{4} x_i^2 = $ const (independent on the choice of a control $u(t)$). Without loss of generality we can take the three–dimensional sphere*

$$S^3 = \left\{ x \,\Big|\, \sum_{i=1}^{4} x_i^2 = 1 \right\}$$

as the state space of the problem.

Thus, the conditions of Theorem 3.1 are satisfied at each point p of the singular circle $x_3^2 + x_4^2 = 1$, $x_1^2 + x_2^2 = 0$ except for $x_4 = 0$. Therefore, there exists a two–dimensional fibre of chattering arcs joining p. In view of Theorems 3.3–3.4, the corresponding fibration of the phase space consists of trajectories which are locally optimal for problem (6.17)–(6.19) outside the points $x_4 = 0$.

Note that it is unprofitable to enter into the singular curve at points $x_4 = 0$ because the transfer of energy from one oscillator to another is minimal at this points. This is the physical reason why Theorem 3.1 is not applicable to these points. Our hypothesis is that the phase portrait of the optimal synthesis in S^3 is similar to that in Fig. 18.

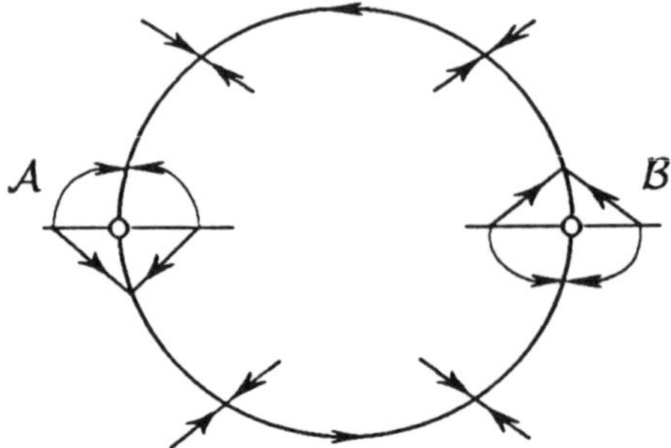

FIG. 18: CONTROL OF TWO INTERDEPENDENT
OSCILLATORS

We see that there should be a dispersal surface through points $\mathcal{A}$ and $\mathcal{B}$ where $x_4 = 0$. From one side of $\mathcal{A}$, we use the fibres joining the singular circle before $\mathcal{A}$, from the other side, we use the fibres joining the singular circle after $\mathcal{A}$. The same is true for the point $\mathcal{B}$. Two–dimensional fibres of the chattering arcs are shown in Fig. 18 schematically as one–dimensional arrows.

6.5 Lowden's Problem

For the first time, the problem of the minimization of fuel expenditure for a space rocket moving in the Newtonian gravitational field under restricted traction was investigated by D. Lowden [1963] in the early sixties.

Let us begin with an exposition of the history of the study of Lowden's problem. It is necessary to mention that the history is intricate to a certain extent. Many results on Lowden's problem have been discovered more than once in view of both the deficiency of information and dissatisfaction with the rigor of some of the mathematical articles on this theme. We are ready to discuss various different versions of events.

The most important result of D. Lowden consisted of the calculation of formulae of Pontryagin's extremals where traction has some intermediate value, not equal to zero, and the maximum possible value. The rocket's motion occurs in a two–dimensional plane; the time of hitting the target manifold is free. Lowden's trajectories are singular in the sense of optimal control theory and have been called "intermediate thrust arcs". It is worthwhile to note that Pontryagin's system for the time optimal problem coincides with that for the fuel–optimal problem with the same differential constraints. All the differences between these problems consist in transversality conditions and the condition of the extremals normality.

Later C. Marchal [1968] investigated the explicit equation of intermediate thrust arcs for the plane problem where time is fixed. The most interesting peculiarity, found by C.Marchal, is the existence of *reversible arcs*, where the intermediate thrust control leads first to an arrest of the rocket at a certain instant and then to motion along the same trajectory in configuration space but in the opposite direction.

The question of optimality of the intermediate thrust arcs is significant from the viewpoint of space navigation and has aroused a wave of attention. It would not be an exaggeration to say that this very problem has given birth to the theory of higher order necessary conditions of optimality of singular extremals.

The first result in this direction belongs to H.J. Kelley [1963, 1964]. However, Kelley's condition on Lowden's spiral is degenerate and does not give any information on optimality. The higher order conditions incited by Lowden's problem were obtained by R.E. Kopp-H.G. Moyer [1965] and H.M. Robbins [1965]. These conditions were more informative. In particular, H.M. Robbins as well as a number of other authors (who have used some different approaches, see [B. Veubeke, J. Geerts, 1964] and [C. Marchal, 1989]), proved that Lowden's spirals (i.e., the singular solutions in the plane with intermediate thrust) are not optimal in the problem with free time. They have proven the optimality of some short arcs of intermediate thrust trajectories for the problem with the fixed time.

Later, the theory of higher order necessary conditions of optimality being generated by cosmonautics began to develop in an independent way, although a number of unsettled questions on Lowden's problem still remain. For example, the optimality of three–dimensional intermediate thrust arcs has not been explored. Another problem, more important from our viewpoint, is the structure of optimal synthesis in Lowden's problem. The first step in this direction was taken by J.V. Breakwell, J.F. Dixon [1975]. They ascertained that the associated problem of minimization of the second variation of Lowden's problem on the singular trajectory such as the reversible arc coincide with the Fuller–Marchal problem (see Problem 3.2). Therefore, J.V. Breakwell, J.F. Dixon suggested using the optimal synthesis in the Fuller-Marchal problem for the correction of small deviations from the singular trajectory in Lowden's problem.

Below, the optimal synthesis for Lowden's problem itself (the time–optimal model) is designed for some class of terminal manifolds in some open domain.

Let us consider the control system that describes the motion of a spacecraft in the central gravity field:

$$\ddot{\mathbf{r}} = \frac{1}{m}\,\mathbf{T} - \mu\,\frac{\mathbf{r}}{|\mathbf{r}|^3},$$

$$\dot{m} = -\frac{1}{c}\,T.$$

Here $\mathbf{r} = (x, y, z) \in \mathbb{R}^3$ is the radius vector, m is the mass of the spacecraft. The control is realized by the emission of a part of the mass. The direction of the thrust $\mathbf{T} = (T_x, T_y, T_z)$ is arbitrary, the magnitude is bounded, $T \overset{\text{def}}{=} |\mathbf{T}| \in [0, 1]$. The numbers μ, c are some positive constants. We look for the trajectories which achieve the transfer from an initial point $\mathbf{r}(0)$, $\dot{\mathbf{r}}(0)$, $m(0)$ to a final point $\mathbf{r}(\tau)$, $\dot{\mathbf{r}}(\tau)$, $m(\tau)$ on some terminal manifold in $\mathbb{R}^7$. In order to state the control problem we can take as the performance index either the time of flight $\tau \to \inf$ or the value of the fuel's expenditure $\int_0^\tau T\, ds \to \inf$. Let us consider the particular case of the problem where the motion occurs in the ecliptic plane and the performance index is the time of the flight.

PROBLEM 6.7. **Lowden's problem.**

Minimize τ *subject to*

$$\dot{X} = U, \qquad \dot{U} = \frac{T}{m}\cos\Theta - \mu\frac{X}{\sqrt{(X^2 + Y^2)^3}},$$

$$\dot{Y} = V, \qquad \dot{V} = \frac{T}{m}\sin\Theta - \mu\frac{Y}{\sqrt{(X^2 + Y^2)^3}}, \qquad (6.23)$$

$$\dot{m} = -\frac{T}{c},$$

with boundary conditions

$$X(0) = X_0, \quad Y(0) = Y_0,$$

$$U(0) = U_0, \quad V(0) = V_0, \quad m(0) = m_0,$$

$$\big(X(\tau), U(\tau), Y(\tau), V(\tau), m(\tau)\big) \in M \subset \mathbb{R}^5.$$

Here the control variables are the thrust magnitude $T \in [0, 1]$ and the direction Θ of the thrust. The variable Θ can be arbitrarily chosen. The goal is to design a family of locally optimal synthesis with the chattering arcs, matched with singular extremals like the reversible arcs of C. Marchal.

Reduction of the Problem

There is a one–parameter symmetry group of system (6.23) induced by the rotations around the origin in (X, Y)–plane. It follows from the Noether theorem that there is the first integral of system (6.17) and hence the order of the system can be decreased by 1. To achieve that, let us put

$$r = \sqrt{X^2 + Y^2}, \quad x = \frac{XU + YV}{\sqrt{X^2 + Y^2}}, \quad y = \frac{XV - YU}{\sqrt{X^2 + Y^2}},$$

$$\phi = \Theta - \arccos\frac{X}{\sqrt{X^2 + Y^2}}.$$

Since Θ is an arbitrary angle, the values of the new variable ϕ can be taken as the new independent control.

For target manifolds $\mathcal{M}$, that are invariant under the corresponding symmetry group, Problem 6.7 is equivalent in the new coordinates to the following one.

PROBLEM 6.8.   *Minimize* τ *subject to*

$$\dot{r} = x, \qquad \dot{x} = \frac{u}{m}\cos\phi - \frac{\mu}{r^2} + \frac{y^2}{r},$$

$$\dot{m} = -\frac{u}{c}, \quad \dot{y} = \frac{u}{m}\sin\phi - \frac{xy}{r}, \tag{6.25.a}$$

with boundary conditions

$$r(0) = r_0, \quad x(0) = x_0, \quad y(0) = y_0, \quad m(0) = m_0$$

$$(r, x, y, m)(\tau) \in \mathcal{M} \subset \mathbb{R}^4.$$

Here u and ϕ are the independent control variables such that $u \in [0, 1]$ and $\phi \in [0, 2\pi]$, μ, c being some positive constants. The system of Pontryagin's maximum principle can be written in the form

$$H = \psi_r x + \psi_x \left(\frac{u}{m}\cos\phi - \frac{\mu}{r^2} + \frac{y^2}{r} \right)$$

$$+ \psi_y \left(\frac{u}{m}\sin\phi - \frac{xy}{r} \right) + \psi_m \left(-\frac{u}{c} \right)$$

(the Pontryagin's function of the problem),

$$\dot{\psi}_r = \left(-\frac{2\mu}{r^3} + \frac{y^2}{r^2} \right)\psi_x - \frac{xy}{r^2}\psi_y,$$

$$\dot{\psi}_x = -\psi_r + \frac{y}{r}\psi_y,$$

$$\dot{\psi}_y = -\frac{2y}{r}\psi_x + \frac{x}{r}\psi_y, \tag{6.25.b}$$

$$\dot{\psi}_m = \frac{\widehat{u}}{m^2}\left(\psi_x \cos\widehat{\phi} + \psi_y \sin\widehat{\phi} \right)$$

(the adjoint system),

$$\max_{\substack{u \in [0,1] \\ \phi \in [0,2\pi]}} u \left(\frac{\psi_x(t)\cos\phi + \psi_y(t)\sin\phi}{m(t)} - \frac{\psi_m(t)}{c} \right)$$

$$= \widehat{u}(t) \left(\frac{\psi_x(t)\cos\widehat{\phi}(t) + \psi_y(t)\sin\widehat{\phi}(t)}{m(t)} - \frac{\psi_m(t)}{c} \right) \tag{6.25.c}$$

(the maximum condition).

Singular Solutions

Let us look for solutions of (6.25) which are not singular with respect to the ϕ–component of the control, i.e., $\psi_x^2 + \psi_y^2 \neq 0$. It follows from (6.25.c) that

$$\cos \widehat{\phi} = \frac{\psi_x}{\sqrt{\psi_x^2 + \psi_y^2}}, \quad \sin \widehat{\phi} = \frac{\psi_y}{\sqrt{\psi_x^2 + \psi_y^2}},$$

hence

$$\widehat{u}(t) = \begin{cases} 1, & \text{if} \quad \dfrac{1}{m}\sqrt{\psi_x^2 + \psi_y^2} - \dfrac{\psi_m}{c} > 0, \\[2ex] 0, & \text{if} \quad \dfrac{1}{m}\sqrt{\psi_x^2 + \psi_y^2} - \dfrac{\psi_m}{c} < 0. \end{cases}$$

Let us calculate u–singular solutions of (6.25) where

$$\frac{1}{m}\sqrt{\psi_x^2 + \psi_y^2} \equiv \frac{\psi_m}{c}.$$

We have

$$H = H_0 + u\,H_1,$$

$$H_0 = x\psi_r + \left(-\frac{\mu}{r^2} + \frac{y^2}{r}\right)\psi_x - \frac{xy}{r}\psi_y,$$

$$H_1 = \frac{1}{m}\sqrt{\psi_x^2 + \psi_y^2} - \frac{\psi_m}{c}.$$

Set $z_1 = H_1$. The derivative of z_1 along the solution of (6.25) is

$$\dot{z}_1 = \frac{z_2}{m^2\left(z_1 + \dfrac{\psi_m}{c}\right)},$$

where

$$z_2 = -\psi_r\psi_x - \frac{y}{r}\psi_x\psi_y + \frac{x}{r}\psi_y^2. \tag{6.26}$$

Also,

$$\dot{z}_2 = z_3,$$

$$\dot{z}_3 = z_4,$$

$$\dot{z}_4 = A + \widehat{u}\,B,$$

where

$$z_3 = \psi_r^2 + \left(\frac{2\mu}{r^3} + \frac{y^2}{r^2}\right)\psi_x^2 - \frac{2xy}{r^2}\psi_x\psi_y + \left(\frac{x^2}{r^2} - \frac{\mu}{r^3}\right)\psi_y^2,$$

$$z_4 = \frac{\mu}{r^4}\left(-8r\psi_r\psi_x - 6x\psi_x^2 + 10y\psi_x\psi_y - x\psi_y^2\right),$$

$$A = -\frac{4x}{r}z_4 + \frac{\mu}{r^4}\left(4x\psi_r\psi_x - 18y\psi_r\psi_y + 8r\psi_r^2\right.$$
$$\left. + \left(\frac{22\mu}{r^2} - \frac{34y^2}{r}\right)\psi_x^2 + \left(\frac{\mu}{r^2} + \frac{9y^2}{r} - \frac{2x^2}{r}\right)\psi_y^2\right), \tag{6.27}$$

$$B = \frac{3\mu\psi_x(3\psi_y^2 - 2\psi_x^2)}{mr^4\sqrt{\psi_x^2 + \psi_y^2}}.$$

If $\psi_x(3\psi_y^2 - 2\psi_x^2) \neq 0$, then the u–singular solution is of second (local) order. Let us find the parametric representation of the surface

$$S = \{z_1 = z_2 = z_3 = z_4 = 0\}.$$

It follows from the equation $z_1 = 0$ that

$$\psi_m = \frac{c}{m}\sqrt{\psi_x^2 + \psi_y^2}.$$

The relation $rz_2 + \dfrac{r^4}{\mu}z_4 = 0$ gives

$$-9r\psi_x\psi_r - 6x\psi_x^2 + 9y\psi_x\psi_y = 0.$$

It follows that either

$$\psi_x = 0$$

or

$$\psi_r = \frac{-2x\psi_x + 3y\psi_y}{3r}. \tag{6.28}$$

Let us consider the u–singular solutions in the domain $\psi_x \neq 0$. If we replace the value ψ_r in the equations $z_2 = 0$ and $z_3 = 0$ by (6.28), we obtain

$$\frac{2x}{3r}\psi_x^2 - \frac{2y}{r}\psi_x\psi_y + \frac{x}{r}\psi_y^2 = 0,$$

$$\left(\frac{4x^2}{9r^2} + \frac{y^2}{r^2} + \frac{2\mu}{r^3}\right)\psi_x^2 - \frac{10xy}{3r^2}\psi_x\psi_y + \left(\frac{x^2}{r^2} + \frac{y^2}{r^2} - \frac{\mu}{r^3}\right)\psi_y^2 = 0.$$

Hence,

$$2x\,\psi_x^2 - 6y\,\psi_x\psi_y + 3x\,\psi_y^2 = 0, \tag{6.29}$$

$$\left(4x^2 + 9y^2 + \frac{18\mu}{r}\right)\psi_x^2 - 30xy\,\psi_x\psi_y$$
$$+ \left(9x^2 + 9y^2 - \frac{9\mu}{r}\right)\psi_y^2 = 0. \tag{6.30}$$

If we multiply (6.29) by $-5x$ and add to (6.30), we obtain

$$\left(-6x^2 + 9y^2 + \frac{18\mu}{r}\right)\psi_x^2 + \left(-6x^2 + 9y^2 - \frac{9\mu}{r}\right)\psi_y^2 = 0,$$

hence

$$3y^2 - 2x^2 = \frac{3\mu(\psi_y^2 - 2\psi_x^2)}{r(\psi_x^2 + \psi_y^2)}. \tag{6.31}$$

It follows from (6.29) that

$$y = \frac{x(2\psi_x^2 + 3\psi_y^2)}{6\psi_x\psi_y}. \tag{6.32}$$

By substituting of (6.32) into (6.31), we obtain

$$x^2\,\frac{4\psi_x^4 - 12\psi_x^2\psi_y^2 + 9\psi_y^4}{12\psi_x^2\psi_y^2} = \frac{3\mu(\psi_y^2 - 2\psi_x^2)}{r(\psi_x^2 + \psi_y^2)},$$

hence

$$x^2 = \frac{36\mu\,\psi_x^2\psi_y^2\,(\psi_y^2 - 2\psi_x^2)}{r(3\psi_y^2 - 2\psi_x^2)^2(\psi_x^2 + \psi_y^2)}.$$

The last relation implies that u–singular trajectories of system (6.25) lie in the domain $\psi_y^2 \geqslant 2\psi_x^2$. The singular manifold of second order can be parameterized by the values $\psi_x,\,\psi_y,\,r,\,m$ as follows:

$$x = \pm\frac{6\psi_x\psi_y}{3\psi_y^2 - 2\psi_x^2}\sqrt{\frac{\psi_y^2 - 2\psi_x^2}{\psi_x^2 + \psi_y^2}}\sqrt{\frac{\mu}{r}}, \tag{6.33}$$

$$y = \pm\frac{2\psi_x^2 + 3\psi_y^2}{3\psi_y^2 - 2\psi_x^2}\sqrt{\frac{\psi_y^2 - 2\psi_x^2}{\psi_x^2 + \psi_y^2}}\sqrt{\frac{\mu}{r}},$$

$$\psi_r = \pm\frac{\psi_y}{r}\sqrt{\frac{\psi_y^2 - 2\psi_x^2}{\psi_x^2 + \psi_y^2}}\sqrt{\frac{\mu}{r}},$$

$$\psi_m = \frac{c}{m}\sqrt{\psi_x^2 + \psi_y^2},$$

At points of the singular surface we have

$$\widehat{u} = -\frac{3\mu m\psi_x(40\psi_x^6 + 36\psi_x^4\psi_y^2 + 22\psi_x^2\psi_y^4 - 9\psi_y^6)}{r^2\sqrt{\psi_x^2 + \psi_y^2}\,(3\psi_y^2 - 2\psi_x^2)^3},$$

$$H_0 = -\frac{3\mu\psi_x^3}{r^2(\psi_x^2 + \psi_y^2)}.$$

Kelley's condition determines the domain

$$\psi_x\left(3\psi_y^2 - 2\psi_x^2\right) < 0.$$

Since $3\psi_y^2 - 2\psi_x^2 \geqslant \psi_y^2 - 2\psi_x^2 \geqslant 0,$ we have

$$\psi_x < 0.$$

The Optimality of the Chattering Arcs

As usual, let us complement the variables $z_1, \ldots, z_4$ by the functions $w_1 = \psi_x$, $w_2 = \psi_y$, $w_3 = r$, $w_4 = m$. To check that (z, w) constitute a basis in $\mathbb{R}^8$, let us calculate the Jacobian $\mathcal{J}$ of the transformation

$$(\psi_m, \psi_r, x, y) \rightarrow (z_1, z_2, z_3, z_4)$$

at points of S_0. We have

$$\mathcal{J} = \begin{vmatrix} -\dfrac{1}{c} & 0 & 0 & 0 \\[2mm] 0 & -\psi_x & \dfrac{\psi_y^2}{r} & -\dfrac{\psi_x\psi_y}{r} \\[2mm] 0 & 2\psi_r & -\dfrac{2y}{r^2}\psi_x\psi_y + \dfrac{2x}{r^2}\psi_y^2 & \dfrac{2y}{r^2}\psi_x^2 - \dfrac{2x}{r^2}\psi_x\psi_y \\[2mm] 0 & -\dfrac{8\mu}{r^3}\psi_x & -\dfrac{6\mu}{r^4}\psi_x^2 - \dfrac{\mu}{r^4}\psi_y^2 & \dfrac{10\mu}{r^4}\psi_x\psi_y \end{vmatrix}$$

$$= -\frac{2\mu}{cr^7}\begin{vmatrix} -r\psi_x & \psi_y^2 & -\psi_x\psi_y \\[2mm] r^2\psi_r & -y\psi_x\psi_y + x\psi_y^2 & y\psi_x^2 - x\psi_x\psi_y \\[2mm] -8r\psi_x & -6\psi_x^2 - \psi_y^2 & 10\psi_x\psi_y \end{vmatrix}.$$

If we replace ψ_r by the expression (6.28), we obtain

$$\mathcal{J} =$$

$$-\frac{2\mu}{3cr^6}\begin{vmatrix} -\psi_x & \psi_y^2 & -\psi_x\psi_y \\[2mm] -2x\psi_x + 3y\psi_y & -3y\psi_x\psi_y + 3x\psi_y^2 & 3y\psi_x^2 - 3x\psi_x\psi_y \\[2mm] -8\psi_x & -6\psi_x^2 - \psi_y^2 & 10\psi_x\psi_y \end{vmatrix}$$

$$= \frac{2\mu\psi_x}{cr^6}\left(3\psi_y^2 - 2\psi_x^2\right)\left(x\psi_x\psi_y + 3y(\psi_y^2 - \psi_x^2)\right).$$

Taking into account (6.32), we obtain

$$\mathcal{J} = \frac{\mu}{cr^6}(3\psi_y^2 - 2\psi_x^2)^2 \frac{x(\psi_x^2 + \psi_y^2)}{\psi_y}.$$

We have, therefore,

$$\mathcal{J} = \pm\frac{6\mu}{cr^6}\sqrt{\frac{\mu}{r}}\psi_x(3\psi_x^2 - 2\psi_y^2)\sqrt{(\psi_x^2 + \psi_y^2)(\psi_y^2 - 2\psi_x^2)} \neq 0$$

at points of the singular manifold S_0. Thus, S_0 is a four–dimensional manifold. Since the multiplication of ψ_x, ψ_y, ψ_r, ψ_m by any real $\lambda \neq 0$ preserves (6.33), we have that the projection πS_0 of S_0 to (x, y, r, m)–space is a three–dimensional manifold (π being the natural projection of $\mathbb{R}^8$ to (x, y, r, m)–space).

Up to now in our examples, the projection of the singular manifold S_0 into the state space had the codimension 2. In Lowden's problem, the projection has the codimension 1. Hence, the singular manifold of the optimal chattering synthesis in the state space depends on the choice of the Lagrangian submanifold in S_0. It follows that only a part of the chattering extremals can be used in generating the optimal synthesis in (x, y, r, m)–space. By choice of the boundary conditions we shall determine some two–dimensional submanifold $\mathcal{X}$ on the singular three–dimensional manifold πS_0 whose open neighborhood is fibred by two–dimensional chattering fibres. First of all, let us find the part of S_0 where the singular control is bounded by $[0, 1]$. Let us show that this set is not empty for all values of constants μ, c. In view of (6.33), we have that $u \geqslant 0$ if

$$\mathcal{P}(\psi_x, \psi_y) \overset{\text{def}}{=} 40\psi_x^6 + 36\psi_x^4\psi_y^2 + 22\psi_x^2\psi_y^4 - 9\psi_y^6 \geqslant 0 \quad \text{and} \quad u \leqslant 1 \quad \text{if}$$

$$-\frac{3\mu}{r^2}m\psi_x \frac{\mathcal{P}(\psi_x, \psi_y)}{\sqrt{(\psi_x^2 + \psi_y^2)(3\psi_y^2 - 2\psi_x^2)^3}} \leqslant 1.$$

The latter inequality is automatically fulfilled if the value of the variable r is large enough.

Let us denote $\left(\psi_x/\psi_y\right)^2 = \alpha$. Since $\psi_y^2 - 2\psi_x^2 \geqslant 0$, we have $0 \leqslant \alpha < 0.5$ in a proper region. The bound $\mathcal{P}(\psi_x, \psi_y) \geqslant 0$ implies that

$$\mathcal{R}(\alpha) \overset{\text{def}}{=} 40\alpha^3 + 36\alpha^2 + 22\alpha - 4 > 0.$$

The cubic polynomial $\mathcal{R}(\alpha)$ is monotonically increasing at $\alpha \geqslant 0$. Its unique root is $\alpha_0 \approx 0.2629\dots$. Therefore, $\mathcal{P}(\psi_x, \psi_y) \geqslant 0$ provides that $\left(\psi_x/\psi_y\right)^2 \in (\alpha_0, 0.5)$.

Now we choose an appropriate target manifold in Lowden's problem to obtain the field of chattering extremals. Let us consider an arbitrary curve Γ in S_0 which

is transversal to the singular trajectories of system (6.25);

belongs to the surface $H = 1$;

meets the transversality condition $\psi\, dx\,\big|_\Gamma = 0$.

Assume that the curve is given by the equations

$$\psi_x = \psi_x(s), \quad \psi_y = \psi_y(s), \quad r = r(s), \quad m = m(s).$$

The values x, y, ψ_r, ψ_m are also functions of s, specified by equations (6.33). The condition $H = 1$ gives

$$r^2(\psi_x^2 + \psi_y^2) = -3\mu\psi_x^3.$$

Therefore, the given values $\psi_x(s)$, $r(s)$ specify the value $\psi_y(s)$. Furthermore, given the functions $r(s)$, $m(s)$, the condition $\psi\, dx\,\big|_\Gamma = 0$ allows us to express $d\psi_x(s)/ds$ as some ordinary differential equation

$$d\psi_x = F_1(\psi_x, r, m)\, dr + F_2(\psi_x, r, m)dm.$$

In the general situation the initial condition $\psi_x(0)$ and functions $r(s)$, $m(s)$ specify the curve Γ as a whole.

The most complicated task is checking the conditions of Theorem 3.4 on the regular projection of the bundles with chattering arcs. To obtain the equation of πS_0, one must eliminate the variables ψ_x, ψ_y between (6.31) and (6.32) (these equations are homogeneous with respect to ψ_x, ψ_y, i.e., they can be expressed as an equation over the single parameter ψ_x/ψ_y). Straightforward algebraic manipulations result in the following equation of πS_0:

$$
\begin{aligned}
(x^2 - 36y^2)(3y^2 - 2x^2)^2 \frac{r^2}{9\mu^2} + (36y^2 - 20x^2)\times \\
\times (3y^2 - 2x^2)\frac{r}{3\mu} + 72y^2 - 62x^2 = 0.
\end{aligned}
\tag{6.34}
$$

Now let us find the projection of the vector $\partial/\partial z_3$ to x–space,

$$D\pi\cdot\left(\frac{\partial}{\partial z_3}\right) = \frac{\partial r}{\partial z_3}\frac{\partial}{\partial r} + \frac{\partial x}{\partial z_3}\frac{\partial}{\partial x} + \frac{\partial y}{\partial z_3}\frac{\partial}{\partial y} + \frac{\partial m}{\partial z_3}\frac{\partial}{\partial m}.$$

It follows from the equations $w_1 = \psi_x$, $w_2 = \psi_y$, $w_3 = r$, $w_4 = m$ that

$$\frac{\partial\psi_x}{\partial z_3} = \frac{\partial\psi_y}{\partial z_3} = \frac{\partial r}{\partial z_3} = \frac{\partial m}{\partial z_3} = 0.$$

Differentiating the expressions (6.26)–(6.27) for z_2, z_3, z_4 with respect to z_3, one can check that

$$0 = -r\psi_x\frac{\partial\psi_r}{\partial z_3} + \psi_y^2\frac{\partial x}{\partial z_3} - \psi_x\psi_y\frac{\partial y}{\partial z_3},
\tag{6.35}$$

$$\frac{r^2}{2} = r^2\psi_r\frac{\partial\psi_r}{\partial z_3} + x\psi_y^2\frac{\partial x}{\partial z_3} - \psi_x\psi_y\left(y\frac{\partial x}{\partial z_3} + x\frac{\partial y}{\partial z_3}\right) + y\psi_x^2\frac{\partial y}{\partial z_3},$$

$$0 = -8r\psi_x\frac{\partial\psi_x}{\partial z_3} - 6\psi_x^2\frac{\partial x}{\partial z_3} + 10\psi_x\psi_y\frac{\partial y}{\partial z_3} - \psi_y^2\frac{\partial x}{\partial z_3}.$$

Differentiation of (6.32) with respect to z_3 yields

$$\frac{\partial y}{\partial z_3} = \frac{\partial x}{\partial z_3} \cdot \frac{2\psi_x^2 + 3\psi_y^2}{6\psi_x\psi_y} = \frac{\partial x}{\partial z_3} \cdot \frac{y}{x}.$$

If $\partial x/\partial z_3 = 0$, then $\partial y/\partial z_3 = 0$ and the first equation of (6.35) gives $\partial\psi_r/\partial z_3 = 0$, which contradicts the second equation of (6.35). Hence, $\partial x/\partial z_3 \neq 0$, and up to a nonzero multiplier, the projection of $\partial/\partial z_3$ is collinear to the vector $\frac{\partial}{\partial x} + \frac{y}{x}\frac{\partial}{\partial y}$,

$$D\pi \cdot \left(\frac{\partial}{\partial z_3}\right) = \frac{\partial x}{\partial z_3}\left(\frac{\partial}{\partial x} + \frac{y}{x}\frac{\partial}{\partial y}\right).$$

It can readily be seen that, in general, the vector $(0,1,y/x,0)$ is not tangent to the singular surface (6.34).

By the condition of Theorem 3.4, the projection of the vector $\partial/\partial z_3$ is required not to belong to the projection of the tangent plane to the switching surface of the chattering family generated by the curve Γ. Recall that at points of Γ this tangent plane is constituted by three types of vectors: the singular velocity vector, say v_1, the tangent vectors to Γ, and the nonsingular velocity vector, say v_2. If the projections of vectors v_1 and v_2 together with the projection of vector $\partial/\partial z_3$ turn out to be linear independent, then the curve Γ can be chosen transversally to this three–plane in x–space.

The singular velocity can be written as

$$v_1 = \left(x, \; \frac{u\psi_x}{\sqrt{\psi_x^2 + \psi_y^2}} - \frac{\mu}{r^2} + \frac{y^2}{r}, \; \frac{u\psi_y}{\sqrt{\psi_x^2 + \psi_y^2}} - \frac{xy}{r}, \; -\frac{u}{c}\right),$$

the singular control u being given by (6.33); the nonsingular velocity (related to $u = 0$) is the following:

$$v_2 = \left(x, \; -\frac{\mu}{r^2} + \frac{y^2}{r}, \; -\frac{xy}{r}, \; 0\right).$$

The projection of $\partial/\partial z_3$ (up to a nonlinear multiplier) equals

$$v_3 = \left(0, \; 1, \; \frac{y}{x}, \; 0\right).$$

It is readily seen that, in general, the vectors v_1, v_2, v_3 are independent and the curve Γ on πS_0 can be chosen in the proper way.

In view of the chattering optimality theorem (Theorem 3.3), the curve Γ generates the field of locally optimal trajectories in Lowden's problem with chattering arcs.

6.6 Robot Control

We consider the following version of controlling a manipulator (robot). In this version, a mobile link is fastened to a massive vertical cylinder which rotates around its fixed vertical axis. The mobile link takes the form of a horizontal advancing arrow–bar (see Fig. 19).

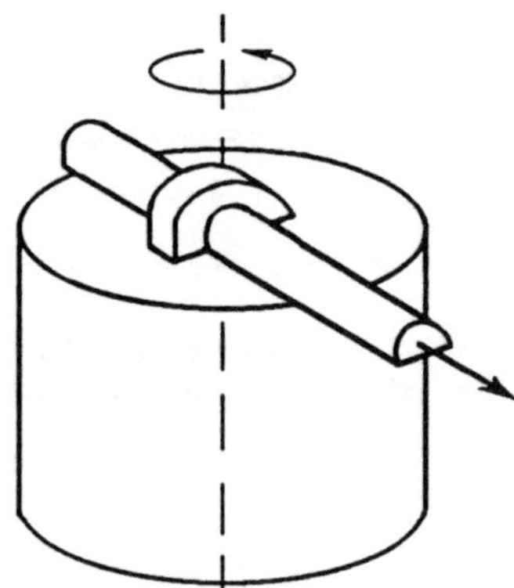

FIG. 19: MANIPULATOR WITH AN ADVANCING BAR

There are two control parameters in the system, $v(t)$ and $u(t)$. The control $v(t)$ is a torque acting on the vertical cylinder, the control $u(t)$ is a force acting on the advancing bar. Both controls are bounded in absolute value.

Let x_1 correspond to the angle of rotation of the vertical cylinder, and x_2 be its angular velocity. Let also x_3 correspond to a position of the gravity center of the advancing arrow–bar, and x_4 be its velocity. Then the movement of the robot is described by the system

$$\dot{x}_1 = \frac{x_2}{1 + x_3^2}, \quad \dot{x}_3 = x_4,$$

$$\dot{x}_2 = v, \qquad \dot{x}_4 = u + \frac{x_2^2 x_3}{(1 + x_3^2)^2} \tag{6.36}$$

Here the denominator $1 + x_3^2$ plays the part of the manipulator's moment of inertia, and the term $(x_2^2 x_3)/(1 + x_3^2)^2$ corresponds to the centrifugal force acting on the bar. Admissible controls in (6.36), (6.37) need to be measurable functions such that

$$|u| \leqslant u_0, \quad |v| \leqslant 1. \tag{6.37}$$

Admissible trajectories $X(\cdot) = (x_1(\cdot), x_2(\cdot), x_3(\cdot), x_4(\cdot))$ are absolutely continuous. We shall deal with the following boundary–value problem for system (6.36).

PROBLEM 6.9. ***Robot control.***

$$T \to \min, \quad X(0) = X_0, \quad X(T) = 0$$

subject to (6.36)–(6.37).

Remark 6.2 *A similar problem can be formulated for the manipulator with another type of the mobile link when the latter is represented by a bar rotating in a vertical plane. The bar is fastened to the vertical cylinder by a hinged joint (see Fig. 20)*

The control system in this case takes the form

$$\dot{x}_1 = x_2, \quad \dot{x}_2 = \frac{v - x_2 x_4 \sin 2x_3}{1 + \sin^2 x_3},$$

$$\dot{x}_3 = x_4, \quad \dot{x}_4 = u + \sin x_3 + C x_2^2 \sin 2x_3. \tag{6.38}$$

If we consider not optimal but locally optimal trajectories, then all the results of this section are also valid for the problem in which $X(t)$ meets (6.37), (6.38).

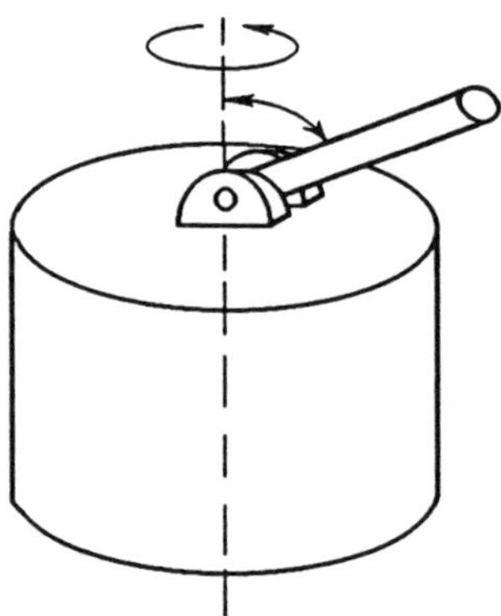

FIG. 20: MANIPULATOR WITH A ROTATING LINK

Note that the problems of manipulator (robot) control are actively treated. The given boundary value problem has been derived from [H.P Geering et al., 1986]. As a rule, the earlier works on robot control treated the controls with one or two switches. Moreover, it was mistakenly stated in [H.P. Geering et al., 1986] that an optimal solution to Problem 6.9 can be directly matched to a singular arc by means of a finite number of switches (see also [S.N. Osipov, A.N. Formalskii, 1988] in this connection). Indeed, the following statement is valid. Consider the region

$$\mathcal{D}_\epsilon \overset{\text{def}}{=} \left\{ X \,\Big|\, \max \left(\sqrt{|x_4|}, |x_3|, |x_1 - A|, |x_2 - B| \right) < \epsilon \right\},$$

$$A > 0, \ B < 0, \ A > B^2/2.$$

For any constants A, B and a sufficiently small $\epsilon > 0$, the optimal solution to Problem 6.9 starting at $X_0 \in \mathcal{D}_\epsilon$ is joined with a singular arc by the chattering control. The optimal feedback control is the following.

All the optimal trajectories with $X_0 \in \mathcal{D}_\epsilon$ first attain the manifold $S_0 \overset{\text{def}}{=} \{X \mid x_3 = x_4 = 0\}$ in finite time with an infinite number of switches of u–components of the control and with a constant control $v = -1$. The manifold S_0 is singular with respect to the u–component of the control. The optimal trajectories in S_0 lead directly to the origin with a constant control $u \equiv 0$ and a single switch of the v–component of the control. There is a one–parameter family of u–chattering extremals through each $(\alpha, \beta, 0, 0) \in \mathcal{D}_\epsilon \cap S_0$. All of them fill a two–dimensional surface, call it $\Sigma_{\alpha,\beta}$, which is smooth outside $\Sigma_{\alpha,\beta} \cap S_0$. The switching surface is a three–dimensional C^1–manifold partitioning $\mathcal{D}_\epsilon$ into two subregions $\mathcal{D}_\epsilon^+, \mathcal{D}_\epsilon^-$, where $u = u_0$ and $u = -u_0$, respectively.

Problem Reduction

Consider the system

$$\dot{x}_1 = \sigma x_2, \quad \dot{x}_2 = v, \quad \dot{x}_3 = x_4,$$
$$\dot{x}_4 = u + x_2^2 x_3 / (1 + x_3^2)^2, \tag{6.39}$$

where u, v, σ are the control variables. Denote by $\mathcal{M}_R$ the set of absolutely continuous functions $X(\cdot)$, such that $X(\cdot)$ is a solution of system (6.36) at the interval $t \in [0, \epsilon R]$ and is a solution of system (6.39) at the interval $t > \epsilon R$ where $\sigma \in [0, 1]$ and (u, v) meet restrictions (6.37). The constant R will be chosen below. Let us consider the following problem.

PROBLEM 6.10. $T \to \min$

subject to constraints

$$X(\cdot) \in \mathcal{M}_R, \quad X(0) = X_0, \quad X(T) = 0.$$

The Filippov theorem [A.F. Filippov, 1961] provides that there exists a solution to both Problems 6.9 and 6.10 at arbitrary initial conditions.

Remark 6.3. *It is easy to see that the optimal value of the terminal time T in Problem 6.10 does not exceed that in Problem 6.9. Hence, if the solution to Problem 6.10 is admissible to Problem 6.9 for some ϵ, R, then it is optimal to Problem 6.19 also.*

We show in Lemma 6.3 below that Problems 6.9 and 6.10 are equivalent indeed. All considerations below are valid provided that $0 < \epsilon \leqslant \epsilon_0$, where $\epsilon_0 = \epsilon(A, B)$ is tacitly assumed to be small enough.

LEMMA 6.3. *Assume that $X^*(\cdot)$ is a solution to Problem 6.10 with an initial point $X_0 \in \mathcal{D}_\epsilon$. Assume also that the triple of functions $(\sigma^*(\cdot),\, u^*(\cdot),\, v^*(\cdot))$ is the corresponding optimal control. Then $\sigma^*(t) \equiv 1$ at $t > \epsilon R$ and $v^*(t) \equiv -1$ at $0 \leqslant t \leqslant \epsilon R$.*

Proof. If $t > \epsilon R$, then the motion in the plane (x_1, x_2) does not depend on the (x_3, x_4)–variation. It follows from (6.39) that

$$\left| x_2(t) - x_2(0) \right| < t, \quad \left| x_1(t) - x_1(0) \right| < \left| x_2(0) \right| \cdot t + \frac{t^2}{2},$$

for any $t > 0$. Hence, for any $X_0 \in \mathcal{D}_\epsilon$ at $t \in [0, \alpha_1 \epsilon]$, the trajectory $X(\cdot)$ of system (6.39) satisfies the estimation

$$x_2^2 \, |x_3| \, (1 + x_3^2)^{-2} < \alpha_2 \epsilon,$$

where the constant α_2 depends on α_1 only.

To prove the lemma we consider two auxiliary problems.

AUXILIARY PROBLEM 6.11. $T \to \min$

subject to

$$\dot{x}_3 = x_4, \quad \dot{x}_4 = u, \quad |u| \leqslant \tilde{u}_0$$
$$(x_3, x_4)(0) = (x_{30}, x_{40}), \quad (x_3, x_4)(T) = (0, 0).$$

where $\tilde{u}_0 = u_0 - \epsilon \alpha_2$.

The restriction on u provides that any admissible trajectory of Problem 6.11 is admissible for system (6.39). It is well–known (see [V. Boltyanskii, 1969]) that the optimal synthesis for Problem 6.11 is given by the switching curve $x_3 = \frac{1}{2u_0} x_4^2 \operatorname{sgn} x_4$. One can check that the optimal time in Problem 6.11 equals

$$T_{\mathrm{opt}} = \kappa x_{40} + 2\sqrt{\kappa \tilde{u}_0 x_{30} + \frac{1}{2} x_{40}^2}$$

where $\kappa = \operatorname{sgn}\left(x_{30} - (x_{40}^2 \operatorname{sgn} x_{40})/(2u_0)\right)$.

The control u being bounded to the segment $\left[-(u_0 - \epsilon \alpha_2),\, u_0 - \epsilon \alpha_2\right]$, one can transfer an initial point along the solution $X(\cdot)$ of system (6.36) to the plane $x_3 = x_4 = 0$ in a time not greater than $L\epsilon$ (where L is some constant). We see that the optimal time T^* in Problem 6.10 is determined by the instant at which $x_1(T^*) = x_2(T^*) = 0$.

Denote $x_1(\epsilon R) = \alpha$, $x_2(\epsilon R) = \beta$. Since $A > \frac{1}{2} B^2$ and $B < 0$, we have $\alpha > \frac{1}{2}\beta^2$, $\beta < 0$. To find the dependence of the function T^* on the parameters α, β let us consider the following auxiliary problem:

AUXILIARY PROBLEM 6.12. $T \to \min$

subject to

$$\dot{x}_1 = \sigma x_2, \quad \dot{x}_2 = v, \quad \sigma \in [0,1], \quad v \in [-1,1]$$

with boundary condition:

$$(x_1, x_2)(0) = (x_{10}, x_{20}); \quad (x_1, x_2)(T) = (0,0).$$

A solution to Problem 6.12 exists by the Filippov existence theorem. Assume that $(x_1, x_2)(\cdot)$ is an optimal solution to Problem 6.12 and $(\sigma, v)(\cdot)$ is a control function. It follows from Pontryagin's maximum principle that there exist absolutely continuous functions $(\psi_1(\cdot), \psi_2(\cdot))$ (nonzero simultaneously) meeting the adjoint system

$$\dot{\psi}_1 = 0, \quad \dot{\psi}_2 = -\sigma \psi_1$$

and the maximum condition

$$\max_{\sigma \in [0,1]} \sigma \psi_1(t) x_2(t) = \sigma(t) \psi_1(t) x_2(t) \quad \text{(a.e.),}$$

$$\max_{v \in [-1,1]} v \psi_2(t) = v(t) \psi_2(t) \quad \text{(a.e.).}$$

Since $\dot{\psi}_1 = 0$, we have $\psi_1(t) \equiv \psi_{10}$. If $\psi_{10} = 0$, then the equation $\dot{\psi}_2 = -\sigma \psi_1$ implies $\psi_2(t) \equiv \psi_{20} \neq 0$. In this case, the v–component of the control is constant along $(x_1, x_2)(\cdot)$. Passing through the origin, the admissible solutions of the system

$$\dot{x}_1 = \sigma x_2, \quad \dot{x}_2 \equiv v$$

are filling the region, denote it G_0, which is constrained by the straight line $x_1 = 0$ and the branch of the parabola $x_1 = -\frac{1}{2} x_2^2 \, \text{sgn} \, x_2$ (see Fig. 21).

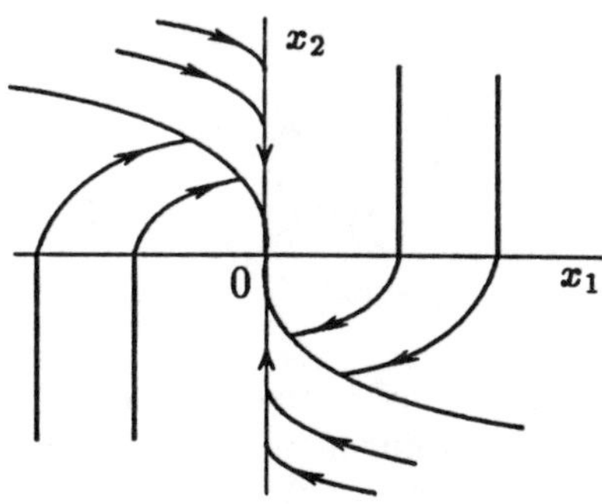

FIG. 21: SOLUTIONS TO AN AUXILIARY PROBLEM

The component v of the control equals 1 at $x_2 < 0$ and -1 at $x_2 > 0$. The component σ is not defined uniquely. The interior of G_0 is an indifference region with respect to σ; on the boundary of G_0 we have $\sigma = 0$ at $x_1 = 0$ and $\sigma = 1$ at $x_1 = -\frac{1}{2}x_2^2 \operatorname{sgn} x_2$.

Let us consider the case $\psi_{10} \neq 0$. If $\psi_{10} > 0$, then the control σ equals 1 at $x_2 > 0$ and 0 at $x_2 < 0$. If $\psi_{10} < 0$, we have $\sigma = 1$ at $x_2 < 0$ and $\sigma = 0$ at $x_2 > 0$. If $\sigma(t) = 0$ at some interval, then $\psi_2(t) \equiv \text{const}$. Therefore, $v \equiv 1$ or $v \equiv -1$. If $\sigma(t) = 1$ on some interval, then the function $\psi_2(t) = -\psi_{10}t + \psi_{20}$ is linear and the optimal trajectory is to enter into the origin after not greater than one switching of v. It is left to calculate the optimal time T_{opt} in Problem 6.12 for an arbitrary initial point $x_{10} = \alpha$, $x_{20} = \beta$ ($\alpha > \beta^2/2$, $\beta < 0$). If we integrate the equation $\dot{x}_2 = v$, then we obtain

$$T_{\text{opt}} = (\beta - x_2^*) + (0 - x_2^*),$$

x_2^* being an ordinate of the intersection point of the curves

$$\begin{cases} x_1 + \frac{1}{2}x_2^2 = \alpha + \frac{1}{2}\beta^2, \\ x_1 = \frac{1}{2}x_2^2, \end{cases}$$

i.e., $x_2^* = -(\alpha + \frac{1}{2}\beta^2)^{1/2}$. This gives, for Problem 6.10, that $\sigma \equiv 1$ at $t > \epsilon R$ and the optimal value of the terminal time is

$$T^* = 2\left(\alpha + \frac{1}{2}\beta^2\right)^{1/2} + \beta + \epsilon R. \tag{6.40}$$

Thus, Problem 6.10 has been reduced to that of minimization of function (6.40) along solutions of system (6.36)–(6.37) at $0 \leqslant t \leqslant \epsilon R$. We assert that if $v^* \neq -1$ on a set of a positive Lebesgue measure in $[0, \epsilon R]$, then the trajectory $X^*(\cdot)$ is not optimal for Problem 6.10.

Consider the solution $X^0(\cdot)$ of system (6.36) with the control $u^0(t) = u^*(T)$, $v^0(t) \equiv -1$. Set $y(t) = x_3^*(t) - x_3^0(t)$, $z(t) = x_2^*(t) - x_2^0(t)$. We have

$$z(t) = \int_0^t \left(1 + v^*(\tau)\right) d\tau \geqslant 0, \tag{6.41}$$

$$\ddot{y} = \frac{(x_2^*)^2 x_3^*}{\left(1 + (x_3^*)^2\right)^2} - \frac{(x_2^0)^2 x_3^0}{\left(1 + (x_3^0)^2\right)^2}. \tag{6.42}$$

Let us rearrange the terms on the right–hand side of (6.42) to separate the functions $y(t)$, $z(t)$ as linear multipliers, denoting the coefficients at $y(t)$, $z(t)$ by $a(X^*, X^0)$, $b(X^*, X^0)$:

$$\ddot{y} = a(t)y + b(t)z,$$

where

$$a = \frac{(x_2^*)^2}{(1+(x_3^*)^2)^2} - \frac{x_3^0(x_2^0)^2\left(2+(x_3^*)^2+(x_3^0)^2\right)(x_3^*+x_3^0)}{\left(1+(x_3^*)^2\right)^2\left(1+(x_3^0)^2\right)^2},$$

$$b = \frac{x_3^0(x_2^*+x_2^0)}{\left(1+(x_3^*)^2\right)^2}.$$

If we take into account that $y(0) = 0$, $\dot{y}(0) = 0$, we can rewrite system (6.42) in the form of an integral equation

$$K\big(y(\cdot)\big) = Z(\cdot),$$

where K is a functional operator

$$(Kw)(s) = w(s) - \int_0^s (s-\tau)a(\tau)w(\tau)\,d\tau,$$

and

$$Z(s) = \int_0^s (s-\tau)b(\tau)z(\tau)\,d\tau.$$

Since $\max\limits_{s\in[0,t]}\left|(Kw)(s)\right| \geqslant \max\limits_{s\in[0,t]}\left|w(s)\right| \cdot \left(1-O(\epsilon)\right)$, the operator K is invertible and

$$y(\cdot)\Big|_{[0,t]} = K^{-1}Z(\cdot)\Big|_{[0,t]}.$$

Hence, there exists a constant $C > 0$ such that

$$\left|y(t)\right| < C\epsilon\left\|z(\cdot)\right\|_{C[0,t]} = C\epsilon\,z(t). \tag{6.43}$$

Let us estimate the difference:

$$\Delta x_1 = x_1^*(\epsilon R) - x_1^0(\epsilon R) = \int_0^{\epsilon R}\left(\frac{x_2^*}{1+(x_3^*)^2} - \frac{x_2^0}{1+(x_3^0)^2}\right)dt.$$

Separate the terms under the integral as follows:

$$\frac{x_2^*}{1+(x_3^*)^2} - \frac{x_2^0}{1+(x_3^0)^2} = \frac{z(t)}{1+(x_3^*)^2} + \frac{x_2^0(x_3^*+x_3^0)y(t)}{\left(1+(x_3^*)^2\right)\left(1+(x_3^0)^2\right)}.$$

Because of (6.43) and the estimation $\left|x_3^0(t)\right| < O(\epsilon)$, we obtain

$$\Delta x_1 \geqslant \left(1-O(\epsilon)\right)\int_0^{\epsilon R} z(t)\,dt.$$

Now (6.41) implies that

$$x_2^*(\epsilon R) > x_2^0(\epsilon R), \quad x_1^*(\epsilon R) > x_1^0(\epsilon R).$$

Calculating the partial derivatives of function (6.41) with respect to α, β, we arrive at the inequalities $\partial T^*/\partial \alpha > 0$, $\partial T^*/\partial \beta > 0$ at $\beta < 0$, $\alpha > \beta^2/2$, i.e., at fixed α the function T^* monotonically decreases as long as β decreases until it intersects parabola $\alpha = \beta^2/2$ and, respectively, at fixed β the function T^* monotonically decreases as long as α decreases until it intersects $\alpha = \frac{1}{2}\beta^2$. This contradicts the optimality of $X^*(\cdot)$. **Q.E.D.**

Set

$$F(x) = \frac{-x_2 x_3^2}{\left(1 + x_3^2\right)\sqrt{x_1 + \frac{1}{2}x_2^2}}.$$

The function $F(x)$ equals exactly the derivative of the function $t + x_2 + 2\sqrt{x_1 + \frac{1}{2}x_2^2}$ with respect to t along solutions of the system

$$\dot{x}_1 = \frac{x_2}{1 + x_3^2}, \quad \dot{x}_2 = -1,$$

so

$$\mathcal{I} \stackrel{\text{def}}{=} \int_0^{\epsilon R} F(x)\,dt = \left(t + x_2 + 2\sqrt{x_1 + \frac{1}{2}x_2^2}\right)\Bigg|_{t=0}^{t=\epsilon R}.$$

It follows from Lemma 6.3 that Problem 6.10 is equivalent to the following one:

$$\mathcal{I} \to \min$$

subject to

$$\dot{x}_1 = \frac{x_2}{1 + x_3^2}, \quad \dot{x}_3 = x_4,$$

$$\dot{x}_2 = -1, \quad \dot{x}_4 = u + \frac{x_2^2 x_3}{(1 + x_3^2)^2},$$

$$|u| \leqslant u_0, \quad x(0) = x_0.$$

The Main Result

THEOREM 6.1. *The optimal solution to Problem 6.10 starting at $X_0 \in \mathcal{D}_\epsilon$ contains a chattering arc. In a finite time, which is continuous in X_0 and does not exceed ϵR, the optimal solution $X^*(\cdot)$ attains the singular manifold $x_3 = x_4 = 0$ with an infinite number of switches of u–component of the control. The optimal motion is followed by a singular arc where $u \equiv 0$.*

Proof.

The upper estimation of the functional of Problem 6.10.

If $t \in [0, \epsilon R]$, then for an arbitrary admissible trajectory $X(\cdot)$ of Problem 6.10 we have $x(t) \subset \mathcal{D}_{C_1\epsilon}$. Hence,

$$\max_{X(\cdot)\in\mathcal{M}_R} \left(\frac{x_2^2 x_3}{1 + x_3^2} : 0 \leqslant t \leqslant \epsilon R \right) \leqslant \gamma_0 < C_2\epsilon.$$

Assume that $x^0(\cdot) = (x_3^0, x_4^0)$ is an optimal solution to Problem 6.11 where $\widetilde{u}_0 = u_1 \overset{\text{def}}{=} u_0 - \gamma_0$. Let

$$K(r) \overset{\text{def}}{=} \left\{ x_3, x_4 \,\Big|\, |x_3| + \frac{x_4^2}{2u_1} \leqslant r \right\}.$$

If $X_0 \in D_\epsilon$, then $(x_{30}, x_{40}) \in K(\lambda\epsilon^2)$, $\lambda = 1 + 1/(2u_1)$. Assume that $(x_{30}, x_{40}) \in K(r)$, $r \leqslant \lambda\epsilon^2$. It follows from (6.42) that the trajectory $x^0(t) = (x_3^0(t), x_4^0(t))$ enters into $(0,0)$ at the instant $\tau^0 \leqslant C_3\sqrt{r}$ (where the constant C_3 does not depend on r). Furthermore, $x^0(t) \subset K(r)$ at $t \in [0, \tau^0]$. If we denote by $X^0(t)$ an admissible trajectory of Problem 6.10 whose projection onto (x_3, x_4)–space coincides with $x^0(t)$, then we have

$$\inf_{X\in\mathcal{M}_R} \mathcal{I} \leqslant \left(r^2 + O(\epsilon) \right) \cdot \left(A + \frac{1}{2}B^2 \right)^{-1/2} \int_0^{\tau^0} (-x_2^0)\, dt \leqslant sr^{5/2}, \quad (6.44)$$

where s does not depend on R.

The lower estimation of the functional of Problem 6.10.

Let

$$\Omega_1 = \left\{ x_3, x_4 \,\big|\, |x_3| \leqslant r/2 \right\}\backslash K(r).$$

Then the x_3–component of the velocity $\dot{x}_3 = x_4$ is separated from zero: $|\dot{x}_3| \geqslant \sqrt{ru_1}$. Let us upper estimate the duration of an arc of an admissible trajectory inside of the region Ω_1:

$$\sup_{X(\cdot)\in\mathcal{M}_R} \left(\tau > 0 : (x_3(t), x_4(t)) \subset \Omega_1, t \in [0, \tau] \right)$$

$$\leqslant \sqrt{\frac{r}{u_1}} < \frac{5}{4}\sqrt{\frac{r}{u_0}}.$$

Assume $X(\cdot)$ to be an arbitrary admissible trajectory of Problem 6.10 such that $x_3(0) = \frac{r}{2}\operatorname{sgn} x_4(0)$. Set $\Omega_2 = \left\{ x_3, x_4 \,\big|\, |x_3| \geqslant r/2 \right\}\backslash K(r)$. We have

$$\inf_{X(\cdot)\in\mathcal{M}_R} \left(\tau > 0 : (x_3(\tau), x_4(\tau)) \notin \Omega_2 \right) \qquad (6.45)$$

$$\geqslant 2\sqrt{\frac{r}{u_1}} \frac{1}{u_0 + \gamma_0} > \frac{3}{2}\sqrt{\frac{r}{u_0}}.$$

Assume $X^*(\cdot)$ to be an optimal solution to Problem 6.10 such that $X^*(0) \in \mathcal{D}_\epsilon$, $(x_3^*, x_4^*)(0) \in K(r)$. Let us estimate the time τ, which is required for $X^*(\cdot)$ to attain the region $K(r/2)$. Suppose that $\big(x_3^*(t), x_4^*(t)\big) \notin K(r/2)$ while $t \in [0, \tau]$, $\tau \leqslant \epsilon R$. Let n be a number of intersections of $X^*(\cdot)$ with the set $x_3 = r/2 \cdot \operatorname{sgn} x_4$ at $t \in [0, \tau]$. Let $\mu = \big\{ t \in [0, \tau] : |x_3^*(t)| \leqslant r/2 \big\}$ and $\nu = [0, \tau]\backslash\mu$. In view of (6.45) we have $\tau > 3/2 \cdot (n-1)\sqrt{r/u_0}$. Hence, $n < 1 + 2/3 \cdot \tau\sqrt{u_0/r}$ and

$$\operatorname{mes} \mu < \frac{5}{4}\sqrt{\frac{r}{u_0}}\, n \leqslant \frac{5}{4}\sqrt{\frac{r}{u_0}} + \frac{5}{6}\tau, \quad \operatorname{mes} \nu \geqslant \frac{1}{6}\tau - \frac{5}{4}\sqrt{\frac{r}{u_0}}.$$

On account of (6.44) we have

$$\int_0^\tau F^* \, dt \leqslant sr^{5/2}, \tag{6.46}$$

where $F^* = F\big(X^*(t)\big)$. Hence,

$$\int_0^\tau F^* \, dt \geqslant \int_\nu F^* \, dt \geqslant Q_1 r^2 \operatorname{mes} \nu \geqslant Q_2 r^2 \tau - Q_3 r^{5/2}, \tag{6.47}$$

for some constants $Q_i > 0$ which do not depend on R. Now (6.46) and (6.47) imply $\tau < Q\sqrt{r}$.

Thus, in a time not larger than $Q\sqrt{r}$, the value of r becomes half as much, in a time $Q\sqrt{r/2}$, it becomes half as much once again, etc. Hence, for $r < \lambda\epsilon^2$, the trajectory $X^*(\cdot)$ starting at $X^* \in \mathcal{D}_\epsilon$ attains the plane $x_3 = x_4 = 0$ in a time not larger than the sum of the following converging geometrical progression:

$$\tau < Q\left(\sqrt{r} + \sqrt{\frac{r}{2}} + \ldots + \sqrt{\frac{r}{2^n}} + \ldots\right) = Q\frac{\sqrt{r}}{1 - \sqrt{1/2}} \leqslant Q^*\epsilon,$$

where $Q^* = Q\sqrt{\lambda}/(1 - \sqrt{1/2})$. Recall that the constant R in the statement of Problem 6.10 was unspecified, and take $R > Q^*$. In this case, we have $\tau < \epsilon R$, and it follows from Remark 6.3 that optimal solutions to Problem 6.10 are also optimal to Problem 6.9. This means that optimal solutions to Problem 6.9 attain the plane S_0 in a time less than R, and this time is continuous in the initial point $X_0 \in \mathcal{D}_\epsilon$.

The order of the singular arcs.

Let us show that the optimal solution to Problem 6.9 starting at $X_0 \in \mathcal{D}_\epsilon$ contains a chattering arc. To achieve that we shall demonstrate that u–singular extremals of Problem 6.9 have second intrinsic order in the region

$\mathcal{D}_\epsilon$. Consider the equations of Pontryagin's maximum principle for Problem 6.9:

$$
\begin{aligned}
\dot\psi_1 &= 0, & \dot x_1 &= \frac{x_2}{1+x_3^2}, \\[2mm]
\dot\psi_2 &= -\frac{\psi_1}{1+x_3^2} - \frac{2x_2 x_3 \psi_4}{(1+x_3^2)^2}, & \dot x_2 &= v^*, \\[2mm]
\dot\psi_3 &= \frac{2x_2 x_3 \psi_1}{(1+x_3^2)^2} - \frac{x_2^2(1-3x_3^2)\psi_4}{(1+x_3^2)^3}, & \dot x_3 &= x_4, \\[2mm]
\dot\psi_4 &= -\psi_3, & \dot x_4 &= u^* + \frac{x_2^2 x_3}{(1+x_3^2)^2},
\end{aligned}
\tag{6.48}
$$

where

$$
\max_{|u| \leqslant u_0} u\psi_4(t) = u^*(t)\,\psi_4(t) \quad \text{a.e.,}
$$

$$
\max_{|v| \leqslant 1} v\psi_2(t) = v^*(t)\,\psi_2(t) \quad \text{a.e.}
$$

We look for u–singular solutions of system (6.48) in the region $\mathcal{D}_\epsilon$. Assume that $\psi_4(t) \equiv 0$ at some interval (t_0, t_1). Then $\psi_3(t) \equiv 0$ at (t_1, t_2). Differentiating the relation $\psi_3 = 0$ along solutions of system (6.48) we obtain $x_2 x_3 = 0$. It follows from Lemma 6.3 that $x_2 \neq 0$ in $\mathcal{D}_\epsilon$–region. If $\psi_1 = 0$ at (t_0, t_1), then $\psi_2(t) \equiv \text{const} \neq 0$ and hence the optimal control is constant in spite of Lemma 6.3. So $\psi_1(t) \neq 0$ and hence $x_3(t) = x_4(t) = u^*(t) = 0$ at (t_0, t_1). Differentiating the function $H_1 = \psi_4$ along solutions of (6.48) leads to the following relations at u–singular extremals:

$$
\frac{\partial}{\partial u}\frac{d^k}{dt^k} H_1 = 0, \quad k = 0,1,2,3;
$$

$$
\frac{\partial}{\partial u}\frac{d^4}{dt^4} H_1 = 2x_2\psi_1 \neq 0.
$$

Thus, u–singular extremals in $\mathcal{D}_\epsilon$ have second intrinsic order. The statement being proved is a consequence of the theorem on conjugation in Chapter 3 (a second–order singular extremal cannot be matched directly with a piecewise continuous nonsingular arc).

Q.E.D.

Optimality of the Chattering Arcs

THEOREM 6.2. *There exists a two–parameter set $\mathfrak{N}_{\alpha,\beta}$ of extremals of Problem 6.9 with the following properties. For any fixed values of parameters (α,β) the $\mathfrak{N}_{\alpha,\beta}$–trajectories pass through the point $X_{\alpha,\beta} \stackrel{\text{def}}{=} (\alpha,\beta,0,0) \in \mathcal{D}_\epsilon$ and contain a chattering arc with respect to the u–component of the control. The switching points of $\mathfrak{N}_{\alpha,\beta}$–trajectories constitute a curve, smooth outside the point of intersection with the S_0–surface. All $\mathfrak{N}_{\alpha,\beta}$–trajectories are optimal for Problem 6.9.*

Proof. Without loss of generality, suppose $u_0 = -1$. We look for the solutions of system (6.48) through the point with the following coordinates:

$$x_1 = \alpha, \quad x_2 = \beta, \tag{6.49}$$
$$x_3 = x_4 = \psi_3 = \psi_4 = 0,$$
$$\psi_1 = -1, \quad \psi_2 = -\beta - \sqrt{\alpha + \beta^2/2}.$$

It follows form (6.49) that, up to a rate factor, the (ψ_1,ψ_2)–field on the S_0–surface coincides with the field of adjoint variables in the time–optimal Problem 6.11 on the (x_3,x_4)–plane. That is, if we integrate the equations

$$\dot{\psi}_1 = 0, \quad \dot{\psi}_2 = -\psi_1, \quad \dot{x}_1 = x_2, \quad \dot{x}_2 = \operatorname{sgn} \psi_2, \tag{6.50}$$

we obtain that a solution of system (6.50) starting at the point (6.49) with the constant control $v = -1$ intersects the switching curve $x_1 = x_2^2/2$ where $\psi_2(t) = 0$. Then the solution of (6.50) with constant control $v = +1$ leads to the origin in the time $\sqrt{\alpha + \beta^2/2}$.

Now let us set $v^* = -1$ in (6.48). We assert that in this case system (6.48) can be reduced to the (3.5), (4.5)–form. Indeed, system (6.48) implies

$$\dot{\psi}_4 = -\psi_3, \quad \dot{\psi}_3 = -2\beta x_3 + O(\epsilon)\Theta_1,$$
$$\dot{x}_3 = x_4, \quad \dot{x}_4 = \operatorname{sgn} \psi_4 + \Theta_2,$$

where Θ_1, Θ_2 denote functions complying with the relations

$$\varlimsup_{\kappa \to +0} \Theta_i\big(\mathfrak{g}_\kappa(\psi_4,\psi_3,x_3,x_4)\big)\kappa^{-2} < \infty,$$
$$\mathfrak{g}_\kappa(\psi_4,\psi_3,x_3,x_4) = (\kappa^4\psi_4, \kappa^3\psi_3, \kappa^2 x_3, \kappa x_4).$$

If we set $z_1 = \psi_4$, $z_2 = -\psi_3$, $z_3 = 2\beta x_3$, $z_4 = 2\beta x_4$, we arrive at the conditions of Theorem 3.1 in the modification of Proposition 4.2. Hence, the existence of $\mathfrak{N}_{\alpha,\beta}$–trajectories has been proved, as required.

The last step is to prove the optimality of $\mathfrak{N}_{\alpha,\beta}$–trajectories for Problem 6.9.

On account of (6.49), we have $H = \sqrt{\alpha + \beta^2/2}$ on $\mathfrak{N}_{\alpha,\beta}$–trajectories. It follows from Theorem 3.2 that the field of adjoint variables $\psi/H = (\psi_1/H, \psi_2/H, \psi_3/H, \psi_4/H)$ corresponding to the $\mathfrak{N}_{\alpha,\beta}$–solutions is a potential one in the $\mathcal{D}_\epsilon$–region.

Since

$$\det \left. \frac{D(z_3, z_4)}{D(x_3, x_4)} \right|_{S_0} = 4\beta^2 \neq 0,$$

it follows from Corollary 3.1 that x–projections of $\mathfrak{N}_{\alpha,\beta}$–trajectories are mutually exclusive and fill some open neighborhood of the surface $S_0 \cap \mathcal{D}_\epsilon$.

However, we cannot use Theorems 3.3 and 3.4 immediately to prove the optimality of the $\mathfrak{N}_{\alpha\beta}$–trajectories because their endpoints do not belong to an open set covered by the extremals field. So we are to join the open set $\mathfrak{O} \subset \mathcal{D}_\epsilon$ covered by the chattering arcs of $\mathfrak{N}_{\alpha,\beta}$–extremals with the two–dimensional surface S_0. Theorem 6.1 implies that optimal solutions to Problem 6.8 exist and lie exactly in $\mathfrak{O} \bigcup S_0$. But being restricted on $\mathfrak{O} \bigcup S_0$, the field ψ/H is potential. From here on, optimality is proved by the same way as in Theorem 3.3.

Indeed, assume that $X^*(\cdot)$ is an optimal solution to Problem 6.9 meeting the initial condition $X^*(0) \in \mathfrak{O}$ and $X^0(\cdot)$ is an $\mathfrak{N}_{\alpha,\beta}$–trajectory starting at $X^0(0) = X^*(0)$. Let us consider the curve in $\mathfrak{O} \bigcup S_0$ produced by the joining of $X^*(\cdot)$ and $X^0(\cdot)$. In view of potentiality, we have

$$\oint_{X^*} \frac{\psi}{H}\, dx = \oint_{X^0} \frac{\psi}{H}\, dx.$$

Let T^0 and T^* be the instants of hitting the origin by the trajectories X^0, X^* respectively. It follows from the maximum condition that the right–hand side of the equation equals T^0 and the left–hand side is not larger than T^*. Thus, $T^0 \leqslant T^*$, which implies the optimality of $X^0(\cdot)$.
Q.E.D.

Chapter 7

MULTIDIMENSIONAL CONTROL WITH CHATTERING

In the previous chapters the main attention was paid to problems with one–dimensional controls, i.e., to problems with a single input. Though Problem 6.3 of stabilization of a rigid body and Problem 6.9 of robot control have several inputs, each input was bounded independently and could been considered as a single one.

This chapter is devoted to essentially *multidimensional controls*. For a general affine in control problem existence of singular solutions is due to flat pieces of the indicatrix of velocities, so we have the following two extreme possibilities: a polyhedral and a smooth indicatrix. Some wide class of problems with a polyhedral (simplicial) indicatrix is considered in the first section. Problems with a smooth indicatrix are represented in the second section by an example with a spherical indicatrix (the dimension of the sphere is less than the dimension of the state space).

7.1 Multidimensional Problems with a Polyhedral Indicatrix

We consider a wide class of affine in control problems with multidimensional control variables.

Let $x = (x_1, x_2, \ldots, x_n) \in \mathbb{R}^n$, $u = (u_1, \ldots, u_n) \in U = \left\{ u \in \mathbb{R}^n \mid \sum_{i=1}^n u_i = 1, u_i \geqslant 0 \right\}$, and $Q : \mathbb{R}^n \to \mathbb{R}^1$ be a smooth positive function increasing with each x_i.

Consider the time–optimal control problem with a target $\mathcal{M}$:

PROBLEM 7.1. *Multidimensional control.*

$$T \to \min$$

subject to

$$x_i^{(k)} = u_i Q(x) \quad (i = 1, \ldots, n),$$

with boundary conditions

$$x(0) = x_0, \ldots, x^{(k-1)}(0) = x_0^{(k-1)};$$
$$\left(x(T), \ldots, x^{(k-1)}(T) \right) \in \mathcal{M}.$$

Let us find singular solutions to Problem 7.1. Set

$$\dot{x} = y_1, \ \ddot{x} = y_2, \ \ldots, \ x^{(k-1)} = y_{k-1} \quad (y_i \in \mathbb{R}^n).$$

Then Pontryagin's maximum principle yields

$$H = \psi_1 y_1 + \psi_2 y_2 + \cdots + \psi_{k-1} y_{k-1} + \langle \psi_k, u \rangle Q(x),$$

$$
\begin{aligned}
\dot{x} &= y_1, & \dot{\psi}_1 &= -\langle \psi_k, u \rangle \frac{\partial Q}{\partial x}, \\
\dot{y}_1 &= y_2, & \dot{\psi}_2 &= -\psi_1, \\
&\cdots & &\cdots \\
\dot{y}_{k-1} &= uQ(x), & \dot{\psi}_k &= -\psi_{k-1}.
\end{aligned}
$$

The maximum condition is

$$\max_{u \in U} \ \langle \psi_k, u \rangle = \langle \psi_k, u^* \rangle, \tag{7.1}$$

u^* being the optimal control.

Condition (7.1) means that u^* is to be chosen in such a way that $\langle \psi_k, u \rangle$ equals the maximal component of the vector ψ_k. If $(\psi_k)_\alpha > (\psi_k)_i$

for the given α and each $i \neq \alpha$, i.e., there is exactly one maximal component (the α–th one), then $u_\alpha = 1$, $u_i = 0$ $(i \neq \alpha)$. If there are several maximal components

$$(\psi_k)_{\alpha_1} = (\psi_k)_{\alpha_2} = \ldots = (\psi_k)_{\alpha_s}, \tag{7.2}$$

then $\sum_{j=1}^{s} u_{\alpha_j} = 1$; $u_i = 0$ $(i \neq \alpha_1, \ldots, \alpha_s)$. If relations (7.2) are fulfilled at some nonzero time–interval, then the corresponding arc is called *singular with respect to components* $\alpha_1 \ldots, \alpha_s$ or, in short, $(\alpha_1, \ldots, \alpha_s)$–*singular*. To find its order, consider first the (1,2)–singular arc

$$(\psi_k)_1 = (\psi_k)_2 > (\psi_k)_i. \tag{7.3}$$

Set $Q_i = \partial Q / \partial x_i$. The multi–index will be used for partial derivatives of higher order, for instance, $Q_{12} = \partial^2 Q / \partial x_1 \partial x_2$. The common value of the maximal components of some vector, say $(\psi_k)_1 = (\psi_k)_2 = \ldots = (\psi_k)_s \geq (\psi_k)_i$, $i > s$, will be denoted by $\overline{\psi}_k$.

Successive differentiation of (7.3) on the singular solution gives

$$(\psi_{k-1})_1 = (\psi_{k-1})_2, \ldots, (\psi_0)_1 = (\psi_0)_2, \quad \overline{\psi}_k Q_1 = \overline{\psi}_k Q_2.$$

Further on, we will consider only the "regular" case $\overline{\psi}_k \neq 0$. In this case, the last equation becomes free from the adjoint variables:

$$Q_1 - Q_2 = 0. \tag{7.4}$$

Let us continue the differentiation of (7.4):

$$\begin{cases} \sum_{i=1}^{n} (Q_{1i} - Q_{2i})(y_1)_i = 0, \\[2mm] \sum_{i=1}^{n} \sum_{j=1}^{n} (Q_{1ij} - Q_{2ij})(y_1)_i (y_1)_j \\[2mm] \qquad + \sum_{i=1}^{n} (Q_{1i} - Q_{2j})(y_2)_i = 0, \\[2mm] \qquad\qquad \vdots \\[2mm] P_{k-1}(x, y) + \sum_{i=1}^{n} (Q_{1i} - Q_{2i})(y_{k-1})_i = 0. \end{cases} \tag{7.5}$$

Finally, the derivative of order $2k$ gives

$$P_k(x, y) + \sum_{i=1}^{n} (Q_{1i} - Q_{2i}) u_i Q(x) = 0.$$

Here $P_{k-1}(x, y)$ and $P_k(x, y)$ denote some functions of x, $y_1, \ldots, y_{k-1}$, polynomial in y.

Equations (7.4), (7.5) determine a $(n-1) \cdot k$–dimensional submanifold S_{12} of the state space $\mathbb{R}^{nk}$ and all (1,2)–singular arcs are to lie in S_{12}.

All $(\alpha_1, \ldots, \alpha_s)$–singular arcs lie in the manifold $S_{\alpha_1, \ldots, \alpha_s}$ which is the intersection of all manifolds

$$S_{\alpha_1, \ldots, \alpha_s} = \bigcap_{i,j=1}^{s} S_{\alpha_i \alpha_j}. \tag{7.6}$$

It suffices to intersect in (7.6) only $(s-1)$–manifolds with independent pairs of indices, for instance, (α_1, α_2), (α_1, α_3), $\ldots$, (α_1, α_s), because all other equations of $(\alpha_1, \ldots, \alpha_s)$–singular arcs follow from these $k(s-1)$.

The control u appears at the $(2k)$–th step of differentiating, so the (intrinsic) order of singular arcs equals k.

To justify the picture of joining singular and nonsingular arcs in the case $k > 2$, one needs Conjectures 5.1–5.5. We are not concerned with this case. The case $k = 1$ (first order singular arcs) has been explored in [L. Zelikina, 1975]. Here we consider the case $k = 2$ (second order singular arcs). This problem seems a natural generalization (for the case of several factors) of the resource allocation problem ($k = 2$, $n = 2$) that was considered in Section 6.3, where the economical motivation of the statement was given. To investigate this generalization in more detail, we limit ourselves to the case $k = 2$, $n = 3$, which includes all principal difficulties of the cases $n > 3$. Take for simplicity $Q(x) = x_1 x_2 x_3$ and set $\dot{x} = y$. Consider the domain $\mathbb{R}_+^6 = \{x_i > 0,\ y_i > 0,\ i = 1, 2, 3\}$. It is readily seen that all trajectories of the system

$$\begin{cases} \dot{x} = y, \\ \dot{y} = u\, x_1 x_2 x_3, \end{cases} \qquad u \in U = \left\{ \sum_{i=1}^{3} u_i = 1,\, u_i \geqslant 0 \right\}, \tag{7.7}$$

starting in $\mathbb{R}_+^6$, remain in $\mathbb{R}_+^6$ for all $t > 0$. As a target we take the line

$$\begin{cases} x_1 = x_2 = x_3 = c_0 > 0, \\ y_1 = y_2 = y_3. \end{cases}$$

Pontryagin's maximum principle yields

$$H = \langle \phi, y \rangle + \langle \psi, u \rangle\, x_1 x_2 x_3.$$

The adjoint system has the form

$$\begin{aligned} \dot{\phi}_1 &= -\langle \psi, u \rangle x_2 x_3, & \dot{\phi}_2 &= -\langle \psi, u \rangle x_1 x_3, & \dot{\phi}_3 &= -\langle \psi, u \rangle x_1 x_2, \\ \dot{\psi}_1 &= -\phi_1, & \dot{\psi}_2 &= -\phi_2, & \dot{\psi}_3 &= -\phi_3. \end{aligned} \tag{7.8}$$

The maximum condition has the form

$$\max_{u \in U} \langle \psi, u \rangle = \langle \psi, u^* \rangle, \tag{7.9}$$

where u^* is the optimal control.

Equations (7.5), specifying the (1,2)–singular manifold S_{12}, are given as follows:

$$S_{12} = \begin{cases} x_1 = x_2, \\ y_1 = y_2. \end{cases}$$

The (1,2)–singular control is $u_1 = u_2 = 1/2$, $u_3 = 0$. We can apply the theory and describe the junction of singular and nonsingular arcs in a neighborhood V_{12} of S_{12}. The same can proceed in a neighborhood V_{13} of the manifold

$$S_{13} = \{\, x_1 = x_3, y_1 = y_3 \,\}$$

(relative to the components (1,3)) and in a neighborhood V_{23} of the manifold

$$S_{23} = \{\, x_2 = x_3, y_2 = y_3 \,\}$$

(relative to the components (2,3)).

These three manifolds mutually intersect at the manifold

$$S_{123} = \{\, x_1 = x_2 = x_3, y_2 = y_2 = y_3 \,\},$$

which is (1,2,3)–singular, the corresponding control being $u_1 = u_2 = u_3 = 1/3$. The disposition of the manifolds S and of the corresponding solutions is shown in Fig. 22.

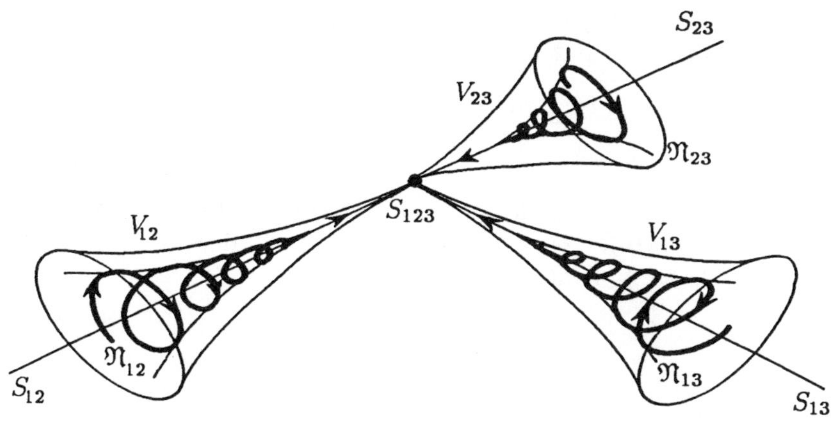

FIG. 22: SOLUTIONS TO THE MULTIDIMENSIONAL
CONTROL PROBLEM

In Fig. 22, the lines S_{12}, S_{13}, S_{23} represent four–dimensional manifolds, the point S_{123} represents a two–dimensional one. The codimension of S_{123} in S_{12} is two (we are unable to show this in Fig. 22 where the corresponding dimension is 1). The size of the neighborhood V_{ij} is decreased while approaching S_{123} and is reduced to zero at points of S_{123} itself.

This example allows us to demonstrate a new phenomenon.

Chattering of Singular and Nonsingular Arcs

The restriction of the function $x_1 x_2 x_3$ to the plane $x_1 + x_2 + x_3 = \text{const}$ achieves its maximum value at the point $x_1 = x_2 = x_3$. Hence the most desirable way to increase x and y is to move along S_{123}. Consider the situation when a trajectory first enters into a manifold $S_{i,j}$ and then follows the (i,j)–singular arc until it attains the manifold S_{123}. Suppose that starting at a point of V_{12} a chattering trajectory attains S_{12} with the alternation of $u_1 = 1$ and $u_2 = 1$. After that, using (1,2)–singular control we wish to hit S_{123}.

Just as in the case of the scalar control, the following proposition for the system (7.7), (7.8), (7.9) is valid.

PROPOSITION 7.1. *Any (1,2,3)–singular arc cannot be matched directly with a piecewise–smooth one (both nonsingular and singular relative to any pair of indices).*

The proof is completely analogous to that in Chapter 2. We leave it for the reader as an exercise.

It is inevitable to hit S_{123} using the chattering, and only the possibility $u_3 = 1$ remains as an alternative to (1,2)–singular control.

PROPOSITION 7.2. S_{12} *is the integral variety of system (7.7), (7.8) with* $u_3 = 1$.

Proof. Let $y_1(0) = y_2(0)$, $x_1(0) = x_2(0)$, $\phi_1(0) = \phi_2(0), \psi_1(0) = \psi_2(0)$, $x_3(0)$, $y_3(0)$, $\psi_3(0)$, $\psi_4(0)$ be initial values of a trajectory, call it γ, with $u_3 = 1$ at $t \in (0, \tau)$. Then

$$y_1(t) = y_2(t) = y_1(0);$$
$$x_1(t) = x_2(t) = x_1(0) + y_1(0)t.$$

From the equations

$$\dot{\phi}_1 = -\psi_3(t)\big(y_1(0)t + x_1(0)\big)x_3(t),$$
$$\dot{\phi}_2 = -\psi_3(t)\big(y_1(0)t + x_1(0)\big)x_3(t),$$

it follows that $\phi_1(t) = \phi_2(t)$. The equations

$$\begin{cases} \dot{\psi}_1 = -\phi_1 \\ \dot{\psi}_2 = -\phi_2 \end{cases}$$
$$\psi_1(0) = \psi_2(0)$$

yield that $\psi_1(t) = \psi_2(t)$. **Q.E.D.**

Proposition 7.2 shows that a control, which alternates $u_3 = 1$, $u_1 = u_2 = 0$ and $u_1 = u_2 = 1/2$, $u_3 = 0$, keeps points inside S_{12}. It is easy to reduce the search of corresponding arcs in S_{12} near S_{123} to the problem with a one–dimensional control. So the conditions of Theorem 3.1 hold.

To design the optimal synthesis we begin with the transversality condition: the vector $(0,0,0,1,1,1)$, tangential to $\mathcal{M}$, is orthogonal to the vector $(\phi_1, \phi_2, \phi_3, \psi_1, \psi_2, \psi_3)$. At points of the (1,2,3)–singular trajectory we have $\phi_1 = \phi_2 = \phi_3$, $\psi_1 = \psi_2 = \psi_3$, and the transversality condition implies $\psi_i = 0$ $(i = 1, 2, 3)$. The coordinates in S_{123} are $\bar{x}$ and $\bar{y}$. System (7.7) has the form

$$\dot{\bar{x}} = \bar{y},$$
$$\dot{\bar{y}} = \frac{1}{3}\bar{x}^3,$$

and its solutions $\bar{y} = \sqrt{\bar{x}^4/6 + C}$ are shown in Fig. 23, see the next page.

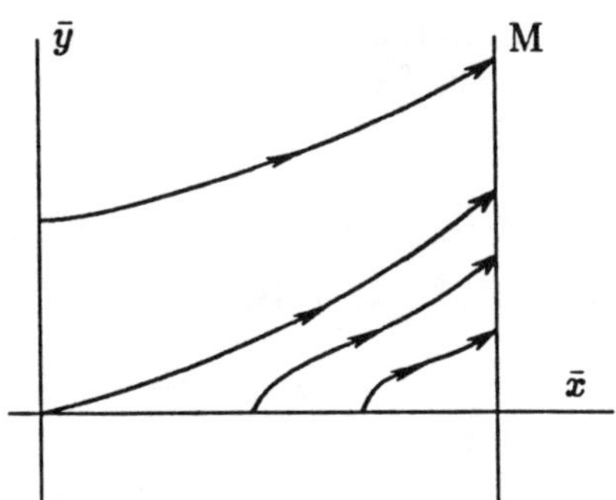

FIG. 23: SINGULAR SOLUTIONS

We can find the values $\bar{x}(t), \bar{y}(t), \bar{\phi}(t), \bar{\psi}(t)$ on S_{123}. We have to emphasize that a field $(\bar{\phi}, \bar{\psi})$ on S_{123} depends on a target $\mathcal{M}$ and determines a fibering $S_{12} \to S_{123}$ by two–dimensional fibres $\mathfrak{M}_{\sigma}^{+}$, $\sigma \in S_{123}$, with the chattering mode which alternates $u_3 = 1$ and $u_1 = u_2 = 1/2$ in each fibre $\mathfrak{M}_{\sigma}^{+}$. Taking any point $\left(\xi \in \mathfrak{M}_{\sigma}^{+}, \phi_{\xi}, \psi_{\xi}\right)$ as a terminal one, we can get the leaf of the chattering mode which alternates $u_1 = 1$ and $u_2 = 1$. This furnishes the fibering $V_{12} \to S_{12}$ by $\mathfrak{N}_{\xi}^{+}$.

All aforesaid is valid mutatis mutandis for any substitution of indices 1, 2, 3.

Now we can summarize. The domain $V_{12} \subset \mathbb{R}_{+}^{6}$ is fibred by two–dimensional mutually exclusive leaves $\mathfrak{N}_{\xi}^{+}$. Trajectories of $\mathfrak{N}_{\xi}^{+}$ attain the four–dimensional manifold S_{12} in finite time with an infinite number of switches from $u_1 = 1$ to $u_2 = 1$, and conversely, and then continue to remain in it. The manifold S_{12} is also fibred by two–dimensional mutually exclusive leaves $\mathfrak{M}_{\sigma}^{+}$. Trajectories of $\mathfrak{M}_{\sigma}^{+}$ attain the two–dimensional manifold S_{123} in finite time with an infinite number of switches from $u_1 = u_2 = \frac{1}{2}$ to $u_3 = 1$, and conversely. Then trajectories follow S_{123} with the singular control $u_1 = u_2 = u_3 = 1/3$ up to the target $\mathcal{M}$.

The proof of optimality of this synthesis in V_{ij} we leave for the reader as an exercise. In case of difficulties, one may consult the similar proof in Section 6.6.

Research problem. We leave unexplored the most intriguing question — the behavior of trajectories that start at points

$$(x, y) \notin V_{12} \bigcup V_{13} \bigcup V_{23}$$

and arrive directly at S_{123}, by passing S_{12}, S_{13}, S_{23} when we have to use successively all vertices of the simplex switching from one to another in some order with the increasing frequency (the order of the switches depends on initial positions).

7.2 Multidimensional Problems with a Smooth Indicatrix

As a typical example of a problem with a smooth indicatrix, we propose here some analog of the Fuller problem (Problem 2.1). Just as in the Fuller problem, it is required to arrest a massive point at the origin using an arbitrary force that is constrained in its absolute value. With this in mind, one looks for a minimum of the mean square deviation of the point from the origin. Unlike the usual Fuller problem, the motion takes place not on a line, but in an n–dimensional configuration space. The admissible controls are the measurable n–dimensional vector–functions whose norms are less than or equal to 1.

Let the range of vector–functions $x(t), y(t), u(t)$ be $\mathbb{R}^{n}$. In this section, the scalar product of the n–dimensional vectors is denoted as the usual multiplication.

PROBLEM 7.2.   *Minimize*

$$\int_0^\infty x^2(t)\, dt$$

subject to

$$\dot{x} = y, \qquad u^2 \leqslant 1, \quad x(0) = x_0, \quad y(0) = y_0.$$
$$\dot{y} = u,$$

Here x, y, u are three n–vectors.

PROPOSITION 7.3. *There exists a unique solution to Problem 7.2.*

The proof is completely similar to that for Problem 5.2 (see lemmas 5.1 and 5.2).

Pontryagin's Maximum Principle

Introduce the adjoint variables $\phi, \psi \in \mathbb{R}^n$, then

$$H = \phi y + \psi u - \frac{x^2}{2}$$

(it can readily be shown that $\lambda_0 \neq 0$ as well as in Problem 2.1).   The adjoint system is

$$\begin{cases} \dot{\phi} = \quad x, \\ \dot{\psi} = -\phi. \end{cases}$$

The maximum condition gives

$$\text{If} \quad \psi \neq 0, \quad \text{then} \quad u = \psi/|\psi|;$$
$$\text{if} \quad \psi = 0, \quad \text{then any control } u \text{ meets}$$
$$\text{the maximum condition.}$$

After the usual redenotation, we obtain the following Hamiltonian system with discontinuous right–hand side (the discontinuity takes place at points of a manifold of the codimension n):

$$\begin{cases} \dot{z}_1 = z_2, \\ \dot{z}_2 = z_3, \qquad\qquad z_i \in \mathbb{R}^n, \\ \dot{z}_3 = z_4, \qquad\qquad i = 1, 2, 3, 4. \\ \dot{z}_4 = -z_1/|z_1|, \end{cases} \qquad (7.10)$$

The $4n$–dimensional vector (z_1, z_2, z_3, z_4) is denoted by z.

DEFINITION 7.1. *A solution $z(t)$ of system (7.10) is called singular on an interval (t_0, t_1) if $z_1(t) = 0$ for all $t \in (t_0, t_1)$.*

It is evident that a unique singular solution of system (7.10) is the zero solution, $z(t) \equiv 0$.

PROPOSITION 7.4. *For any point* (x_0, y_0), *the optimal solution to Problem 7.2 hits the origin in finite time.*

The proof of the proposition is a multidimensional version of Theorem 6.1. Here we indicate only the main stages of the proof and leave the details as an exercise for the readers.

Just as in the proof of Theorem 6.1, we define the following subsets of $\mathbb{R}^{2n}$:

$$K_r = \left\{ (x, y) \in \mathbb{R}^{2n} \mid |x| + \frac{y^2}{2} \leqslant r \right\},$$

$$\Omega_1 = \left\{ (x, y) \in \mathbb{R}^{2n} \mid |x| < \frac{r}{2};\ (x, y) \notin K_r \right\},$$

$$\Omega_2 = \left\{ (x, y) \in \mathbb{R}^{2n} \mid |x| \geqslant \frac{r}{2};\ (x, y) \notin K_r \right\}.$$

i. The Upper Estimation of the Functional

(a) For any $(x_0, y_0) \in K_r$, there exists a control $u(t)$ (which does not need to be optimal) such that the corresponding trajectory remains in K_r and attains the origin at an instant $T \leqslant C_1 \sqrt{r}$, C_1 being a positive constant.

(b) For this trajectory, we have:

$$\int_0^T x^2(t)\, dt \leqslant C_1 r^{5/2}.$$

ii. The Lower Estimation of the Functional

(a) Let $\big(x(t), y(t)\big)$ be any admissible trajectory with initial conditions $|x(0)| = r/2$, $x(0)y(0) > 0$; and T be an instant such that $\big(x(t), y(t)\big) \notin \Omega_2$. Then $T \geqslant 2\sqrt{r}$ (i.e., one cannot leave the region Ω_2 in a time less than $2\sqrt{r}$).

(b) Let $\big(x(t), y(t)\big)$ be any admissible trajectory, and T be an instant such that $\big(x(t), y(t)\big) \in \Omega_1$ for all $t \in [0, T]$. Then $T \geqslant \sqrt{r}$ (i.e., one cannot remain in Ω_1 for a time larger than $\sqrt{r}$).

(c) Let $\big(x(t), y(t)\big)$ be any admissible trajectory with initial conditions $\big(x(0), y(0)\big) \in K_r$ and τ be an instant such that $\big(x(t), y(t)\big) \notin K_{r/2}$ for all $t \in [0, \tau]$. We fix the following notation

$$\mu = \big\{ t \in [0, \tau] \mid \big(x(t), y(t)\big) \in \Omega_1 \big\},$$

$$\nu = \big\{ t \in [0, \tau] \mid \big(x(t), y(t)\big) \in \Omega_2 \big\},$$

n being a number of invasions of the trajectory into the set Ω_1. In view of (b) we have

$$\tau \geqslant (n - 1)2\sqrt{r},$$

and, hence,

$$n \leqslant \frac{\tau}{2\sqrt{r}} + 1,$$

$$\operatorname{mes} \mu \leqslant n\sqrt{r} \leqslant \frac{\tau}{2} + \sqrt{r},$$

$$\operatorname{mes} \nu \geqslant \tau - \left(\frac{\tau}{2} + \sqrt{r}\right) = \frac{\tau}{2} - \sqrt{r}.$$

For $t \in \nu$, we have $x^2(t) \geqslant (r/2)^2$ and, consequently,

$$\int_0^\tau x^2(t)\,dt \;\geqslant\; \frac{r^2}{4} \operatorname{mes} \nu \;\geqslant\; \frac{\tau r^2}{8} - \frac{r^{5/2}}{4}.$$

iii. *Estimation of the Time of Hitting the Origin*

Combining the upper and the lower estimation, one obtains

$$C_1 r^{5/2} \;\geqslant\; \int_0^\infty x^2(t)\,dt \;\geqslant\; \frac{r^2 \tau}{8} - \frac{r^{5/2}}{4}.$$

Hence, $\tau \leqslant Q\sqrt{r}$, where Q is some constant.

Thus, the optimal trajectory emanating from any point $A_1 \in K_r$ attains the set $K_{r/2}$ at some point A_2 in a time, that does not exceed $Q\sqrt{r}$. Starting from A_2, the trajectory attains the set $K_{r/4}$ in a time, that does not exceed $Q\sqrt{\dfrac{r}{2}}$, and so on. Hence, the total time of hitting the origin does not exceed

$$Q\sqrt{r} + Q\sqrt{\frac{r}{2}} + \cdots = \frac{Q\sqrt{r}}{1 - 1/\sqrt{2}}.$$

Q.E.D.

PROPOSITION 7.5. *Let T be an instant of junction of a nonsingular solution $\big(x(t), y(t), u(t)\big)$, $t \in (0,T)$, with the singular one, $z \equiv 0$. Then $u(t)$ does not have a limit as $t \to T - 0$.*

Proof. Denote $\tau = T - t$. Suppose that $u(\tau) \to v$ as $\tau \to +0$. Then $u(\tau) = v + o(1)$;

$$\frac{dy}{d\tau} = -u(\tau); \quad y(0) = 0;$$

so

$$y(\tau) = -v\tau + o(\tau).$$

It follows that

$$\frac{dx}{d\tau} = -y = v\tau + o(\tau); \quad x(0) = 0;$$

so

$$x(\tau) = \int_0^\tau \big(vt + o(t)\big)\, dt = \frac{v}{2}\tau^2 + o(\tau^2).$$

In addition,

$$\frac{d\phi}{d\tau} = -x = -\frac{v}{2}\tau^2 + o(\tau^2); \quad \phi(0) = 0,$$

so

$$\phi(\tau) = \int_0^\tau \Big(-\frac{v}{2}\tau^2 + o(\tau^2)\Big)\, d\tau = -\frac{v}{6}\tau^3 + o(\tau^3).$$

It follows that

$$\frac{d\psi}{d\tau} = \phi = -v\frac{\tau^3}{6} + o(\tau^3), \quad \psi(0) = 0,$$

so

$$\psi(\tau) = \int_0^\tau \Big(-\frac{v}{6}\tau^3 + o(\tau^3)\Big)\, d\tau = -\frac{v}{24}\tau^4 + o(\tau^4).$$

Multiplication of the last relation by v yields

$$v\psi(\tau) = -\frac{\tau^4}{24} + o(\tau^4),$$

hence $v\psi(\tau) < 0$ for sufficiently small τ. On the other hand, the maximum condition yields

$$u(\tau)\psi(\tau) = \big|\psi(\tau)\big| > 0 \quad \text{and} \quad v\psi(\tau) > 0,$$

so $v\psi(\tau) \geqslant 0$ for sufficiently small τ.

The contradiction proves the proposition. **Q.E.D.**

Let us define the action of the group $\mathcal{O}(n) \times \mathbb{R}_+$ on a tuple $\xi \overset{\text{def}}{=} \big(u(t), y(t), x(t), \phi(t), \psi(t)\big)$ as follows. Let $\zeta \in \mathcal{O}(n)$. Then

$$\zeta \circ \xi \overset{\text{def}}{=} (\zeta u, \zeta y, \zeta x, \zeta \phi, \zeta \psi).$$

Let $\lambda \in \mathbb{R}_+$. Then

$$\mathfrak{g}_\lambda(\xi) \overset{\text{def}}{=} \big(u(t/\lambda), \lambda y(t/\lambda), \lambda^2 x(t/\lambda), \lambda^3 \phi(t/\lambda), \lambda^4 \psi(t/\lambda)\big).$$

PROPOSITION 7.6. *The action of the group $\mathcal{O}(n) \times R_+$ respects the set of solutions to Problem 7.2 as well as the set of solutions of equation (7.10).*

Proof is evident.

Thereafter, we consider the velocity vectors y as free vectors lying in the configuration space $X = (x_1, x_2, \ldots, x_n)$.

Let $L_0 \subset X$ be a two–dimensional subspace of $\mathbb{R}^n$ spanned on x_0 and y_0.

PROPOSITION 7.7. L_0 *is an integral variety of the flow of optimal solutions to Problem 7.2.*

Let the initial conditions x_0 and y_0 belong to some two-dimensional subspace L_0. It follows that for any $t > 0$ the corresponding optimal trajectory $x(t)$ and its velocity $y(t)$ belong to L_0.

Proof. Consider the following *auxiliary problem* specified as the restriction of Problem 7.2 to the plane L_0. It means that the admissible controls are measurable functions $u(t)$ such that $u(t) \in L_0$, $|u(t)| \leqslant 1$; the admissible trajectories $x(t)$ and $y(t)$ lie in the plain L_0; and the functional is $\int_0^\infty x^2(t)\,dt$.

Denote by π the operator of the orthogonal projection of the configuration space X into L_0 and by Id the identity mapping of $\mathbb{R}^n$. Let $\big(x(t), y(t)\big)$ be any admissible trajectory of Problem 7.2 with initial conditions x_0 and y_0 in the plain L_0. Then the trajectory $\big(\pi x(t), \pi y(t)\big)$ corresponds to the control $\pi u(t)$ which is admissible for the *auxiliary problem*. We have

$$\int_0^\infty x^2(t)\,dt = \int_0^\infty \big(\pi x(t)\big)^2 dt + \int_0^\infty \big((Id - \pi)x(t)\big)^2 dt$$

$$\geqslant \int_0^\infty \big(\pi x(t)\big)^2 dt \geqslant J_0,$$

where J_0 is an optimal value of the *auxiliary problem*.

Existence of solutions to the *auxiliary problem* and uniqueness of the solution to problem 7.2 provide that the optimal control $u(t)$ lie in L_0 and the corresponding trajectory $x(t)$ and its velocity $y(t)$ lie in the plain L_0 too.   **Q.E.D.**

COROLLARY 7.1. *Problem 7.2 can be reduced to the following problem with a two–dimensional control:*

PROBLEM 7.3.   *Minimize*

$$\int_0^\infty x^2(t)\,dt$$

subject to

$$\dot{x} = y, \qquad |u| \leqslant 1, \quad x(0) = x_0, \quad y(0) = y_0.$$
$$\dot{y} = u$$

Here x, y, u are three two–dimensional vectors.

PROPOSITION 7.8. *If the vectors* $x(0)$ *and* $y(0)$ *are collinear, that is,*

$$rk \left\| \begin{matrix} x_1(0) & y_1(0) \\ x_2(0) & y_2(0) \end{matrix} \right\| = 1,$$

then for all t,

$$rk \left\| \begin{matrix} x_1(t) & y_1(t) \\ x_2(t) & y_2(t) \end{matrix} \right\| = 1,$$

where $x(t), y(t)$ *is the optimal solution with initial conditions* $x(0), y(0)$.

Proof is quite similar to that of Proposition 7.7.

COROLLARY 7.2. *The torical cone*

$$K \stackrel{\text{def}}{=} \left\{ (x, y) \in \mathbb{R}^4 \mid x_1 y_2 - x_2 y_1 = 0, \ (x, y) \neq (0, 0) \right\}$$

is the integral variety of the flow of solutions to Problem 7.3. Hence, if $(x_0, y_0) \notin K$, *then the corresponding solution does not intersect* K.

The cone K divides the four–dimensional state space (x, y) into two congruent parts,

$$V_1 = \left\{ (x, y) \in \mathbb{R}^4 \mid x_1 y_2 - x_2 y_1 > 0 \right\},$$
$$V_2 = \left\{ (x, y) \in \mathbb{R}^4 \mid x_1 y_2 - x_2 y_1 < 0 \right\}.$$

the Hamiltonian system (7.10) turns into the following eight–dimensional discontinuous system:

$$\begin{cases} \dot{z}_1 = z_2, \\ \dot{z}_2 = z_3, \\ \dot{z}_3 = z_4, \\ \dot{z}_4 = -z_1/|z_1|. \end{cases} \tag{7.11}$$

Let us clarify the behavior of the trajectories of system (7.11) on the cone K. We seek the solutions of system (7.11) in the form

$$z_i(t) = p_i(t)V, \quad i = 1, 2, 3, 4, \tag{7.12}$$

where $p_i(t)$ are some scalar functions, and V is a constant two–dimensional vector, $\|V\| \leqslant 1$. Then (7.11) yields

$$\dot{p}_1 = p_2, \quad \dot{p}_2 = p_3, \quad \dot{p}_3 = p_4, \quad \dot{p}_4 = -p_1/|p_1|,$$

i.e., for an arbitrary unit vector V, the relation (7.12) generates a solution of system (7.11) iff $p(t)$ is a solution of the standard four–dimensional Fuller system

$$p_1^{(4)} = -\text{sgn } p_1.$$

The behavior of these solutions is described in Lemmas 2.5 and 2.6.

The next paragraph contains some euristic speculations that help to find solutions to Problem 7.3.

It seems reasonable that in the vicinity of the origin the optimal choice of the control u is designed to quench the velocity and simultaneously to turn it toward the origin. In view of Proposition 7.4 and 7.5 we would get spiral–like trajectories. In the set V_1, the pairs of velocities have a right orientation and this orientation cannot change along an optimal trajectory in view of Corollary 7.2. Hence, the domain V_1 is filled by dextrorse–spirals, which attain the origin twisting counter–clock–wise and undergo an infinite number of rotations in finite time. The picture in V_2 is quite similar, but the trajectories are sinisrorse–spirals and rotate clock–wise. When initial position (x_0, y_0) approach a generating line l of the cone $K = \{x_1 y_2 = x_2 y_1\}$, these spirals become more and more stretched and, in the limit, they lie in l and coincide with solutions to Problem 2.1. Since switching points of these solutions constitute geometric progressions, it is natural to conjecture that solutions to problem 7.3 lie on logarithmic spirals. But unlike trajectories of a focus for linear differential equations, these solutions attain the origin in finite time. The relation $y \sim \sqrt{x}$ for optimal trajectories of Problem 2.1. helps us to guess the parameterization of spirals.

These euristic speculations allow us to find some explicit solutions of equation (7.11) and hence a solution to Problem 7.3.

We will use the complex notation for the two–dimensional vectors, namely,

$$(R\cos\,\alpha, R\sin\,\alpha) \,=\, Re^{i\alpha}.$$

PROPOSITION 7.9. *The function* $z_1^*(\tau) \stackrel{\text{def}}{=} A_0\tau^4 e^{i\alpha\ln|\tau|}$ *and its successive derivatives are solutions of system (7.11) when* $A_0 = \dfrac{1}{126}$, $\alpha = \pm\sqrt{5}$.

Proof. We differentiate z_1 and denote its successive derivatives by
$$z_1^{(j)} \,=\, z_{1+j}.$$

We have

$$z_2 \,=\, \frac{dz_1}{d\tau} \,=\, A_0(4 + i\alpha)\tau^3 e^{i\alpha\ln|\tau|},$$

$$z_3 \,=\, \frac{dz_2}{d\tau} \,=\, A_0(4 + i\alpha)(3 + i\alpha)\tau^2 e^{i\alpha\ln|\tau|},$$

$$z_4 \,=\, \frac{dz_3}{d\tau} \,=\, A_0(4 + i\alpha)(3 + i\alpha)(2 + i\alpha)\tau e^{i\alpha\ln|\tau|},$$

$$\frac{dz_4}{d\tau} \,=\, A_0(4 + i\alpha)(3 + i\alpha)(2 + i\alpha)(1 + i\alpha)e^{i\alpha\ln|\tau|}.$$

Equation (7.11) implies

$$A_0(4 + i\alpha)(3 + i\alpha)(2 + i\alpha)(1 + i\alpha) = -\operatorname{sgn} A_0,$$

or, equivalently,

$$A_0 = \frac{1}{126}, \qquad \alpha = \pm\sqrt{5}.$$

Q.E.D.

The function

$$z(t) = z^*(T - t)$$

is a solution of system (7.11), defined for $t < T$, and it tends to zero as $t \to T$. The function

$$z(t) = z^*(t - T)$$

is the solution to system (7.11), defined for $t > T$, and it escapes from the origin at the instant T.

The action of the group $\mathcal{O}(2)$ generates the following one–parameter family of solutions:

$$(\kappa \circ z)_{j+1}(\tau) = A_j \tau^{4-j} e^{i\alpha(\ln|\tau|+\kappa)},$$
$$A_0 = 1/126, \quad \alpha = \pm\sqrt{5},$$
$$A_{j+1} = A_j(5 - j + i\alpha), \quad j = 1, 2, 3,$$
$$\kappa \in \mathcal{O}(2).$$

The action of the group $\mathbb{R}_+$ gives us nothing new because this family appears automodelling relative to the action of $\mathbb{R}_+$. Indeed,

$$(g_\lambda(z))_{j+1}(\tau) = \lambda^{4-j} A_j \left(\frac{\tau}{\lambda}\right)^{4-j} e^{i\alpha \ln\,(|\tau/\lambda|)}$$
$$= A_j \tau^{4-j} e^{i\alpha(\ln\,|\tau|+\ln\,\lambda)}.$$

COROLLARY 7.3. *The functions*

$$x(t) = \kappa^2 z_3^*(T - t),$$
$$y(t) = \kappa z_4^*(T - t),$$
$$u(t) = \frac{\kappa^4 z_1^*(T - t)}{|\kappa^4 z_1^*(T - t)|}$$

are solutions to Problem 7.3.

The angle between vectors $x(t)$ *and* $y(t)$ *is constant and equals* $\text{arctg } \sqrt{(5/2)}$, *the angle between vectors* $y(t)$ *and* $u(t)$ *is constant and equals* $\pi - \text{arctg } \sqrt{5}$.

Corollary 7.3 gives the possibility to design effectively an optimal strategy for the following class of initial conditions

(i) the angle between $x(0)$ and $y(0)$ equals $\text{arctg } \sqrt{5}$;

(ii) $|y(0)|^2 = |x(0)A_0|\sqrt{(16 + \alpha^2)(9 + \alpha^2)}(4 + \alpha^2) = |x(0)|\sqrt{6}/2$.

Namely, the optimal strategy $u(t)$ is such that $|u(t)| = 1$ and the angle between the reference vector $y(t)$ and $u(t)$ is constant and equals $\pi - \text{arctg } \sqrt{5}$.

Remark 7.1 *We have found only the "circular" spirals as solutions of equation (7.11) but we did not find the "elliptic" spirals.*

Research Problem

Find all solutions of the equation (7.11) which tend to zero as $t \to T$.

Research Problem

Investigate the class of problems with the criterion $\int_0^\infty x^2\, dt$ and the control system $\ddot{x} = u$, where $x, u \in \mathbb{R}^n$, and $u(t) \in U \subset \mathbb{R}^n$ for various shapes of U.

EPILOGUE

The investigation of particular examples receives much emphasis in the book. It is a striking fact that so many specific applied problems fit into the framework of Chattering Theory. This may be quite understandable for metaphysical reasons. One of the dogmas of the Russian Orthodox Church runs as follows: Christ is God and the genuine man simultaneously. These two natures of Christ, the divine one and the human one, coexist being inmergeable and yet inseparable. Similarly, one may say that Example and Theory are unmergeable and yet inseparable. It is the anticipation of the existence of a nice general theory that leads us to study one or another type of examples since examples furnish useful insight into the theory. Meanwhile, Example (being considered in a somewhat generalized, theoretical sense) is the heart of Theory and the latter merely reflects specific features inherent in examples. Moreover, Example and Theory presuppose each other and mathematical discovery is an act of breaking this close circle.

Let us recall the old questions about the source of mathematical discoveries. Does a mathematician create a new reality or does he remember only, by Plato's theory, eternal immutable ideas? Does mathematical discovery appear as a result of the explorer's intellectual efforts or does God himself show him solution?

For instance, after many years of hard and vain work, Carl Gauss saw the unlooked–for proof of the reciprocity law in an instant, as if it had been shown to him. Likewise, having climbed into an omnibus, Henry Poincaré suddenly understood the genuine nature of automorphic functions (its invariance under the action of a discrete group of Lobachevski's plane) as though God himself had shown him this idea. On the other hand, Leopold Kroneker thought that only integers were created by our Lord and all the rest was made by human hands.

Note that the Russian philosopher Vladimir Soloviev (1853–1900) considered ideas to be changeable like living things. This opinion correlates with a strange and deep Plotin's remark that ideas have bodies and souls.

We are inclined to the point of view of Charles Hermite who reasoned that mathematical discovery is a result of close inspection. This means that mathematical reality does exist. Communicating with an idea, mathematician sees some perfect image (in the mind's eye, by Hamlet's expression) and he must inspect it with utter scrutiny to realize (may be, it is better to say, to put into it) its genuine sense.

LIST OF FIGURES

Bibliography

1. Alexeev, V.M., Tikhomirov, V.M., Fomin, S.V., *Optimal Control*, Consultants bureau (a Division of Plenum Publishing Corpor.), New York, 1987.

2. Bershchanckii, Ya.M., *Paths of linear systems with relay type nonlinearity*, Automation and Remote Control **43** (1982), no. 7, 853–860.

3. Boltyanskii, V.G., *Sufficient Conditions for Optimality and the Justification of Dynamic Programming Methods*, SIAM J. Control **4** (1966), no. 2, 326–361.

4. Boltyanskii, V.G., *Mathematical Methods of Optimal Control*, 2nd rev. aug. ed. (in Russian), "Nauka", Moscow, 1969, English transl. of 1st ed., Holt, Reinhart and Winston, 1971.

5. Borisov, V.F., *Structural stability of the synthesis of optimal trajectories in the Fuller problem* (in Russian), Vestn. MGU, Ser. Matematika, Mekhanika (1987), no. 4, 64–66, English transl. Moscow University Math. Bulletin, **42** (1987), no. 4, 71–74.

6. Borisov, V.F., *The design of the optimal synthesis with chattering mode* (in Russian), Dokl. Akad. Nauk **302** (1988), no. 4, 785–789, English transl. Soviet. Math. Dokl. **38** (1989).

7. Borisov, V.F., Zelikin, M.I., *Modes with switching of increasing frequency in the problem of controlling a robot* (in Russian), PMM **52** (1988), no. 6, 731–738, English transl. in Applied Math. and Mech. **52** (1988), no. 6, 731–738.

8. Borshchevskii, M.Z., Ioslovich, I.V., *The problem of the optimum rapid braking of an axisymmetric solid rotating around its center of mass* (in Russian), PMM **49** (1985), no. 1, 35–42, English transl in Applied Math. and Mech. **49** (1985), no. 1, 24–30.

9. Breakwell, J.V., Dixon, J.F., *Minimum–fuel rocket trajectories involving intermediate thrust arcs*, J. Optimiz. Theory and Appl. **17** (1975), no. 5, 465–479.

10. Brunovskii, P., *Existence of regular synthesis for general control problems*, J. Different. Equat. **38** (1980), no. 3, 317–343.

11. Brunovskii, P., *Regular synthesis for the linear–quadratic optimal control problem with singular control constraints*, J. Different. Equat. **38** (1980), no. 3, 344–360.

12. Brunovskii, P., Mallet–Paret, J., *Switchings of optimal controls and the equation $y^{(4)} + y^{\alpha}\mathrm{sgn}\, y, \quad 0 < \alpha < 1.$*, Časopes pro pěstovane matematiky roč. **110** (1985), no. 3, 302–313.

13. Coddington, E.A., Levinson, N., *Theory of Ordinary Differential Equations*, McGraw–Hill Book Company, Inc., New York, Toronto, London, 1955.

14. Cohn, P.M., *Algebra*, vol. 1, J. Wiley & Sons, London, 1974.

15. Dicoussar, V.V., Milutin, A.A., *Qualitative and Numerical Methods of the Maximum Principle* (in Russian), "Nauka", Moscow, 1989.

16. Dorling, C.M., Ryan, E.P., *Minimization of non–quadratic cost functional for third order saturating system*, Intern. J. Control **34** (1981), no. 2, 231–258.

17. Dunford, N., Schwartz, J.T., *Linear Operators*, part 1: General Theory, Interscience Publishers, N,Y., London, 1958.

18. Filippov, A.F., *On some problems of theory of optimal control*, SIAM J. Control **1** (1962), no. 1, 76–84.

19. Filippov, A.F., *Differential Equations with Discontinuous Right–Hand Sides* (in Russian), "Nauka", Moscow, 1985, English transl. Reidel, 1989.

20. Fuller, A.T., *Relay Control Systems Optimized for Various Performance Criteria*, Automatic and Remote Control, (Proc. First World Congress IFAC, Moscow, 1960), vol. 1, Butterworths, London, 1961, pp. 510–519.

21. Fuller, A.T., *Minimization of various performance indices for a system with bounded control*, Intern. J. Control 41 (1985), no. 1, 1–37.

22. Fuller, A.T., Grensted, P.E., *Minimization of integral–square error for non–linear control system of third and higher order*, Intern. J. Control 2 (1965), no. 2, 33–73.

23. Geering, H.G., Guzzela, L., Hepner, S.A., Onder, C.H., *Time–optimal motions of robots in assembly tasks*, IEEE Trans. Automat. Control 9 (1971), no. 2, 161–172.

24. Hartman, Ph., *Ordinary Differential equations*, J. Wiley & Sons, New York, London, Sydney, 1964.

25. Hirsh, M.W., Pugh, C.C., Shub, M., *Invariant manifolds*, Springer–Verlag, Berlin, Heidelberg, N.Y., 1977.

26. Kelley, H.J., *Singular extremals in Lawden problem of optimal rocket flight*, AIAA J. 1 (1963), 1578–1588.

27. Kelley, H.J., *A second variation test for singular extremals*, AIAA J. 2 (1964), no. 8, 1380–1382.

28. Kelley, H.J., Kopp, R.E., Moyer, H.G., *Singular extremals*, Topics in Optimization, (ed. by G.Leitmann), Acad. Press, N.Y., 1967, pp. 63–103.

29. Kopp, R.E., Moyer, H.G., *Necessary conditions for singular extremals*, AIAA J. 3 (1965), no. 5, 1439–1444.

30. Kupka, I., *Geometric Theory of extremals. Fuller phenomenon*, Proc. XXIV Conf. Decision and control, Lauderdale (Fl.), 1985, pp. 711–713.

31. Kupka, I., *Fuller's phenomena*, Perspectives in Control Theory (Sielpia, 1988), Progr. Systems Control Theory, Birkhäser, Boston, 1990, pp. 129–142.

32. Kupka, I., *The ubiquity of Fuller's phenomenon*, Nonlinear controllability and optimal control, Monograph textbooks Pure Appl. Math. N 133 (ed. by H.Z. Sussman), Dekker, N.Y., 1990, pp. 313–350.

33. Lawden, D.F., *Optimal Trajectories for Space Navigation*, Butterworth, London, 1963.

34. Lee, E.B., Marcus, L., *Foundations of Optimal Control Theory*, J.Wiley & Sons, New York, London, Sydney, 1967.

35. Lewis, R.M., *Definition of order and junction condition in singular control problems*, SIAM J. Control and Optimiz. 18 (1980), no. 1, 21–32.

36. Magaril–Il'yaev, G.G., *Kolmogorov inequalities on the half–line* (in Russian), Vestnik MGU, ser. Matematica, Mekhanika (1976), no. 5, 33–41, English transl. in Moscow University Mathematical Bulletin 31 (1976), no. 5–6, 25–32.

37. Marchal, C., *Chattering arcs and chattering controls*, J. Optimiz. Theory and Appl. 11 (1973), no. 5, 441–468.

38. Malgrange, B., *Ideals of differential functions*, Oxford University Press, England, 1966.

39. Marchal, C., *Second–order tests in optimization theory*, J. Optimiz. Theory and Appl. 15 (1975), no. 5, 633–666.

40. Marchal, C. , *Theoretical research in deterministic optimization*, ONERA publ. (1971), no. 139, 1–159.

41. Marchal, C., *Généralization tridimensionelle et étude de l'optimalité des arcs a poussé intermédiaire de Lowden (dans un champ Newtonien)*, Rech. Aerospat. (1968), no. 123, 3–13.

42. Nitezky, Z., *Differential Dynamics*, Introduction to the orbit structure, The MIT Press, Cambridge, Massachusetts, and London, England, 1971.

43. Osipov, S.N., Formalskii, A.M., *The problem of the time–optimal turning of manipulator* (in Russian), PMM **52** (1988), no. 6, 929–946, English transl. in Applied Math. and Mech. **52** (1988), no. 6, 725–731.

44. Pontryagin, L.S., Boltyanskii, V.G., Gamkrelidze, R.V., Mishchenko, E.F., *The Mathematical Theory of Optimal Processes*, 3rd ed. (in Russian), "Nauka", Moscow, 1976, English transl. of 2nd ed. Wiley, 1962, and MacMillan, 1964.

45. Robbins, H.M., *Optimality of Intermediate–Thrust arcs of Rockets Trajectories*, AIAA J. **3** (1965), no. 6, 1094–1098.

46. Rockafellar, R.T., *Convex Analysis*, Princeton University Press, Princeton, New Jersey, 1970.

47. Ryan, E.P., *Optimal Control for Second Order saturating System*, Intern J. Control **30** (1979), no. 4, 549–564.

48. Shoshitaishvili, A.M., *On bifurcation of topological type of singular points of vector fields depending on parameters*, Amer. Math. Sov. Transl. **118** (1982), no. 2, 91–121.

49. Sternberg, S., *Lectures on Differential Geometry*, Prentice Hall, Inc. Englewood Cliffs, N.Y., 1964.

50. Sussman, H., *Synthesis, Presynthesis, Sufficient conditions for optimality abs. subanalytic sets*, Nonliner Contollability and Optimal Control, Monograph Textbook Pure Appl. Math., 133, Tekker, N.Y., 1990, pp. 1–19, (Summary).

51. Telesnin, V.R., *On a problem of optimizing transition processes*, Proc. of Steklov Institute of Math. **166** (1986), no. 1, 261–271.

52. Veubeke, B.M.F., Geerts, J., *Optimization of multiple impulse orbital transfers by the maximum principle*, Rept. OA–4 presented at XV Astronautic Congress (1963).

53. Zelikin, M.I., Borisov, V.F., *Fields of Optimal Trajectories containing singular second order extremals and extremals with frequent swithchings* (in Russian), Dokl. Acad. Nauk SSSR **304** (1989), no. 5, 1050–1053, Engl. transl. in Soviet Math. Dokl. **39** (1989), no. 1, 188–191.

54. Zelikin, M.I., Borisov, V.F., *Synthesis in Problems of Optimal Control Containing a trajectory with switchings whose frequency increases, and second–order singular trajectories* (in Russian), Matemat. Zametki **47** (1990), no. 1, 62–73, English transl. in Mathematical Notes of Academy of Sciences of USSR, **47** (1990), iss. 1–2, 41–49.

55. Zelikin, M.I., Borisov, V.F., *Regimes with increasingly more frequent switchings in optimal problems* (in Russian), Trudy Mat. Inst. Steklova **197** (1991), 85–167, Engl. transl. in Proc. of the Steklov Inst. of Math., 1993, iss. 1, 95–186.

56. Zelikina, L.F., *On the question of regular synthesis* (in Russian), Dokl. Akad. Nauk SSSR **267** (1982), no. 3, 532–535, English transl. in Soviet Math. Dokl. **26** (1982), no. 3, 635–639.

57. Zelikina, L.F., *Universal manifolds and turnpike theorems for a class of optimal control problems* (in Russian), Dokl. Akad. Nauk SSSR **224** (1975), no. 1, 31–34, English transl. in Soviet Math. Dokl. **16** (1975), no. 5, 1136–1140.

INDEX

Systems & Control: Foundations & Applications

Series Editor
Christopher I. Byrnes
School of Engineering and Applied Science
Washington University
Campus P.O. 1040
One Brookings Drive
St. Louis, MO 63130-4899
U.S.A.

Systems & Control: Foundations & Applications publishes research monographs and advanced graduate texts dealing with areas of current research in all areas of systems and control theory and its applications to a wide variety of scientific disciplines.

We encourage the preparation of manuscripts in TeX, preferably in Plain or AMS TeX— LaTeX is also acceptable—for delivery as camera-ready hard copy which leads to rapid publication, or on a diskette that can interface with laser printers or typesetters.

Proposals should be sent directly to the editor or to: Birkhäuser Boston, 675 Massachusetts Avenue, Cambridge, MA 02139, U.S.A.

Estimation Techniques for Distributed Parameter Systems
H.T. Banks and K. Kunisch

Set-Valued Analysis
Jean-Pierre Aubin and Hélène Frankowska

Weak Convergence Methods and Singularly Perturbed
Stochastic Control and Filtering Problems
Harold J. Kushner

Methods of Algebraic Geometry in Control Theory: Part I
Scalar Linear Systems and Affine Algebraic Geometry
Peter Falb

H - Optimal Control and Related Minimax Design Problems
Tamer Başar and Pierre Bernhard